DEVELOPING SOCIOLOGICAL KNOWLEDGE

Theory and Method

However, you are just not sure what your committee members want and whether or not they will "buy" your idea. This could be an issue even with regard to some technical issues. For instance, how much literature review is considered enough, and is not considered overdone? It is sometimes really hard for you to decide. The best you can do is to keep good communication with your committee and ask for feedback on your drafts. Sometimes a brief discussion will be extremely helpful since not only you can adapt to their expectations but your committee may get a better understanding of your ideas as well as good answers to their questions and concerns.

Final approval of your thesis/dissertation not only means required signatures from your committee members but also refers to the acceptance by the library. Colleges and universities usually have very strict requirements for formatting and binding a thesis and a dissertation. You should carefully read and follow those guidelines before you produce the final copies. Those regulations can be very detailed and are usually contained in a student manual available from appropriate divisions of your college or university. You should also watch for the deadline of each term/quarter; otherwise you may not be able to graduate in time.

Once you have made sure that your work has met the requirements, it is time for you to reorganize your books and other reference materials, your hand-written notes, your computer files, your data, and all other things used and resources mobilized for completing the project. While taking the inventory and filing away the items, you may take the opportunity to think about your future research direction and some immediate new work plans. Probably it is also the best time for you to formulate your thoughts regarding scholarly publication.

DEVELOPING SOCIOLOGICAL KNOWLEDGE

Theory and Method

Bernard P. Cohen
Stanford University

Second Edition

NELSON–HALL 𝑛ℎ CHICAGO

Editor: Dorothy Anderson
Copy Editor: Jean Scott Berry
Designer: Claudia von Hendricks
Cover Design: Richard Meade, Claudia von Hendricks

Cover Illustration:
One of the drawings from *The Choreographer Dream Series*
by Nancy Carrigan

The cover drawing is one of a series which uses the idiom of dance to point out that even when parameters (color, direction, and shape) are very limited, the number of possible combinations may be in the trillions. The variables of human behavior, the parameters of which are so difficult to draw, are infinite...an interesting challenge for the social scientist. Dance-trained artist Nancy Carrigan used pencils, ink, and graph-paper to complete the work.

LIBRARY OF CONGRESS CATALOGING-IN-PUBLICATION DATA

Cohen, Bernard P.
    Developing sociological knowledge: theory and method / Bernard P. Cohen.—[2nd ed.]
        p.   cm.
    Bibliography: p.
    Includes index.
    ISBN 0-8304-1123-2
    1. Sociology—Methodology.   2. Sociology—Research—Evaluation.
I. Title.
    HM24.C617 1989   88-18864
    301'.01'8—dc19   CIP

Manufactured in the United States of America

10   9   8   7   6   5   4   3   2   1

 ™ The paper used in this book meets the minimum requirements of American National Standard for Information Sciences—Permanence of Paper for Printed Library Materials, ANSI Z39.48-1984.

To the Memory of My Father, Max Cohen

CONTENTS

Preface to Second Edition   xi
Preface to First Edition   xiii

ONE

Introduction   1

Sociology as Science   3
Some Basic Assumptions   13
The Approach of This Book   24

TWO

Actions, Values, and Objective Sociology   27

The Relationship between Sociological Knowledge and Action Decisions   28
The Distinction between Factual Statements and Value Judgments   33
Can Sociology Be as Objective as Natural Science?
*   Should Sociology Be as Objective as Natural Science?   37*
Positivism, Theory-Ladenness, and the Possibility of Objective Science   43
Summary   46

THREE

Two Important Norms of Science   49

The Orientation of Science   51
Intersubjective Testability as a Norm   60

FOUR

Ideas, Observations, and Knowledge Claims   67

Debunking a Myth   69
Ideas—The Raw Material of Science   72
The Error of Reification   73
The Impracticality of Holism   74
Scientific Knowledge Claims: Statements about Events   75
Simple Knowledge Structures   85

FIVE

Simple Knowledge Structures   89

A Study Based on a Simple Knowledge Structure   90
A Simple Knowledge Structure and When, Where, and How to Do a Study   94
Summary   112

SIX

Particularizing and Generalizing Strategies   115

The Orientation to Phenomena   118
Particularizing and Generalizing Strategies and the Problem of Prediction   121

SEVEN

Concepts, Definitions, and Concept Formation   127

Properties of Concepts   128
Definition   136
Types of Definition   138
Types of Definition and the Objectives of Concept Formation   143

EIGHT

Tying Concepts to Observations 147

Operationism: The Case of the IQ 148
The Strategy of Indicators 153
Evaluating Indicators: Reliability and Validity 155

NINE

The Special Problems of Quantitative Concepts and Quantitative Indicators 163

Concepts and Irrelevant Numbers 164
Quantification 166

TEN

From Simple Knowledge Structures to Theories 177

Theory 178
Elements of a Theory 183
What a Theory Does 189
Criteria for the Evaluation of a Theory 195

ELEVEN

A Theory and Its Analysis 199

The Problem 199
Conceptualization of the Problem 203
The Theory 206
Analysis and Evaluation of the Theory 218
Summary 224

TWELVE

The Logical Analysis of a Theory 227

The Calculus of Propositions 228
Using the Calculus of Propositions 230

THIRTEEN

The Empirical Evaluation of Ideas 239

Explanatory Research—Hypothesis Testing 240
What Empirical Research Cannot Do 244
Is Proof Possible in Experimental Research? 248
What Experiments Can Do 258
Limited Value of the Single Study 262
Conclusions 263

FOURTEEN

Making Empirical Research More Informative 265

Constructing the Set of Alternative Explanations 268
Designing an Explanatory Study 273
One Study as Part of a Series 288

FIFTEEN

Cumulative Research Programs 291

CRPs and TRPs 292
Generalizing 297
Integrating Diverse Studies in a CRP 299
Strategies for CRPs 302
General Questions for a CRP 306

SIXTEEN
Stages of a CRP   311

Early Stages of a CRP   312
Intermediate Stages   314
More Advanced Stages   318
A Summary and Two Conclusions   325

SEVENTEEN
Some Final Comments   327

Producing Sociological Knowledge   329
Applying Sociological Knowledge   331
A Consumer's Guide   333

REFERENCES   337

INDEX   343

PREFACE TO THE SECOND EDITION

Since this book appeared, events inside and outside the discipline have affected the climate for developing sociological knowledge. Experiencing the consequences of these events, both the positive and the negative, has reinforced this author's conviction in the positions taken in the preface to the first edition.

Societal support in the form of government and foundation funding of research has declined significantly. Part of this stems from ideological motives, but part represents disillusionment with sociology's ability to contribute to the solution of society's problems. The decline in funding affects not only the amount and quality of research but also, and perhaps of more long-term significance, the amount and quality of training. Fewer and smaller projects mean fewer opportunities for apprenticeship training. Funding constraints mitigate against designing new research that specifically addresses the question at hand, but force researchers, especially those doing doctoral thesis research, to make do with available data even if the data must be stretched to fit the problem. In some cases, sociologists are channeled away from empirical research entirely into other ways of expressing their professional identity.

Producers of sociological knowledge need to be much better versed in, and more self-conscious about, theory and method in order to cope effectively with diminished resources. Reversing the trend of declining resources requires much greater recognition on the consumer's part of what sociology can and cannot do; consumer sophistication is the best antidote to both illusions and disillusionment. An understanding of the fundamental issues discussed in this book will help create more self-conscious producers and more sympathetic consumers.

Controversies concerning the possibility and desirability of a "scientific" sociology have continued in the years since this book first appeared. The ideological debate over "positivism" has become more strident if not more enlightening. Concurrently, some social scientists have claimed that "science has failed" while others, although scientifically oriented, have lost their nerve and have become almost apologetic for their commitment to science. The conclusion that science has failed is, to say the least, premature; even if there were no successes at all—and there are many—we cannot have sufficient perspective on the last half-century, during which almost all the scientific work has been done, to make such judgments. But, more important, it is not clear what the "science" is to which both proponents and opponents refer. A significant part of the debate rests on serious misconceptions of science on both sides. Although there are important issues, they are not well understood and debate rarely focuses on them. Serious analysis of the topics examined in this book may not end the contro-

research interest. You can reorganize the system of knowledge about a central issue or issues in a way that you think is most appropriate. And you can fully develop your scientific curiosity about the problems that remain unresolved. You will have the opportunity to demonstrate your scholarship, analytical integrity, and research methodology by clearly laying out the plan for your thesis or dissertation.

The thesis/dissertation proposal is sometimes called an overview. Once you start writing your proposal, you start framing and phrasing your thesis or dissertation. Usually the first part of your thesis/dissertation will contain the core contents of your proposal, though later revision of the design and frequent modification of the writings may be required.

Students may seek suggestions and opinions from their advisors in preparing their proposal. As a matter of fact, faculty approval of the thesis/dissertation proposal is often a formal requirement of university regulations. For the doctoral students, their proposal is particularly important since they cannot formally embark on their dissertation unless all (or a required majority of) their committee members vote to approve it. This usually requires an oral examination (pre-dissertation defense) before the committee. If you fail, in theory you may fail your entire doctoral career. But if you succeed in the exam with outstanding performance, you may make the committee believe that a final examination will not be needed. In such a case, your final defense may be waived at the discretion of your committee.

The proposal is especially important if you wish to obtain some funding for your thesis/dissertation project. To apply for funding, you need to write your proposal with special care. This demands great patience from you because you must follow a specifically designated writing format.

Starting with chapter five, we will discuss some important items you need to cover in writing your proposal as well as your thesis/dissertation. For those who want to know more about proposal writing in general and thesis/dissertation process in particular, there are now plenty of reference books available. Check with the library staff or search by yourself using such key words as "proposal writing," "thesis guide," or "dissertation guide," and you will be able to find some of the references if your library has them. The problem with those "comprehensive" guides is that they tend to include too many things that the students actually do not need to spend time on. Those books will add a burden of reading on you yet may not speak much to the point, that is, the real frustrations and successes you will experience as a student. Since you can hardly

CHAPTER FOURTEEN

Working on Publication

Publication is the other term for knowledge dissemination. Science would be of little use if its research findings were known only to the researchers who have produced them. Publication is also a personal dream. Probably few researchers, especially those new degree holders who have just disengaged themselves from the thesis and/or dissertation process, have never thought about publishing their research results. More than that, publication has become the norm, or a normal part of the expectation on scholarly life. It has been even so stressed that no matter (indeed!) what kind and quality of research one is doing, "publish or perish" becomes the rule of the game. Since the thesis or dissertation is one of the most important research experiences in one's professional life, it deserves publication at least the author believes. In fact, thesis and dissertation research is well recognized as an important source of scientific information and understanding. In more utilitarian terms, a good start of publication record based on thesis or dissertation research gives one a jumpstart in her career.

Not everyone in this group feels successful in publication, however. In fact, not many thesis and dissertation research projects get fully published, or get published at all. The reasons may be various. In real terms, the research effort is primarily aimed at serving some immediate needs, i.e., to complete the degree, to satisfy the requirements of the funding source, etc. After the long term stress, it may be just too hard to initiate the publication process. Indeed, the completion of the project is often a great turning point in one's life: graduation, job search, adjustment to new life, hunting for new funding sources, so on and so forth. And time flies. When the researcher finally has the time and inclination to think about getting the thesis, dissertation, or some other kind of research report published,

versy but should raise the level of discussion.

It is significant that, despite decreased external support and increased internal attack, scientific sociology has taken significant steps forward. More sociological studies are concerned with developing and testing theory than was the case a decade ago. More sociologists are self-consciously involved in cumulative research programs, and there have been more empirical studies which confront alternative theories with one another. Rigorously formulated, empirically supported sociological theories are alive and well. And, as more students become aware of the challenges, the ranks of the producers of sociological knowledge will increase.

This book is addressed to graduate students who intend to be producers of sociological knowledge and serious undergraduates who most certainly will be consumers of sociology in business, government, education, and so on, in their roles as professionals or in their roles as citizens. Students need to be serious because the issues are complex and difficult and we cannot, even if we wanted to do so, water them down. Students who want their sociology to be intellectually challenging will find much to ponder.

This revised edition considerably expands the coverage of the book. New chapters examine: problems of quantitative measurement (chapter 9); what empirical research can and cannot do (chapter 13); research design (chapter 14); and cumulative research programs (chapters 15 and 16). The expanded discussion of cumulative programs (which was touched only briefly in chapter 12 of the first edition) represents one of the most important additions to the analysis of theory and method. Chapter 2 has a new section dealing with one issue in the Positivism/Anti-Positivism/Post-Positivism debate: whether the fact that observations are "theory-laden," that is, that presuppositions play a role in collecting and interpreting them, makes objectivity impossible. Finally, other minor modifications were made, including some new examples and a change in terminology in Chapter 6.

I again want to acknowledge the continuing support and valuable criticism of my wife and professional colleague, Elizabeth G. Cohen. Chapters 13 to 16 have benefited considerably from the comments and suggestions of Joseph Berger, Ronald Jepperson, Steven Silver, and Julie Rainwater. Dara Kerkorian assisted with the typing of the manuscript, and Anita Cohen–Williams prepared the index. Finally, many students who have worked with this book over the past six years have offered a number of valuable criticisms and recommendations. Many of the changes had their beginnings in questions from some of these students.

PREFACE TO THE FIRST EDITION

This book has been in the back of my mind for nearly twenty years and in the front for more than four. It began when I first taught methods. I felt that a research methods course was important, not only to future sociologists but also as part of the general education of students who would be consumers of research. Yet the way in which such a course was typically approached met the needs of neither type of student. In developing my course over the years, I tried to deal with issues that would be important both to producers of sociology and to consumers, and I resolved to develop a text that would be useful to both types of students.

Social scientists are often their own worst enemies. While many recognize the need for a sophisticated and sympathetic lay public, they do very little to develop such a public. To my mind, one prerequisite for the general education of the lay public is an overview of the sociological research enterprise with an emphasis on its objectives and on the serious intellectual problems that must be confronted. It makes no sense to teach a future lawyer the details of writing a questionnaire or the subtleties of experimental design. On the other hand, a vital part of the general education of the future lawyer or anyone else in society, for that matter, is the development of critical skills for the evaluation of the claims of sociological researchers. But learning these skills requires a deep understanding of some of the fundamental issues of scientific sociological research.

It also makes no sense to teach the future sociologist how to draw a stratified random sample, or any similar technique, before that future sociologist has an appreciation of why he or she needs to know a particular technique. The future producer of knowledge needs the same foundation as the consumer. Only after the future sociologist has developed skills for the critical evaluation of research can this person appreciate the need to build on this foundation and acquire the technical skills necessary to be a producer.

In large part, the teaching of research methods has not been addressed to developing a foundation based on understanding the enterprise and its intellectual challenges. In concentrating on the trees, students typically have no conception of the forest. While research experiences, which often form the core of methods courses, are quite successful teaching devices, the student rarely generalizes what is learned to other aspects of sociology. In part, that is due to the lack of a framework on which to hang those research experiences.

First and foremost then, this book aims to provide an overview of the sociological research enterprise. In dealing with the objectives and the problems inherent in achieving those objectives, it intends to provide a conceptualization that can be helpful to the student in all encounters with sociology.

Two other features of research methods courses motivate this book. The first is the traditional separation of theory from methods. Methods courses rarely mention theory and theory courses typically ignore research methods. The second feature is the traditional concern of methods courses with the procedures for doing a single isolated study. Most such courses focus on the techniques required to plan and execute one piece of research without even perfunctory reference to where that piece of research might fit into the ongoing development of sociological knowledge.

This book has as its fundamental premise that theory and method are inseparable. We believe that a central portion of sociological knowledge is theoretical knowledge and a central use of research methods is to develop and evaluate theory. An appreciation of the importance of method—either particular methods or method in general—depends upon an understanding of theory, how theory develops, and what is required to evaluate theory. An understanding and appreciation of theory, in turn, depends upon the knowledge of empirical methods, their strengths, and their limitations in producing and evaluating theory. Sociology cannot be wholly theoretical nor wholly empirical. Yet it is rare that the student has the opportunity to see how theoretical issues bear on method or how issues of method relate to theory. It is our aim to present a systematic discussion of issues of theory and method and how they relate to each other.

Recently, the problem of the cumulative development of knowledge has begun to receive proper attention from sociologists. While the concept of cumulation has its controversial aspects, there is, nevertheless, a recognition that cumulation is an important characteristic of scientific knowledge. But this recognition has not yet filtered into the curricula for theory courses or methods courses. The incompatibility of belief in cumulation and emphasis on the single isolated study has yet to be recognized. It is our belief that an overview of the sociological enterprise requires us to look beyond the problems and criteria of the single empirical study. Research is a process. It is carried out over time and by a collective. In this work, we have tried to emphasize the implications of viewing research as a collective enterprise through time.

The ideas developed here resulted from sources of influence too numerous to acknowledge. Yet I must express my indebtedness to some of the people who, in a variety of ways, have stimulated the development of the material presented in this volume. First and foremost, I am deeply in debt for many things, intellectual and nonintellectual, to my wife, Dr. Elizabeth G. Cohen, whose own cumulative research enterprise has raised intriguing issues of theory and application. I have also been favored by participating in research over the years with an exciting group of professional colleagues, among them Joseph Berger and Morris Zelditch, Jr. I especially want to thank the people who read and critically evaluated early drafts of this manuscript: Alex Inkeles, Sanford Dornbusch, Jonathan Reider, and Morris Zelditch, Jr.

The typing and preparation of this manuscript were shared by the late Lillian Lipsitz and Nora Schoenfeld. I am deeply grateful for their devotion to this difficult task.

Finally, I am grateful to many generations of students, both graduate and undergraduate, whose provocative questions and thoughtful critiques have contributed to making this formulation better in many ways, not the least of which has been to force me to adhere to my own injunction to communicate explicitly and precisely.

April 2, 1979

ONE

Introduction

This book discusses criteria and strategies for developing and evaluating sociological knowledge. Those who are, or intend to be, professional sociologists have definite reasons for pursuing these topics. While some may think that it is possible to do research guided by intuition alone, most sociologists believe that an understanding of standards and a sense of direction are prerequisites for competent sociological inquiry. They accept the general principle although they may disagree over particular criteria or particular strategies.

Sociologists, however, are not the only ones called on to evaluate sociological research, even though we may wish we were. At one time, the only consumers of a sociological study were other sociologists. Today, the average person frequently has to decide whether or not to believe some sociological claim. Sometimes as voters, taxpayers, and concerned citizens we have to act on matters to which these claims are relevant. Consider a few examples from one day's news. A radio commentator asserts that sociological research finds no evidence for any positive effects of school-based drug education programs; voters may accept this and vote to cut off funds for the school district's drug education project. In one newspaper story, a researcher claims that the divorce rate is much lower than people think and concludes that concerns for the disintegration of the family are overdrawn; the board of directors of a family service agency may use this to justify reducing predivorce counseling programs. Another newspaper reports a recent study which shows that "people are remarkably poor judges of the way they really impress others." According to the article, the researcher cited "the overriding norm of politeness as the major reason for this. People just do not tell you when you are making a fool of yourself; they simply let it pass in silence ... things go more smoothly that way." Some may place confidence in this research and stop worrying about the impressions they make on others.

The last report also poses problems for the professional sociologist because the reporter asserts that the study may shake the edifice of an important social psychological theory which holds that people construct a self-image from their judgment of the impressions they make on others. While not many adherents of the theory are likely to abandon it on the basis of a newspaper article, there are questions about the basis for reacting to the newspaper story and to the study itself.

Sociological claims increasingly impinge on decisions people make in their jobs and occupations; and many organizations, private as well as governmental, now commission studies to assist their decision makers. Market research uses sociological techniques to evaluate the potential for sales of products. Many companies do personnel and organizational studies to improve, for example, recruitment procedures or the flow of information. Lawyers use research to provide information for accepting or rejecting potential jurors. Public health agencies use sociological studies to determine how to disseminate information for disease prevention. While professional researchers conduct most of this research, someone in the organization has to define what information the organization wants and someone has to evaluate whether the results provide a trustworthy basis for action.

Those who produce sociological knowledge and those who consume it—either actively through commissioning research or more passively by way of media exposure—are called upon to evaluate sociological claims. Without some understanding of criteria, a person cannot put good questions for the research to answer and cannot evaluate the answers the research provides. But where does one obtain such understanding? One would assume that professional sociological training in research methods provides it. While methods courses and textbooks provide some basis for evaluating sociological claims, their focus tends to be very specific, sometimes highly technical, and at too concrete a level of analysis. They emphasize the *tactics* rather than the strategy of research.

The consumer does not need to evaluate the specific techniques for collecting the data but does need some basis for assessing whether the findings are reliable. Furthermore, the would-be professional researcher needs some general criteria in order to put the properties of particular techniques into their proper context. Expertise in the advantages and disadvantages of various question forms is wasted in the absence of an appreciation of the general problems of relating concepts to be observed and techniques for observing them. Emphasis on techniques rather than conceptual understanding of issues fosters the development of people who are specialists in a particular technique rather than people who are able to choose their techniques on the basis of the substance of the problem under investigation. In the worst cases, methods courses provide recipes to be followed uncritically rather than foundations for critical analysis and reasoned choices.

Courses in sociological theory sometimes address problems of evaluating sociological claims and sometimes also consider issues surrounding different strategies of investigation. When they do so, however, the approach tends to be highly abstract and not sufficiently technical. Learning various approaches to defining social facts is important, but it does not help the researcher in formulating a research problem nor does it alert the consumer to unsound interpretations of sets of social facts.

Our analysis points to a gap in meeting the needs of both consumers and producers of sociological claims. Theory courses offer a general orientation and a broad conceptual foundation for considering sociological claims; methods courses train people in specific techniques for collecting and analyzing data. Few books and even fewer courses deal with principles that guide choices in research, and it is rare to find more than perfunctory consideration of the standards that inform the evaluation of sociological ideas. Symptomatic of the problem is the almost total separation of theory and method in sociology. Although theory and method are integrally involved in developing and evaluating sociological knowledge, discussions of theory rarely consider the role of empirical research in the generation and testing of ideas, and discussions of method seldom get beyond the cliché that substantive considerations should guide empirical research.

To begin to fill this gap, this book emphasizes the intimate connection between issues of research methods and issues involved in the creation and use of theory. For example, one cannot consider how to measure a sociological phenomenon like racial discrimination without conceptualizing the ideas involved in that phenomenon; yet many writers treat the problem as simply one of thinking up questions for a questionnaire. Or, one cannot evaluate a set of theoretical ideas without thinking about how to observe the empirical implications of the ideas; yet sociologists debate about whether deviant behavior, for example, should be classified into four, five, or thirteen theoretical categories with no thought about the empirical import of the alternative schemes. Nearly forty years ago, Robert K. Merton wrote two essays, "The Bearing of Sociological Theory on Empirical Research" and "The Bearing of Empirical Research on Sociological Theory" (Merton, 1949). Merton recognized the interdependence of theory and empirical research, but many sociologists seem to have forgotten Merton's teaching and new generations of students need to be initiated to the ideas. This book intends to revive this general concern and to pursue the implications of interdependence for doing sociology and evaluating its products.

SOCIOLOGY AS SCIENCE

Although not all sociologists accept the position, this book treats sociology as a science. In discussing theory and method in sociology, in part we will be defining the requirements for a scientific sociology. But using the word *science* immediately involves us in a host of difficult issues. To many persons the word science is an emotional symbol. To some it is a very positive symbol: anything that can be called "science" is held in awe; if something is "scientific" it must be right. But a substantial number of persons have come to look on "science" as a negative symbol. With no more understanding than those who are in awe of science, the denigrators of science blame it for all of society's ills, from atomic bombs to traffic noise.

Although the detractors of science have increased numerically in the past decade, science is still a prestige symbol. Given this fact, it is not surprising that some people attempt to extend the term *science* to cover all knowledge and any kind of knowledge-producing system. While some may sincerely hold variant views of the nature of science, others are simply trying to wrap the mantle of science around their own activities and thus benefit from attaching the aura of science to any claims they want to make. Despite these efforts, there are differences between scientific claims and other kinds of knowledge claims; scientific knowledge is not coextensive with all human knowledge. Perhaps one reason for the increase in the number of critics who condemn all science has been the failure of the worshipers of science to recognize and acknowledge the limitations of scientific knowledge.

Along with the increasing number of critics of science in general has come an increasing number of doubters of the possibility of social science. There have always been significant critics who have argued that, while natural phenomena can be studied scientifically, human phenomena have not been, are not now being, and can never be studied scientifically. In part, the skepticism about scientific social science arises from the same mysterious aura that surrounds the natural sciences. In part, the skepticism arises from the false belief that all natural phenomena are amenable to scientific investigation and from the argument that, unless all human phenomena are similarly amenable, a scientific study of human phenomena is impossible. In part, the skepticism has an ideological base: there are those who believe that human behavior should not be studied scientifically, and this leads them to argue that it cannot be studied scientifically.

The question of whether human phenomena *should be* studied scientifically is very different from the question of whether such phenomena *can be* studied scientifically. The first is a value question, and people with different values will take different positions. The second is a factual question. One purpose of this book is to show what criteria apply to scientific studies and to argue that *some* human phenomena can be studied in ways that meet these criteria.

Sociologists are divided on the issue. Some serious scholars argue that sociology should not be a science, whether or not it can be.[1] Others believe that, while social science is possible, it must be based on a different model from the natural sciences. Even among those sociologists thoroughly committed to science, there are serious differences about what it means to be a science. While value arguments often come down to matters of faith, the current debate in sociology can certainly benefit from clarification of the issues. One of the objectives of this book is to clarify what scientific sociology entails.

Many students find intolerably boring the questions, Is sociology a science? Can sociology be a science? Should sociology be a science? To them the questions simply represent another one of those parochial, academic debates that are inflicted on students who take introductory courses. Students cannot be blamed for asking why the devil they should care about whether sociology should be or can be a science. Although some of us believe that students should care, our belief boils down to a value judgment, and debating the value question would only lead to each side claiming that its values are better, which is hardly a resolution.

---

1. See the list of suggested readings at the end of this chapter.

Many sociology courses often treat the science issue as something to be put out of the way quickly in order to move on to such interesting subjects as crime, delinquency, or social conflict. Elementary texts must make a gesture to sociology as a science, either a bow or a Bronx cheer, but their chapters on theory and method rarely contain more than symbolic gestures. Students get the message and quickly skim over such discussions. But that approach determines the student's attitude toward the "really interesting sociological subjects," whether or not the student or the teacher realizes it. In the extreme cases, the student uninterested in these questions misses the opportunity to acquire distinctive criteria by which to evaluate an interesting sociological idea, and the student either uncritically accepts or uncritically rejects what sociologists say about conflict, the family, stratification, and other social phenomena.

While it is not possible to argue that you, the reader, should care about these questions, it is possible to claim that you cannot avoid caring. As the examples at the beginning of this chapter implied, the results of social science touch so many avenues of modern life that it is virtually impossible not to take a stance toward these sciences, even if the stance is to totally dismiss the results. If, for example, you believe that the findings of a sociological study are not better than your personal opinions about the matter, you have implicitly taken a stance that that particular study is not scientific.[2] If you take the same attitude toward all sociological studies, then your stance is that sociology is not scientific. Furthermore, if you are committed to such an attitude for the foreseeable future, then you have taken the stance that sociology cannot be scientific. Finally, if you believe that sociology cannot be scientific but that sociologists should be encouraged to promote their opinions and value judgments, then your implicit stance is that sociology should not be a science.

These implicit stances have implications for your personal life, for how you do whatever job you will have after you leave school, for how you behave as a citizen in a complex society, and, if you intend to be a professional sociologist, for how you pursue sociology. Unless you plan to be a hermit, the claims and findings of sociological studies will constantly bombard you, if only in the columns of the daily newspaper. As you read these claims you will be called upon to judge them, and your judgments will certainly be colored by the attitude you have taken toward sociology as a science.

Would-be producers of sociology need to understand how to evaluate claims because to produce new sociological knowledge requires building on previous work in a selective and critical manner. Being a sophisticated consumer is prerequisite to making a useful contribution.

---

2. Of course, it is logically possible that you could regard the study as scientific but consider all science as no better than your personal opinion. Since you are unlikely to take such position in confronting the results of a physics experiment, it is highly improbable that you would regard all science as simply opinion, and highly probable that you implicitly distinguish either the particular study or social science in general from "science."

she may feel that things have become somewhat remote or even out-of-date. The more the time elapsed, the harder to make a further commitment, especially when the topic no longer serves as one's focal interest.

There are people, however, who have kept publishing since they first obtained their research results, or shortly after they finished their degree or some other requirements. The pressure of time and other life demands do not seem sufficient to account for the failure of people in publishing their own research results. If they are given the chance to do it again, probably they will do much better. In many cases, maybe they just have to start early and do it. Maybe they have to keep trying, which, of course, means a firm commitment and further investment of personal resources including time, energy, and money. Probably a major factor in causing the hesitation and indecision is that too much uncertainty is involved in the pursuit of publication. Specifically, who knows whether the research project and its results are publishable or not?

What is publishable

To many students, publication is somehow within sight but beyond reach. For the honors and Master's students especially, publishing something based on their thesis research would be such a rare success. For the doctoral students, particularly those whose coursework constitutes a substantial part of the requirements for their degrees, their dissertations are not necessarily considered suitable for publication. Only for those who meet their degree requirements solely or mainly by research, which is the case for so-called research students in some British universities, publishability would be taken as a criterion of the acceptability of their work. Even so, most of the doctoral theses remain unpublished or inadequately published/disseminated.

In real terms, far more theses, dissertations, and other research reports are publishable than are deemed to be. Ask any person who has been publishing to compare those serious but unpublished works with some of the published "junk articles." For the coverage of new issues and the depth of inquiry, including the effort in a systematic literature review, the former often carry the weight that warrants publication or being treated as actual publications (if not in terms of the range of circulation). Few accomplishments would have so far-reaching impact as a thesis, dissertation, or funded major research project on academic and professional life. After many years investigators may still highly appreciate

One major purpose of this book is to provide criteria that producers and consumers can use to judge sociological claims, whether they occur in the mass media or in technical journals. The criteria we will examine arise from a scientific frame of reference; hence, an understanding of them involves an understanding of the principal features of science in general. Showing that sociological studies measure up to these criteria answers the question of whether some sociology, at least, can be scientific. An appreciation of these criteria can also enlighten us as to those realms of human phenomena which may in the future be amenable to scientific treatment and those which are not likely to be amenable. Although these criteria may be largely irrelevant to people who are firmly committed to the value position that sociology should not be a science, understanding the criteria may have some impact on the value judgments. There are some who believe that sociology cannot be a science and draw the false inference that, because of this, it should not try to be a science. The inference is false because there are no logical ways to draw a value conclusion from a factual premise. Nevertheless, the premise may be the psychological basis for holding the value position. Showing the falsity of the premise (i.e., showing that sociology can be scientific) can undermine the psychological basis, although not the moral basis, for believing that sociology should not be a science.

This book intends to give a clear affirmative answer to the question, Can sociology be scientific? In so doing it may produce some value change, but that is not its intention. For example, a demonstration that some kinds of sociological research are fully consistent with the criteria of scientific research does not deal with the position that the "more important" problems are those which cannot be studied scientifically. Since all science has limitations, it will always be possible to argue that the really important values concern phenomena that are beyond the limits of science. What the really important phenomena are will always be matters of faith; but knowing when we must go on faith and when we can use other criteria for judgment is an important step in the process of clarification.

People have had opinions about human phenomena since the beginning of language. We know that the earliest written documents deal with describing and explaining human behavior. Since most of us are human, it is only natural for us to claim expertise in understanding what humans do, how they do it, why they do it, and what they are likely to do in the future. Ask someone a question about some human phenomenon, and you almost always will get an answer. Furthermore, there is a tacit agreement that everyone is entitled to his opinion.

But the word *opinion* is ambiguous. Some opinions are preferences; so, George's opinion that licorice ice cream is better than chocolate is indeed something he is entitled to, even if the thought repels me. On the other hand, opinion also means a belief about a matter of fact; hence, George may or may not be entitled to the opinion that prisons rehabilitate criminals. Those of us who believe that human behavior can be studied scientifically are concerned with how we decide that George's opinion about prisons is, or is not, correct. We believe there is a better way than merely confronting one personal opinion with another. Furthermore, we believe that there can be a resolution of differences of opinion by means of agreed-upon procedures that would be acceptable to people of widely different value positions. For example, it is possible to develop procedures to resolve the question of whether prisons rehabilitate criminals, whereby the resolution would be acceptable both to people who believe that prisons should rehabilitate criminals and to people who believe that prisons should punish criminals.

The concern of a science of human behavior, then, is to resolve differences of opinion about matters of fact through the use of procedures whose validity does not depend upon the prejudices of the user. Not all disputes about factual matters can presently be resolved. Indeed, there are probably some factual issues which are unlikely to be amenable to collective resolution even in the far-distant future. The important point is that there are some disputes that can be resolved right now and many more that will be resolvable as our knowledge expands and our procedures become more refined.

In the first instance, then, we can look upon science as a set of agreed-upon procedures for collectively resolving differences of opinion about matters of fact, and we can look upon a scientific sociology as a set of agreed-upon procedures for collectively resolving differences of opinion about matters of fact in the area of human social behavior. The objective of this book, then, is to spell out these procedures and the criteria they represent, and to indicate in general terms how they are used and how they are improved. An understanding of these procedures should demonstrate that in many areas there is a better way to resolve differences than simply entitling people to their personal opinion.

An Example Where Differences Need to Be Resolved

One of the interesting phenomena of the past few years centers around the activities of the women's movement for equal rights. The rhetoric of the proponents and opponents of equal rights for women is filled with claims about the causes and consequences of discrimination against women. Indeed, there are even arguments about the "fact" that there is discrimination against women. For example, one can compile an impressive set of statistics which show that employed women earn considerably less than men in comparable occupations. Those who argue discrimination point to these statistics as definitive evidence; but those who argue that there is no discrimination claim that the statistics merely reflect the choices of women to move in and out of the labor market (instead of remaining continuously in the labor market to develop a career and obtain the salary increases that come with career advancement). Some even argue that women are more content in lower salaried positions; that is, striving may produce higher salaries, but at the sacrifice of other important values like contentment. Even this simple example, however, has two different facets: one is the question of the "fact" of discrimination and the second concerns an explanation of the possible causes of discrimination. To resolve the question of fact, we must agree on a definition of *discrimination*. But since our definition has implications for action, it is not a simple matter to formulate such a definition without considering the issue of causes.

Strong advocates of women's rights are unlikely to accept any definition of discrimination that results in the claim that women are not discriminated against. Male chauvinists, on the other hand, are not likely to accept any definition that would result in the claim that women are discriminated against. Isn't the situation therefore hopeless? Won't opinions about this "fact" always depend upon what axe is being ground, or whose ox is being gored? Is it not inevitable that the only resolution is to allow the male chauvinists and militant feminists to be entitled to their own opinions? Are we not forced to the position that this conflict will always be resolved by the side that has the most power? The very history of developments of the women's movement suggests reasons for optimism. Even the strongest male chauvinist is willing to admit that statistics do show some salary discrimination against women, although he may believe that that is the way it ought to be. And even the hardiest feminist is willing to grant that not all of the salary differentials in comparable occupations indicate active conspiring to keep women's salaries down. In short, there are potentials for agreement on some matters of fact, even among groups whose interests are as opposed as those of the male chauvinist and the militant feminist.

In principle, the fact that a set of statistics, such as the average salaries for men and women in various occupations, can provide some basis for agreement is a hopeful sign. In principle, at least, it is possible to totally dismiss these statistics as inaccurate, irrelevant, or distorted. Yet the procedures by which the data are gathered, analyzed, and presented provide a basis for evaluating them that does not depend upon group interests, prejudices, or values. Despite the saying, "There are lies, damned lies, and statistics," the gathering of data by a government agency, utilizing elaborate procedures which themselves must meet rigorous technical standards, promotes collective trust of the "facts."

While it is possible to obtain collective agreement on the "facts," it is more difficult to collectively agree on explanations or interpretations of whether a given fact is relevant to the issue under consideration. If, for example, occupational discrimination were a by-product of other kinds of sex discrimination, it might be possible to claim that the salary statistics for men and women were only incidentally relevant to the question of sex discrimination.

Sometimes, deciding whether or not a given set of facts are relevant requires an understanding of the social processes which led to those facts. Suppose, for example, the male chauvinist argued that lower salaries for women were not due to any male discrimination, but were the result of women being their own worst enemies. Women, our chauvinist argues, do not strive for high salaries because, from earliest childhood, a girl who strives to achieve is punished by other women. With a little imagination, it is possible to explain away our salary statistics, that is, to construct an argument that claims that these statistics result from processes totally unconnected with acts of sex discrimination. It is also possible to construct an explanation which places these data directly in the center of sex discrimination.

It may be true that women are prejudiced against women, but the link connecting that prejudice to the salary differentials between men and women has yet to be established. To put the matter directly, an apparently simple issue of whether salary statistics are relevant to the question of sex discrimination depends upon a theory of occupational socialization which links events in childhood and adolescence to occupational outcomes for adults. It is no wonder that these matters are highly controversial, since the theories that we have available are only in a primitive state and cannot provide the definitive linkages necessary for collective decisions.

The general problem is one of choosing between these explanations. What standards must an explanation meet before it is preferred to alternatives? This is one of the most crucial questions that an analysis of theory and method in sociology, or in any other science, must address.

Choosing among Alternative Claims through Dispassionate Analysis

Advocates of a position like our male chauvinist and our militant feminist often evaluate explanations according to how well they agree with their preconceived notions—those that fit must be true, those that don't are obviously false—or how useful they are to advancing their cause. In general, evaluating research results or theoretical principles on the basis of what we already believe is not good practice; such a strategy protects us from ever being wrong and hence prevents us from ever learning anything.

Sometimes we feel so passionately about something that any question about it represents an attack on our fundamental being. We react by defending against the question, rather than by analyzing and attempting to answer it. Sociological claims frequently deal with issues that rouse passions. Is it possible to analyze these claims at all? Is it possible to evaluate them independently of passionately held preconceptions? Some ideologies deny the possibility of dispassionate analysis in areas of strongly held values. Such a position, however, can become a self-fulfilling prophecy; if you hold such a view, then you never attempt to analyze issues about which you feel strongly. Thus you rule out the possibility of discovering that you are wrong, that is, discovering that you *can* analyze issues tied to important values and can analyze them dispassionately. An approach that rules out the possibility of being wrong is remarkably convenient for some ideologies.

Dispassionate analysis is difficult. The capacity for dispassionate analysis is not an inborn skill. Rather, the techniques must be learned and other passions must be cultivated: the passion for truth, the passion for intellectual honesty, and the passion for avoiding self-delusion (and collective delusion). Like other learned techniques, these skills benefit from practice; we need to master basic principles but, more important, we need to apply them continually. In this book, we will develop principles and we will also apply them. Nothing is off-limits to analysis, not even the analytic principles themselves. Asking searching questions promotes the development of techniques, although we must recognize that not all questions have a ready answer.

The first principle for analyzing sociological claims is:

Analysis requires criteria of evaluation that are independent of the content of the claim.

The reason that newspapers often describe the sampling procedures used in their polls is that the truth of the claim depends in some sense on the sample being "fair" or "unbiased." Considerable effort has been devoted to developing procedures for obtaining "fair" samples, and these procedures are applicable *regardless of the content of the study.* Hence, when we ask how well the poll conformed to the procedures for obtaining "fair" samples, we are applying a criterion that is independent of the content of the claim. Someone who is not an expert in sampling may not be able to judge whether the particular procedure used is sufficient to obtain a "fair" sample but, once the importance of "fair" samples as a criterion is established, the lay person can judge whether or not the poll taker has shown concern for meeting the criterion.

While we assert the necessity for *some* criteria that are independent of the content of the claim, we do not require all analytic standards to be independent. Rules of logic and statistical procedures assist in the evaluation of evidence for a claim, but criteria for assessing the importance of a particular set of evidence will usually incorporate aspects of the content of the claim. In other words, the dispassionate analysis we describe does not require the total objectivity of a completely detached and disinterested outsider; if such objectivity were possible, it would probably not be very helpful, for such detachment and disinterest usually accompany total ignorance of the matter at hand. What is essential is the willingness to ask searching questions and to change one's position when the answers are compelling. Furthermore, many people mistakenly equate dispassionate analysis with coldness, aloofness, insensitivity and, in the extreme, inhumanity. The analysis we advocate is not passionless but involves many passions; the passions for truth and understanding can operate in harmony with passionately held ideas.

Another Example

Explaining male-female salary differentials provides an excellent case for developing our ability to analyze a social phenomenon dispassionately. Most of us have feelings about the issue even though they may not be as strong as those of the male chauvinist or the militant feminist. Can we put these feelings aside temporarily and analyze what one social psychological study may contribute to our understanding?

Several years ago, Phillip Goldberg did an experiment with students at an all-women's college (Goldberg, 1968). He chose six professional fields: law, city planning, elementary school teaching, dietetics, linguistics, and art history. From each of these fields he took one article from the professional literature. The articles were edited and put together in two sets of booklets. Each article bore a male name as author half of the time and had a female name attached the other half. If the first article bore the name John T. McKay in the first set, the same article in the second set would appear under the name Joan T. McKay. Each booklet contained three "male-author" articles and three "female-author" articles.

The subjects were asked to evaluate each article using nine rating questions. Goldberg found that on all nine questions, "regardless of the author's occupational field, the girls consistently found an article more valuable—and its author more competent—when the article bore a male name" (1968, p.30).

At first sight, this study should make our male chauvinist happy. It seems to confirm his view that women are their own worst enemies; it lends credence to his belief that women's views of themselves and other women are responsible for salary differentials between males and females. And the first reaction of our militant feminist would probably be to dismiss the study as meaningless and irrelevant.

But wait a minute. If we closely examine Goldberg's study, we might find that it was a well-done study that followed all the rules of good scientific procedure. In so doing, we need not join the ranks of our male chauvinist. First of all, notice that the study used only female subjects. Quite conceivably, if a comparable group of male subjects was used, the differential evaluation of "male" and "female" authors would have been even stronger in favor of the males. In other words, males might have "put down" female authors even more than the women students did. In short, the study is not sufficient to answer the question; it may, however, be a useful step that would eventually provide a satisfactory way to explain salary differentials so that the militant feminist's dismissal is at least premature.

It requires a huge inferential leap to use this study to support the opinion that women's prejudice against women causes male-female salary differentials. In casual conversation, however, we often take such "facts" and build an argument to support our personal opinion. Casual acquaintance with the results of social research, then, can be worse than no acquaintance at all, another instance illustrating the old saying that a little knowledge is a dangerous thing. The purpose of developing criteria for evaluating research is to avoid such misuse of social science findings and to provide a set of standards for both the understanding and the appropriate use of sociological theory and research.

This book does not intend to present a theory that explains salary differentials between men and women. It does intend to examine the properties that such a theory should have, to consider general approaches to constructing theories with these properties, and to discuss questions that a consumer must address in dealing with the meaning and significance of data such as our salary statistics. In treating these issues, we will consider the question of what constitutes a scientific explanation and the criteria by which one chooses one explanation over its alternatives. It is possible, for example, to use a theory to explain the results of Goldberg's study, and we will show that such a theory has the virtue of linking this apparently isolated experiment with other research dealing with problems far removed from women's evaluations of men and women. Why this is a virtue will become clear later on. For the present, however, we simply note that agreed-upon collective evaluations depend upon a much broader context than is taken into account in any single study.

The discussion so far has implicitly assumed that people want to evaluate their opinions and that they are eager to discard one position in favor of a better view. In this day and age, that cannot be taken for granted. Many people are very comfortable with their opinions or prejudices; they have little desire to examine them closely, much less change them. For some, the important aspect of an opinion is whether it wins out. Some male chauvinists and some militant feminists are more concerned with putting forward opinions that advance their cause than with attempting to solve a problem. Such people reverse the old saying so that it becomes: It's not how you play the game that counts, but whether you win or lose.

While we do not understand all the reasons an individual may have for clinging to an opinion (reasons of security, reasons of power, and so on), we do know that examination of one's own opinions and beliefs is the exception rather than the rule. Indeed, sociology frequently asks students to analyze their own opinions and to confront these opinions with alternatives, and this may account for a good deal of the resistance that sociology teachers get from their students. But if the student is going to understand sociology, and if sociology is going to help understand social phenomena, questioning and analysis are required. One cannot question everything; there are some points, even in a highly developed science, where matters must be taken on faith. The danger occurs when everything is taken on faith. Even if an opinion is comfortable, is useful in winning arguments, and contributes to the power of the believer, it could represent the ''kidding oneself'' syndrome. This syndrome affects not only individuals but whole societies, as the example of the ''light at the end of the tunnel'' during the Vietnam War illustrates.

When one has to make important decisions, one wants to examine and evaluate the alternatives and avoid kidding oneself. The undergraduate student who is absolutely convinced that medicine is the only career for him or her, and refuses to examine any other options, sometimes wakes up as a very unhappy medical student. The student who has avoided taking a course that has a bad reputation among his friends sometimes discovers (when forced by requirements to take the course) that the course is pretty good. By testing the friends' opinions, he or she might find that they use different criteria for evaluating courses. In short, we are suggesting that both big and little decisions are better decisions when based on knowledge, and that knowledge requires the testing of opinions.

These views seem like common sense. However, not only is common sense quite uncommon, but the view that analysis and knowledge are necessary in decision making has recently had strong opposition. The opposition takes many forms: for example, ''It is better to feel than think'' and ''One should rely on one's deep-seated intuitions rather than objective facts.'' What Charles Reich (1970) called Consciousness Three, a revolutionary level of consciousness, is nothing more than the old worshiping of the irrational. Indeed, Marvin Harris (1978) likens Consciousness Three to the set of attitudes that sustained witchcraft. The partisans of the irrational, who trust their own unanalyzed opinions in preference to critical analysis and external evidence, will find little to interest them in this book. Those who are concerned with analyzing and evaluating opinions, whether sociological opinions or opinions about everyday life, should find useful guides in our examination of ways to analyze sociological opinions.

SOME BASIC ASSUMPTIONS

In analyzing and evaluating sociological claims, the first thing to recognize is that some claims are more than mere opinions. Our language distinguishes between knowledge and opinion, and we recognize intuitively that what we call knowledge rests on firmer foundations than what we call opinion. To be sure, some sociological claims are opinions, perhaps even expert opinions, but what we regard as sociological knowledge provides firmer bases for belief than even expert opinion. In order to distinguish among knowledge, expert opinion, personal opinion, and prejudices, we require some understanding of the foundations that support what we call knowledge. These foundations represent a series of tests which can be applied to sociological beliefs. Those beliefs which pass the more difficult of these tests come closer to knowledge. When we talk about sociological knowledge, we have in mind application of the criteria for scientific knowledge. Although some sociologists believe that there can be sociological knowledge that is not scientific knowledge (some regard empathetic understanding as sociological knowledge), we will focus on the foundations of, and criteria for, scientific sociological knowledge.

The most basic assumption of this book is that *scientific knowledge is theoretical knowledge, and the purpose of methods in science is to enable us to choose among alternative theories.* What constitutes theoretical knowledge and what methods are relevant to evaluate alternative theories comprise the focus of the remaining chapters; hence, it is premature to offer simple definitions of theory and method at this point. Nevertheless, certain very important consequences follow from this first assumption.

Scientific sociological knowledge is more than a collection of facts, more than a body of shared opinions or prejudices, more than a perspective for viewing the world, and more than conventional wisdom. Although some would argue that scientific knowledge is a body of facts, the position will not stand up to close analysis. A telephone directory incorporates a body of facts, and few would confuse telephone company publications with scientific treatises. The ''body of facts'' position, then, would include things that most people normally exclude from science. Although few sociologists would endorse this position, especially when we put it in its bald form, many sociologists behave as if the job of sociology was simply to gather more and more facts.

As we have already noted in examining income statistics of males and females, deciding what is a fact is not a simple matter. Our assumption points to a fundamental distinction between facts and theoretical knowledge. It suggests that, although facts are essential for evaluating theories, collections of facts may or may not lead to theoretical knowledge.

Considerable controversy focuses on what we regard as a fundamental distinction between facts and theoretical knowledge. Some readers may be familiar with the debates over Positivism and Post-Positivism; in part these debates concern what is known as the "theory-ladenness of facts," one view of which denies the importance of our distinction. We will analyze the concept of theory-ladenness in chapter 2.

While we claim that sociological knowledge must be shared and collective, it is easy to see that not every shared, collective belief belongs to the body of scientific knowledge. After all, superstitions are shared and collective, but we recognize that scientific knowledge is more than superstition. Furthermore, the question of who does the sharing becomes a crucial issue, particularly in the social sciences. Most of us are willing to grant the existence of physical science knowledge, even when we do not share that knowledge; but, when it comes to sociological knowledge, many feel that unless they know and share the belief it cannot possibly be called knowledge. Humans find it hard to accept that there can be knowledge about human behavior that is esoteric and inaccessible to them. While we insist that a belief incorporated in sociological knowledge must be collectively held, we reject the idea that the collective must encompass all of the human race.

Much has been written about the sociological perspective as a unique way of viewing the world. The argument maintains that sociologists ask different questions about phenomena and that sociological knowledge consists of an orientation that generates unusual questions. The perspective that suggests that a person's self-conception arises from what other people think of him does raise different questions than the perspective that regards a person's self-conception as being a result of the abilities and traits that are inborn.

While sociology has a distinctive perspective and does raise different questions about phenomena, a perspective by itself cannot be called scientific knowledge. Perspectives may generate both useful and useless questions, both answerable and unanswerable questions, both trivial and significant questions. One sociologist, working in a medical setting, asked a series of questions that his physician colleagues would never have asked, and he was able to show that the incidence of a particular disease correlated highly with social class. But his physician colleagues were unimpressed. As one of them put it, "I can't change the patient's social class." While this physician may have been narrow-minded in not seeing the usefulness of tracing out the implications of social-class background of the particular illness, it does illustrate our point: a perspective is not self-justifying; it is the responsibility of the sociologist to demonstrate both the consequences of the perspective and the usefulness of these consequences.

Conventional wisdom, folk knowledge, and folklore characterize every social group. Indeed, much of our everyday life depends upon employing folk knowledge. Students who come to a sociology class expect, on the basis of folk knowledge, that the teacher will speak to them in their own language. Literally thousands of such well-founded beliefs guiding everyday actions could be collected, organized, and published. Indeed, such a collection might describe a good deal of human behavior. But at best, such a collection would represent compendia of historical fact. The conventional wisdom does not tell us when it does not apply, when it changes, or how it changes; only by assuming that the conventional wisdom of the historical present is eternal can we have useful guides for action. Aspects of conventional wisdom may enter into scientific sociological knowledge—and this may be the basis on which some critics attack sociology as belaboring the obvious—but when it does so, it enters in highly qualified ways, and the qualifications are as important as the folk knowledge itself.

To illustrate the importance of qualifying folk knowledge, consider two examples. "Familiarity breeds contempt" is a widely held maxim; but there is also the folk belief "To know him is to love him." These maxims undoubtedly express useful insights. We have already said that there are important grains of truth in conventional wisdom, or folk knowledge, but these two maxims cannot both be true for the same situation at the same time. To transform these maxims into useful knowledge requires qualifying each maxim so that such qualifications tell us something about the circumstances is which each maxim is likely to apply. To qualify the maxims is to say something about the conditions under which each is likely to be true, or to further specify the properties of the terms in the maxim—as, for example, *how much* familiarity is required to breed contempt. Without such qualifications it is always possible to explain away, after the fact, the circumstances in which the maxim fails. The ability to explain away guarantees that the maxim cannot be wrong; but, if after the fact every failure becomes an exception to the rule, the rule is not a useful guide to action. Before the fact, we do not know if the rule or the "exception" applies.

The second assumption of this book is already contained in the first, but making it explicit emphasizes its importance. We assume that *theoretical knowledge does not consist solely of facts and cannot be generated simply by collecting facts.* It is particularly important for the serious student of sociology to recognize this assumption, because early exposure to sociology seems to suggest that the opposite principle guides sociological research. In many courses, the emphasis is on reading sociological studies and mastering their findings; but the findings of any study, or the collection of findings from a series of studies, typically is a set of facts. We could have fifty studies showing income differentials between employed males and employed females and still not have any theoretical knowledge. The fifty studies might give us confidence that we have a set of historical facts, but we would still be in the dark about what these facts mean. We call these facts historical because the findings of our fifty studies characterize the state of affairs at par-

their work during those periods. It may have served as an important launch pad for their entire careers afterwards.

The problem is, publication of a work is a process different from its creation. It is like the marketing of any goods or service, whose success depends heavily on the means and skills for competition. In reality, intellectual products are also dictated by the demand-supply relationship, and not simply judged in terms of absolute academic value (if judgable at all). Therefore, contrary to the popular view, any rigorously trained student researcher may publish her scientific work that contributes to human knowledge and understanding in a particular field. As a student researcher, however, you must know the "ropes" of getting the work published.

First, from the very beginning of your research, you need to bear in mind that the thesis/dissertation or a funded project is among the best opportunities for deriving some important publications. You should plan for these publications even before you start writing the thesis/dissertation or the research report. Think about some potential publishers and/or journals for communicating your work. Although you are dictated by the school rules or agency guidelines that may be quite different from those of the publishers and journals, it makes a big difference in writing whether or not you have your goals and publication requirements in your mind. If you want your work to be published in a book, you must know the difference between a book and an ordinary thesis/dissertation or research report. Directing yourself toward that goal whenever possible in writing will save a lot of revision later.

Secondly, you need to understand the essence of writing for publication. Writing your report as a book means you are aiming at a higher standard of academic achievement than the ordinary. In that sense, not every thesis, dissertation, or research report can become a book. Successful works tend to deal with topics of greater importance. Nothing is perfect, but your work must show some significant advancement in the understanding of your field. You should, however, anticipate the difficulty and ensure the researchability within the limits of your time and other resources. If writing a book is not feasible, then deriving some journal articles may be the consideration.

Thirdly, you should be prepared for revision or rewriting. In preparing and submitting a thesis, dissertation, or a research report, you are a person who is being supervised, directed, and evaluated. In contributing for publication, you must present yourself, the author, as an independent, knowledgeable, or even authoritative researcher. You will also be evaluated, not as a student but as an

ticular times and in particular places. Unless we interpret these facts, we really have no basis for determining whether a study done ten years from now would show the same facts or, if not, what the direction of change would be.

To state the position as strongly as possible, we argue that observations of social facts, no matter how carefully gathered or rigorously analyzed, do not add up to theoretical knowledge. While many sociologists have adopted the strategy of fact gathering, and many aspects of the structure of the discipline of sociology promote this strategy, we contend that piling study upon study will not generate useful theoretical knowledge, even if each study exemplifies the ideals for conducting empirical research.

We emphasize this assumption to contrast the position of this book with the position known as empiricism. Empiricism emphasizes the importance of observation and of creating knowledge by amassing observations and generalizing from these observations. Later on, we will analyze the difficulties with the strategy of empiricism; for the present, it is sufficient to point out that this book begins from a different set of assumptions from those of the empiricists. Sociology can be *empirical,* that is, can use observations to evaluate knowledge claims, without being empiricist. The reader should be aware, however, that empiricism has many adherents among sociologists. Some sociologists define theory in empiricist terms. For example, Spencer (1976) defines a theory as "a generalization based on observed facts that explains a supposed causal relationship between those facts." Such a definition is wholly consistent with the empiricist tradition, but is quite inconsistent with the approach of this book. To anticipate later chapters, we argue that observed facts are neither necessary nor sufficient for a sociological theory. Furthermore, generalization as an activity properly applies to theories, not to facts. A convincing argument for these claims depends upon a careful analysis of theory, of the role of theory in research, of facts, and of the role of facts in formulating and evaluating theories. While the empiricist position is seriously and sincerely held within sociology, we believe that it has fundamental difficulties and that the position developed in this book presents a viable alternative.

We should also emphasize that just as we reject empiricism we also reject what Frederick Crews has called "theoreticism," which he describes as "frank recourse to unsubstantiated theory, not just as a tool of investigation, but as anti-empirical knowledge in its own right." Crews sees theoreticism as a prominent feature of today's intellectual climate in contrast to 1960 when nearly everyone

> would have concurred with the American critic R. S. Crane's observation that one of the most important marks of the good scholar is "a habitual distrust of the a priori; that is to say, of all ways of arriving at particular conclusions which assume the relevance and authority, prior to the concrete evidence, of theoretical doctrines or other general propositions." (Crews, 1987, p. 37)

We believe that Crane's position is no less valid today than in 1960; while we stress the centrality of theory, theory that is impervious to evidence does not meet the requirements of scientific sociological theory.

The third basic assumption of this book is that *sociology students, both under-graduate and graduate, typically lack a clear conception of scientific sociological theory and need such a conception.*

There are many reasons why undergraduate sociology courses do not provide a systematic and well-developed conception of scientific sociological theory. In the first place, there are many sociological traditions which do not emphasize the development and evaluation of theory. The empiricists, although they are numerically large, represent only one of these traditions. Some sociologists, for example, regard a perspective as equivalent to a theory. Hence, in introductory theory courses, one learns about Conflict Theory, Structural-Functional Theory, and Symbolic Interactionist Theory. Without question, each of these schools of sociology asks distinctive and important questions. But these courses rarely ask what makes each of these perspectives a theory. What conception of theory allows us to treat each as a theory? What are the consequences of such a conception of theory? How do we evaluate those consequences? Again to anticipate, the conception of theory that allows us to call one of these perspectives a theory turns out to be too broad to be useful, and it is not consistent with the conception of theory held in other sciences.

Another reason why undergraduate courses rarely present a clear conception of sociological theory arises from the fact that presenting such a conception involves close analysis of difficult and sometimes technical issues. Some instructors believe, even when they share this author's perspective, that a deep understanding of theory is necessary only for those who would produce sociological knowledge. Since few undergraduate students intend to become professional sociologists, these instructors argue that complex technical issues are better saved for advanced graduate courses. To be sure, the analysis necessary for understanding scientific theory and the role that theory should play in sociological research is more demanding than memorizing the findings of some studies or remembering what Durkheim said about social facts. But the issues are not so complex or technical that they foreclose elementary treatment, providing the student understands the reason for the effort.

The Needs of Would-Be Sociologists

Since we assume that sociological knowledge is primarily theoretical knowledge, we believe that would-be producers of sociology must understand scientific sociological theory in order to contribute effectively to the growth of knowledge. Choosing a research problem, selecting a method (survey, historical, participant observation, experiment, etc.), planning and executing an investigation, making decisions about the relevance or irrelevance of data, picking data analysis techniques, discriminating among various interpretations, and writing up results all are informed by available sociological ideas. Highly developed theory provides sound guidelines for conducting and evaluating sociological research. One major purpose of this book is to explicate how scientific sociological theory can provide these guidelines.

In the absence of highly developed theory, the commitment to pursue such theory itself provides direction for the enterprise. An appreciation of what this objective entails assigns high priority to some types of work and rules out others. The researcher pursuing scientific theory will emphasize attempts to explain some set of social facts and will not engage in collecting unrelated facts. Analyzing issues of theory and method, as we will do in this book, prepares the would-be producer to develop research priorities on a sound foundation, one that recognizes both opportunities and limitations.

The researcher oriented to applying sociological knowledge or solving practical social problems also needs to understand the nature of theory and the process of theory building. Successful applications are more likely when ideas are well-formulated and have empirical support, but the researcher must decide when an idea is sufficiently well-formulated and what is adequate empirical support. A perennial problem for applied work concerns when it is legitimate to apply ideas that have been tested in one setting to a totally different setting. Related to this is the whole topic of how one generalizes from one successful problem solution to another. These matters have close ties to the development of theory so that an understanding of the nature of theory can inform the applied researcher in dealing with them.

We also believe that an understanding of theory and of the relation of theory to method will enable researchers to make better use of their training in research methods. Although it is a cliché, it is nevertheless a fundamental truth that any research method entails assumptions about the phenomenon under investigation. Some sociologists, however, ignore this truism and employ methods that are totally inconsistent with their substantive assumptions. Our analyses of issues of theory and method should sensitize would-be researchers to the dangers of uncritical use of a method or technique and they should be less prone to this type of error.

Every researcher approaching a project faces a large number of critical choices. New researchers—especially students doing their first study—are frequently overwhelmed by the number of important choices they must make because they do not have any basis for deciding or even for assigning importance to the needed decisions. They do not understand the reasons for choosing one alternative over another and they lack the conceptual tools for analyzing the set of decisions. How does one recognize a researchable question without any guiding principles? Why should someone worry about possible biases in the research if the issue of bias has never been considered previously? When does a researcher decide that a conceptualization of the problem is sufficient to proceed with the investigation in the absence of any criteria for evaluating concepts? What does an investigator do to insure that a study will be informative to the set of ideas and questions that generated the project? Who can the new researcher turn to for constructive criticisms, and what is the basis for accepting or rejecting criticism? Student researchers, for the most part, can lean on their faculty advisers in dealing with research decisions and sometimes the experience of working with a mentor engenders transferable models usable when the student moves on to independent work. In many instances, however, choices and decisions occur without any understanding on the student's part of the rationale behind them. While this book will not present recipes for resolving the choice dilemmas of the beginning researcher, it will provide some general principles that can aid both the novice and the experienced researcher in thinking through research decisions.

The Needs of Consumers

While a clear conception of sociological theory is essential to students who would be producers of knowledge, this author considers a clear conception no less essential for students who, as members of society, will be forced into the role of sociological-knowledge consumers. Everyone, directly or indirectly, is a consumer of socio-logical knowledge. And in these days of consumerism, can we doubt that appropriate public consumption requires a high level of consumer sophistication?

One of two attitudes characterizes most consumers today. On the one hand, many believe that if a policy decision is based on sociological research, that policy decision is justified. On the other hand, many regard sociological research as so inadequate or so biased that it cannot inform any policy decision. The changing fortunes of public-opinion polling among politicians illustrate both attitudes. A few years ago, the trend toward increased use of polling in political campaigns was clear and apparently irreversible. The poll was becoming enshrined as an indispensable campaign tool, and there were many candidates who would not make a move without first taking a poll. Then the trend was reversed, with more and more politicians questioning the accuracy and utility of political polling. Some regarded polls as dangerous, and one legislature even launched an investigation of political polling with a view to limiting its use by law. Then the pendulum swung back: in a recent election, one presidential candidate's pollster was perhaps the most prominent figure in his campaign. Unfortunately, both supporters and detractors of polling have lost sight of the central issue: Polls in themselves are neither useful nor useless, neither benign nor dangerous; only the objectives of polls and the uses to which their results are put can be evaluated along these dimensions. The same poll may be useful for one objective and useless for another, accurate for one purpose and misleading for another, dangerous when used in one way and benign when used in another.

Attitudes which sanctify sociological research and those which damn it are part of a larger syndrome of lay attitudes toward science. We might say that these attitudes relate to either an optimist's or a pessimist's view of science in general. The optimist believes that science can solve all human problems, while the pessimist not only plays down the problem-solving capacity of science but also believes that science creates more serious human problems in the very process of attempting to solve other problems. Blaming science for pollution, for example, has become commonplace in recent attacks on science.

A reasoned analysis of both the strengths and limitations of science will show that neither the optimist nor the pessimist can justify his sweeping position. The view that science can save us suffered severe blows with the development of the atomic bomb, the environmental crisis, and the energy shortage. These generated and reinforced dire forebodings that science will destroy us. But science can neither save nor destroy us; we will take care of that ourselves. Other institutions of society, using, misusing, or ignoring science, will determine whether we are saved or destroyed. Both the optimist and the pessimist must understand the limitations of science. Science is a human social institution and is fallible. Despite its limitations science has been, and will continue to be, useful to society because it possesses strengths not central to other social institutions. An understanding of these limitations should give heart to the pessimists, who regard science as Frankenstein's monster, and should protect the optimists from too easy disillusionment.

One argument for the importance to laymen of understanding theory and method arises from the conviction that this will promote an understanding of the strengths and limitations of sociology and of science in general. Unfortunately, scientists themselves do not always promote lay understanding of the limitations of science. Without question, the mid-twentieth century has witnessed the overselling of science, in part by extravagant claims for science advanced by scientists themselves. Furthermore, scientists have used the prestige of science to make pronouncements on subjects that are clearly outside of science, in much the same way that baseball stars promote razors. A recognition of the limitations of science should constrain scientists from extravagant claims and testimonials. But if self-discipline will not provide the proper constraints, then it is up to others to criticize, question, and reject extravagant claims. The ability of others to do this depends on a sophisticated understanding of theory and methods in science.

This discussion has focused on lay attitudes and the layman's role in general. Let us look for a moment at the role of the more active consumer. Typically, this consumer has a problem to solve and wants sociological knowledge to help him or her solve it. The consumer may be a school superintendent wanting to improve the organization of his teaching staff; a national leader wanting to promote her country's economic development; an advocate of the women's movement wanting to advance her cause among the public; or a city planning commissioner desiring to develop a rational urban plan. Such consumers turn regularly to sociologists for advice but often do not know how to evaluate the advice when they receive it. Sometimes they fall back on their own taste or their trust of the person advising them in deciding whether or not to accept advice. Taste and trust will always be a part of evaluation; but taste and trust, while necessary, are not sufficient.

Sound evaluation demands, first of all, a realization that sociologists cannot provide every conceivable kind of advice; it is important to know what kind of advice can be appropriately sought. A clear understanding of theory and method should aid the consumer in formulating questions that can be answered and questions whose answers can be evaluated. In the second place, consumers often feel cheated when the advice given is qualified; at best, they regard such qualifications as incidental and safely ignored; at worst, they treat qualifications as nit-picking pedantry that casts doubt on all of the sociologist's advice. A proper understanding of theory and method will result in an appreciation of these qualifications. It is sad but true that all sociological advice must be qualified, for all scientific knowledge is qualified knowledge. One hope for this book is that it will make consumers suspicious of all unqualified advice.

Laymen, both as consumers and as citizens, have frequent opportunities to evaluate science, and these evaluations have an impact. In an interdependent society, science cannot ignore the judgments of other institutions of society, judgments made by laymen. It is in the interest of both science and society that these judgments be soundly based, for unsound judgments have a cost to society. Enthusiastic overevaluation or disillusioned underevaluation serves neither society nor science. Yet one of the great failings of scientists, particularly social scientists, has been the abdication of the responsibility to provide laymen with the basis for intelligent judgments of science. A general understanding of theory and method provides such a basis.

We have already suggested that there is a better way to understand social phenomena than by entitling everyone to his own personal opinion. Perhaps the personal opinion of experts represents a better way than the personal opinion of laymen; but laymen have become skeptical and distrustful of expert opinion. Some of this skepticism and distrust is justified, for experts are hardly infallible; nor are they immune from bias and prejudice. On the other hand, an uncritical rejection of all expert opinion is just as bad as uncritical acceptance. Avoiding either of these pitfalls depends upon guidelines for evaluating expert opinion. These guidelines derive from at least a rudimentary understanding of the basis on which an expert forms his opinion. The soundest foundation for expert opinion is a good theory which has been tested by methods acceptable to other experts regardless of their prejudices or biases. An understanding of theory and method then becomes a tool guiding lay evaluation of expert opinion.

We can sum up our reasons for arguing the importance of a clear conception of sociological theory and method. Social life constantly demands either individual or collective decisions that deal with social phenomena. The more these decisions are based on knowledge, the better they are likely to be. In making decisions people must evaluate the alternatives, not only in terms of their preferences but in terms of the consequences of choosing each alternative. When there is knowledge available we are able to anticipate the consequences, providing that we can evaluate the knowledge. Since we have said that such knowledge is theoretical knowledge, its evaluation depends upon understanding how to evaluate theory. And one cannot understand how to evaluate theory without a conception of what theory is.

Recent controversy surrounding educational desegregation illustrates our position. Society and individuals concerned with school systems, including government officials, school board members, teachers, parents, and even school children, have been forced to confront the problems of segregated schools and alternatives for dealing with these problems. A prominent government official with responsibilities in this area once called together a group of sociologists and asked them whether most sociologists are of the opinion that segregation is detrimental to education. They answered unanimously that most professional sociologists were of the opinion that racial segregation is detrimental to education. But how was the official to evaulate such an answer? It could be that most sociologists were simply expressing their own liberal prejudices. Had the official had a more sophisticated understanding of sociological theory, he might have asked a more appropriate question. As it was, the wrong answer was given to the wrong question, and he would have been ill-advised to base his decisions on what the group told him. Had they been asked about the state of knowledge in sociology regarding the educational effects of segregation, the group would have given a more useful answer. The answer would have been more complicated and much more detailed, but would have pointed to effects of segregation in which there was a great deal of theoretical confidence, effects of segregation which were possible but not well understood, and so on. In short, answers to the right question would have opened up many more ways to think about the problem.

The Inseparability of Theory and Method

The final fundamental assumption of this book is that *sociological theory and sociological method are inseparable.* Here again, we have a situation in which our assumption runs counter to the experience of most sociology students. Textbooks typically present separate chapters on theory and research method, usually at the beginning of the book, before proceeding to the substantive problems of sociology. These textbooks reflect the state of the field: much of sociology separates theory and method in teaching, research, writing, and advising problem solvers. There are separate courses in theory and in methodology, separate sections of the American Sociological Association on theory and methodology, separate journals, and separate textbooks; it is almost as if theorists and methodologists do not talk to one another. It is rare, for example, that a theory course discusses problems of method in dealing with theories, and even rarer that a methodology course raises theoretical issues.

Yet evaluating theories requires evidence, and the evaluation of evidence demands an evaluation of the methods by which the evidence was obtained. If theories are to be more than the wise sayings of great men, their evaluation depends upon knowing how to evaluate; and knowing how to evaluate unavoidably involves issues of method. Even deciding which of the so-called wise sayings are indeed wise requires methodological sophistication.

Similarly, methods for gathering observations, analyzing data, and making inferences inevitably entail theoretical sophistication. Most methods courses begin with the idea that methods are not independent of substance, that not every method is appropriate for every problem, but that the substantive problems determine the choice of methods. Having begun with this correct and important pronouncement, methods courses then proceed to ignore their own advice and to teach methods in the abstract. But methods themselves depend upon theoretical foundations and must be justified in theoretical terms. While abstract criteria exist for evaluating methods, and these criteria may be applicable to a wide range of theoretical orientations, it is also true that there are particular criteria which depend upon the theory being investigated with the method. We may, for example, be concerned with how reliable the evidence generated by a particular method is, and that concern may be important regardless of the theory employed in the research. Yet judging the reliability of data obtained by a particular interviewing method may depend upon assuming that the interviewees can answer questions about themselves; but another theoretical position might make the opposite assumption. Hence, the same interviewing method would not be appropriate for the two diverse theoretical positions. As we mentioned in the previous section, the use of methods requires the user to make assumptions. Whether or not these assumptions are compatible with the user's theoretical position is an issue that must be examined, and such examination requires an understanding of both theory and method and how they fit together.

Problems related to the divorce of method from substance arise from the widely accepted distinction between qualitative and quantitative methodology. The distinction is unfortunate, for methods are the means to the solution of certain problems of substance; and, in dealing with the substance of sociology, no one can guarantee that problems will be only quantitative or only qualitative. The distinction is defensible only in isolation from substance. Part of the difficulty in sociological research results from the attempt to apply methods because they are quantitative or because they are qualitative, rather than because they are suitable to the problem at hand.

This book will not distinguish between quantitative and qualitative methods; such a distinction puts the emphasis in the wrong place. Here we will treat methods as *proposed solutions to problems of the collection, interpretation, and use of observations in the service of substantive sociological considerations.* From this formulation, two conclusions follow: first, methods are instruments that serve substantive ends, and so different substantive ends require different instrumental solutions; second, the substantive end determines when quantitative solutions are needed and when qualitative solutions are required. The implicit belief among many sociologists that quantitative solutions are always better than qualitative will not stand up to close scrutiny; assigning numbers to some observations where the assumptions underlying these numbers are violated often leads to nonsensical conclusions. On the other hand, the view that a sociologist never needs quantitative methods is not supportable either. The way that training in methods occurs fosters these views; what training obscures is the fact that, while methods may be *studied* in the abstract, methods are never *used* in the abstract but are used to solve particular problems. Our final assumption puts the emphasis back where it belongs—on the problems to be solved.

One principal use of method occurs in the empirical evaluation of theory. As we have already noted, many of the issues about method and the choice of method involve theoretical questions. For these reasons, a sound strategy requires that we develop a conception of theory before dealing with issues of method. By and large, we will follow that plan; occasionally, however, there are issues of method that are particularly pertinent to elements of the present conception of theory. When those issues arise, they will be discussed in context with full development of the issues of method. This should help the reader's understanding and also should illustrate the intimate relation of theory and method.

THE APPROACH OF THIS BOOK

The approach of this book is an analytic one. Sound judgment depends on sound analysis. We will concentrate on broad issues and general principles although our analysis will sometimes involve considering a range of specific details. Except as illustration, however, we will not discuss particular techniques or particular schools of theory or methodology. We view the process of developing sociological knowledge as requiring the solution of a number of problems. We will present these general problems and some proposals for their solution; some of these proposed solutions are widely accepted among sociologists, some are controversial, and some are new. These problems and all proposed solutions merit searching analysis.

What is meant by analysis here is nothing more or less than the application of critical reason to a problem. If asked to name the central element in science, this author would choose without a moment's hesitation the collective application of critical reason as the central element. In developing a conception of theory and method, critical reason will be our most important tool. If we are to achieve our objective of understanding, we cannot read passively, accepting arguments and memorizing slogans or recipes for theory and method. It is less important that we agree with arguments than that we subject them to close critical scrutiny. Applying critical reason will teach us what to judge, how to judge it, and, most important, when to reserve judgment.

No matter how sophisticated a layman may be, there will always be circumstances when he must reserve judgment or delegate judgment to others. Although science is a public activity open to public scrutiny and criticism, the general public is not always the appropriate critic; some critical judgments require a high level of technical competence. Every science, however, must have its relevant public, an audience that is competent to judge the particular scientific activity. The idea of a relevant public will be considered further in chapter 3, but one additional comment is in order now. In judging scientific activities, laymen can be properly concerned with whether or not the relevant public has done its job in evaluating activities that are beyond lay competence.

We have been discussing the process of critical reasoning in the abstract. The time has come to get down to cases, to apply our critical reasons to the analysis of an issue in order to grasp how the process works. As a warm-up exercise, the next chapter examines one issue that has been controversial almost from the beginning of sociology, the issue of the relation of sociological knowledge to social action. It is a profound issue which we are unlikely to resolve, and it is chosen to illustrate not only the application of critical reason but the enormous challenge confronting us in using our critical reason.

The next chapter will provide an illustration of the approach we will take. As we consider an issue, we will attempt to conceptualize and analyze it. The analysis will have several objectives: to sharpen the issue; to explicate its underlying assumptions; to draw implications from these assumptions or to point out logical incompatibilities among them; to determine where evidence is required and what kind of evidence must be brought to bear; to indicate how to judge the issue; and, above all, to indicate where judgment must be reserved.

The strategy here is preeminently an effort to apply critical reason to the methodological problems of sociology. By comment and practice we will underscore the role of reason, because we regard the use of critical reason as the central feature of the scientific enterprise. The demands of reasoned analysis require that the very commitment to reason be made explicit. But there is also another basis for our emphasis; even among sociologists committed to a scientific approach, the role of reason has been underplayed. Our explicitness should heighten everyone's consciousness.

Although reason plays a key role in science, the individual's ability to reason critically is not sufficient. Science is a collective enterprise, and the ability of the collective to apply critical reason is crucial. In looking at scientific activities, one must distinguish between the individual and the collectivity. One person may have brilliant insights; but unless they can be shared by others, the insights are outside the realm of science. Insights do not have to be shared universally, but they cannot be the private possession of one genius. Or, to take another example, one person may be biased but the collective may operate to overcome the biases of its members. The reader should note that the emphasis on science as a collective, public, social enterprise is second only to our emphasis on using critical reason.

The objectives of this book, together with the emphasis on collective critical reason, impose two constraints on the problems and solutions to be presented. The problems and solutions must be understandable to the collective and must stand the test of critical reason. Since the reader belongs to the collective, at least temporarily, he has the obligation to analyze carefully and critically every claim put forward. Problems are not problems unless they are collectively recognized, and solutions are not solutions unless they meet collective tests.

expert in the particular field, against a higher standard.

The task of rewriting

Planning your work for publication at the inception of your project lays an advanced foundation for your progress in writing. For a graduate student, preparing for the thesis/dissertation and publishing from the beginning of the graduate career may eventually lead to an outstanding achievement. The selection of courses will be guided by some initial ideas about your thesis/dissertation project. The selected coursework, in turn, will provide information and insights needed to establish the direction of your research. Your various theoretical, research, an/or term papers can be designed to serve as various parts of your thesis/dissertation, which may themselves be developed into different articles for publication. For doctoral students and even some research-oriented and ambitious master's students, the final product, i.e., the thesis or dissertation, can be written with the intention to be published as a book.

Your thesis/dissertation or any other research report, however, will not come straight out as a book or as a series of journal articles. Those research products are first aimed at satisfying various particular requirements. The thesis or dissertation, for instance, is used as a training tool for you to learn and to show your knowledge and understanding of your field. It is also an opportunity to demonstrate your mastery of the research methodology required of a beginning scholar. For this purpose, you are expected to provide a detailed and relatively complete review of the literature. You also should follow a systematic research procedure embracing all the major components and considerations deemed important in a research course. A book or a journal article, on the other hand, is an instructional and/or research vehicle for you to teach or lead students and researchers to the information, knowledge, and understanding that are new and/or enlightening to them. The purposes and the readers are different, so are the requirements for contents, organization, and presentation. Therefore, if you want to publish your work you will have to face the task of rewriting.

Generally speaking, the very detailed literature review and the "systematic" research procedure frequently seen in theses and dissertations usually do not fit a journal article. Although there are also guidelines that appear to be rigid, the latter prefers the author to look like an older hand with more skillful treatment of the material, such as merging the statement of problem with a succinct review

SUGGESTED READINGS

Goode, William J. *Explorations in Social Theory*, ch. 1. New York: Oxford University Press, 1973.

Goode's chapter describes the various activities of sociologists. It not only presents an overview of the field but also points to some of the controversial issues which need to be addressed.

Lee, Alfred McClung. *Toward Humanist Sociology*. Englewood Cliffs, N.J.: Prentice-Hall, 1973.

It is difficult to find sociologists who explicitly say sociology should not be a science. But in his argument for a science with more humanistic concerns, Lee implicitly denies many of the widely accepted requirements for science.

TWO

Action, Values, and Objective Sociology

In the first chapter we emphasized the importance of critical reason in a scientific discipline. In this chapter we are going to employ our reason in the analysis of two central questions confronting contemporary sociology. These questions are: Can sociology be scientific? Should sociology be scientific?

The current debate often confuses the two questions and generates more heat than light. Only by separating the possibility question from the desirability question can we achieve an understanding of the issues involved and a possible resolution of some of those issues. Furthermore, our ability to analyze depends upon a clearer specification of our two questions. Since the word *scientific* has so many meanings, the present formulation can lead only to confusion. People with diametrically opposed positions can project their own meanings into *scientific* and answer yes to both questions. For our analysis, then, let us reformulate the questions as: Can sociology be an objective science in the same sense that the natural sciences are objective? Should sociology attempt to be objective in the sense that the natural sciences are objective?

One's whole attitude toward both the possibility of an objective sociology and the desirability of an objective sociology depends in part on clear understandings of the relation of knowledge to action and of the distinction between statements of fact and value judgments. Hence, we must first analyze the problem of the relationship between knowledge and action decisions. Can knowledge justify particular action decisions? More specifically, can sociological knowledge justify particular social actions? In addition, we must examine the distinction that we have introduced between factual questions and value questions, or between factual statements and value judgments.

For the consumer, the relationship of knowledge to action obviously represents a central concern. The layman also must understand how knowledge can contribute to action decisions and what the limitations of this contribution are. As we shall show, sociological knowledge by itself cannot determine action decisions; such knowledge, by itself, cannot even determine whether a decision is right or wrong, or whether one decision is better than another. Action decisions can be informed by sociological knowledge, and it is our assumption that informed decisions are better than uninformed decisions. But deciding on action involves considerations that are outside the realm of scientific knowledge, considerations that include values and preferences.

An understanding of the distinction between factual statements and value judgments is also an issue of importance to the layman. Later in this chapter we will show that the process of resolving conflicts about factual statements differs from the process of resolving value conflicts. The ability to make this distinction is a prerequisite to knowing what we are arguing about and how we can go about resolving our arguments and even, in some cases, to recognizing that the only resolution to some value conflicts is an agreement to disagree.

Finally, the central questions of this chapter are crucial to any evaluation of sociological work. Sociologists who believe that sociology can and should be objective must be evaluated by different standards from sociologists who believe either that sociology cannot be objective or that it should not be objective. An understanding of the issues involved here will guide the layman in distinguishing the players on the various teams and in applying appropriate standards to sociologists representing different positions. Laymen who understand the issues involved should be able to distinguish sociological activities which measure up to standards of objectivity from those which represent axe grinding. At present, as Professor Merton has put it so well, "The layman...cannot always discriminate between the 'research' which has all the outward trappings of rigorous investigation...but which is defective in basic respects [and] the genuinely disciplined investigation. The outward appearance is mistaken for the reality: 'All social researches look alike to many laymen' " (Merton, 1973, pp. 75–76).

Let us turn to the problem of the relationship of knowledge to action decisions. Later, we will undertake the analysis justifying the distinction between fact and value. Finally, we will deal with the central questions of this chapter: Can sociology be objective in the same sense that the natural sciences are objective? Should sociology be objective in the same sense that the natural sciences are objective?

THE RELATIONSHIP BETWEEN SOCIOLOGICAL KNOWLEDGE AND ACTION DECISIONS

In order to analyze the relationship between sociological knowledge and action decisions, let us look again at our example of sex discrimination. Suppose that a carefully done study indicated that women were occupationally disadvantaged by the personnel policies of a large organization, such as a university. For the moment, let us also suppose that both our male chauvinists and our militant feminists agreed on the findings of the study and on the interpretation of these findings. Many readers would be unwilling to make this assumption, and from our earlier comments it is clear that such an assumption is not automatically true.

We will return to this problem later. But for the sake of the argument, skeptical readers are asked to make the supposition, if only tentatively. Finally, let us suppose that another study indicates that an affirmative-action program is likely to remove many disadvantages that women face in this organization.

In this situation, given our assumptions and armed with the findings of the two studies, is it not absolutely clear that the organization should institute an affirmative-action program? The answer is no. Before someone from the women's movement rushes out to sue this author, we should examine the reasons why it is possible to answer this question with a direct and unequivocal no.

Even taking our suppositions as true, the facts from our studies cannot in themselves justify any course of action. From the perspective and the values of the male chauvinist, the issue is quite clear. At the very least, instituting an affirmative-action program depends not only on the facts of the case but also on the value judgment that discrimination against women is bad and should be eliminated. By definition, our male chauvinist would not hold such values and therefore would find the facts of the studies an inadequate justification for an affirmative-action program.

From the point of view of a militant feminist, however, the problem is much more subtle. Many of the values we hold are held quite unconsciously and with the belief that all right-thinking people share them. Indeed, one of the consciousness-raising activities of the women's movement is not only to make values explicit but also to make women aware of areas of value conflict. For our feminist, as long as the value judgments that discrimination is bad and that it should be eliminated remain implicit, it would appear that our hypothetical studies directly and immediately justify the organization's instituting an affirmative-action program.

But suppose the shoe were on the other foot. Suppose that the second study indicated that an affirmative-action program would so disrupt the organization that it would be impossible for the organization to achieve its goals. While our male chauvinist would claim that that kind of study justifies the action decision not to implement an affirmative-action program, our feminist would hardly agree.

The point is that action decisions depend not only on the facts or on factual statements but also on the values held by those making the decisions. Furthermore, since the consequences of an action decision must often impinge upon a whole host of values, action decisions require assessing the factual consequences of each course of action for each of the values involved, and require balancing the values themselves according to which values are more, and which values less, important.

Since this issue is so terribly important, let us examine another example. Conventionally, factual statements are referred to as "is" statements and value judgments as "ought" or "should" statements. Deciding about an "is" statement involves ideas of true and false, or correct and incorrect, and represents a different decision process from deciding about "ought" statements, which typically involve "better" or "worse." Consider the following argument as an example:

A. There is a disease.
B. Hospitals cure disease.
C. Therefore, we should build hospitals.

A and *B* are factual and decidable, at least in principle. *A* is fairly easy to decide, given most common definitions of disease; *B* presents more difficulties—there are some who would claim that hospitals aggravate disease; but it, too, is decidable. However, there is no known logic that allows us to deduce *C* from *A* and *B,* and no amount of evidence justifies *C* as a proper course of action. Deciding *C* depends on the content of other values of the decision maker. At the very least, the decision maker must hold the value, "Disease ought to be cured." And someone who held the value that "Reducing taxes is more desirable than curing disease" would not arrive at *C* as a conclusion.

Even if the value that disease ought to be cured is widely shared, the action decision depends on balancing a hierarchy of other values, and the decision-making process becomes subtle and complex. Certainly, action decisions do not require that all decision makers share the same hierarchy of values. But the mechanisms of resolving value conflicts are not well understood. We have a somewhat better understanding of the mechanisms for resolving factual conflicts. One important feature of the way we are now looking at the problem is that it calls attention to two fundamentally different decision-making processes: the essentially political (in the best sense of that word) process of resolving value questions and the essentially cognitive process of deciding factual questions.

We can illustrate the interplay of political and cognitive processes in action decisions by examining several scenarios. Let us conceptualize a typical action decision as follows: A decision maker must choose among a number of alternative courses of action, each of which has consequences, and each of which represents realizing some values at the expense of others. For the sake of simplicity, let us restrict ourselves to an action decision with only two alternatives (although we must recognize that most action decisions are much more complex and involve many alternatives). For example, if the head of an organization must decide whether or not to introduce a particular affirmative-action program for women into his organization, the two alternatives are (1) introduce the program or (2) not introduce the program. Not introducing the program may have as its consequences low morale and alienation of the women in the organization who feel discriminated against, a possible lawsuit from some governmental agency, and an increased sense of security for the men who might feel threatened if such a program were introduced. On the other hand, the decision to go ahead with an affirmative-action program may have as its consequences improvement of the morale of the women, avoidance of a lawsuit, and an increase in insecurity and feelings of threat on the part of the men.

Each of these consequences has value implications for a number of different values. We can assume, for example, that morale of both men and women in the organization are valued. We can also assume that avoiding a lawsuit is a value, while fighting a lawsuit is a negative value. (These may be values because they are means to other ends or because they are valued in and of themselves: avoiding a lawsuit may be a good thing in itself or it may be good because it conserves the valued resources of the organization, which then can be put to other purposes.)

With this simplified model of the elements of an action decision, let us consider some scenarios in which research does, or does not, play a role. Suppose our organizational president says, "Affirmative action is morally right. We must provide greater opportunities for those who have suffered discrimination. The hell with the consequences; we're going to do it." On the other hand, the president could say, "Affirmative action is nothing but reverse discrimination; reverse discrimination is morally wrong. The hell with the consequences; we're not going to do it." In either case, our president has made a political decision that obviously needs no research to justify it. The result of the decision depends upon the relative powers of the decision maker and those affected by the decision. As long as the decision is based on the moral value of the program alone, the process is purely political.

A second scenario focuses on decisions based completely on lawsuit avoidance as a paramount value. In this scenario, legal expertise is the key to the decision. The organizational president consults his legal staff and asks them to estimate the likelihood of litigation and the consequences of litigation for each possible course of action. Lawyers then consult the law and precedents in other cases in order to estimate the consequences. In this scenario, while there is no social science knowledge directly involved, we still have a situation in which cognitive knowledge plays a role in the political process, so the process is no longer purely political.

In a third scenario, our organizational president calls in a social scientist to investigate what the consequences would be to the morale of women if an affirmative-action program were instituted or if such a program were not instituted. The research could show that either decision would not affect women's morale, or it could show that one decision would increase and the other decision would decrease women's morale. (Incidentally, it is not a foregone conclusion that the decision to go ahead with the program would increase women's morale whereas the decision not to go ahead would decrease it; in principle at least, it is conceivable that it might work the other way.) Whatever the research results, our decision maker is not bound to a particular decision. For example, even if the research showed that instituting the program would increase the morale of women, the president could decide that other values were much more important, and hence decide not to introduce the program. Such a decision in no way affects the truth of the research results. It may be perfectly true that the decision alternative has the consequences claimed by the research study. But, independent of any cognitive investigation, the decision maker may decide that these consequences are politically less important, or the decision maker may value them less.

The one-sidedness of this scenario should be obvious to the reader. Yet the analysis would be the same if our organizational president had asked the researcher to investigate the consequences of each decision for both men and women. The research could show that an affirmative action program would increase the morale of women and decrease the morale of men, and the political decision could be either to go ahead or not go ahead with the program. Of course, if the research showed that instituting the program would improve the morale of both sexes, our decision maker might see that he had no problem where he thought he had one, and might go ahead. On the other hand, it is still possible that other values might enter such that, despite favorable morale consequences for both sexes, the decision would be to not institute an affirmative-action program.

These scenarios illustrate several points.

1. Choosing among values is a political process that may be narrowly defined where the decision maker weighs his own values, or may be broadly defined where the values of the affected constituencies enter into the choices to be made.

2. The role of research is to make factual statements about particular consequences of particular alternatives. Since research cannot possibly investigate all the possible consequences of all the possible alternatives, there will always be uncertainty about the possible consequences of a decision alternative, and there may always be unforeseen consequences. No lawyer can guarantee, regardless of law and of precedents, that a lawsuit will, or will not, be brought against the organization. No social scientist can guarantee that only positive consequences, or only negative consequences, will follow from an action decision.

While it might be reasonable to argue that our decision maker would be wise to commission research investigating as many consequences of as many alternatives as possible, action decisions rarely take place in the kind of leisurely atmosphere that allows enough time and provides enough money to conduct such researches.

3. Action decisions, therefore, always involve considerations beyond factual knowledge. For example, a decision maker must decide how much uncertainty can be tolerated, and no amount of knowledge can make that decision.

The discussion may seem pessimistic and could even be interpreted by the lay reader as an argument against doing research at all when one has to make action decisions. But our scenarios show one other thing which must not be overlooked. Our discussion brought up the situation in which research found (to the decision maker's surprise) that instituting an affirmative-action program improved male morale, and this disclosed to the decision maker that what might have been a problem was not a problem at all. On the other hand, the research might have shown that women's morale was decreased by instituting an affirmative-action program, thus suggesting to the decision maker that what was apparently the solution to a problem was, in fact, the creation of a problem. Just these kinds of situations— where unexpected findings point to unexpected problems, or a lack of problems— represent the principal argument for bringing knowledge into the decision-making process. Better decisions can result from the recognition that not all consequences can be known and not all values maximized, combined with the desire to use knowledge, even if limited, as a safeguard against decisions based solely on the decision maker's prejudice.

The decision maker's prejudice can also affect the questions asked of the researcher. We have already mentioned the "kidding oneself" syndrome, a bias against which we must be constantly on guard. Suppose our decision maker values an affirmative-action program as morally right, and commissions a study of the reactions of the men in the organization in order to anticipate problems that might arise in instituting such a program. Our decision maker automatically assumes that only favorable consequences among women would follow from such an action. It is conceivable that the decision maker has been led astray by prejudices. Suppose, for example, the women in the organization looked at the special treatment involved in affirmative action as simply reconfirming their lower status in the organization. Suppose they interpreted such an action as sending the message that they could not meet the same standards as men and, therefore, were inferior to men and needed to be patronized. Unless the decision maker considered that possibility, the action decision could boomerang. And when the decision maker overlooks alternatives because bias and prejudice are blinders, it is the researcher's responsibility to suggest that other consequences of decision alternatives be considered.

One final comment is in order. We have suggested that the relationship between knowledge and action decisions is a complex one, that knowledge is always limited, and that uncertainty always exists. Some readers may take this as a justification for the status quo in any situation. After all, what is, is known, and the known is better than the unknown. But this position does not follow from our argument. First of all, inaction has consequences. Secondly, what is, is not necessarily what will be, and the assumption that the present state will persist is not always cognitively justified. Hence, considering action and inaction as alternatives with consequences means that the status quo is neither justified nor unjustified on cognitive grounds, but again is a position held because of values. It should be clear that the position taken in this discussion is that neither the status quo nor change is always good or always bad; both depend on cognitive claims and political evaluations.

THE DISTINCTION BETWEEN FACTUAL STATEMENTS AND VALUE JUDGMENTS

In the last section we noted that factual statements are typically "is" statements and value judgments are typically "ought" or "should" statements; that the decision about factual statements involves truth or falsity, while the decision about value judgments entails the judgment of good or bad, better or worse; and that deciding truth or falsity is a cognitive process, while deciding better or worse is a political process.

The entire discussion of the relation of knowledge to action decisions in the last section rested on these distinctions. We claimed that it was possible to decide that choosing action alternative A did or did not lead to consequence C. In other words, we could decide that the statement "action alternative A leads to consequence C"

was either true or false. Our argument, however, went one step further: Consequence *C* could itself be evaluated as good or bad, or better or worse, than some other consequence, and this decision would be a value judgment. In order to decide on action alternative *A*, such a value judgment must be made, at least implicitly. The analysis in the last section, therefore, totally depends upon accepting the distinction between factual statements and value judgments.

But there are serious sociologists who deny that such a distinction is meaningful. These sociologists claim that deciding so-called factual questions depends upon the values of the decision maker. They assert that so-called objective answers of truth and falsity depend upon the social and cultural perspective of the questioner, and that values form the core of this perspective. In short, they claim that the distinction is artificial and denies human experience. On the other hand, many sociologists believe that sociology and social science can be objective in the same sense that the natural sciences are objective, and they base their belief to some extent on the validity of the distinction between fact and value. While a full analysis and justification of the distinction between factual statements and value judgments would require more than we can deal with here, we must examine a few of the major points of disagreement between us and our critics.

First of all, our critics argue that the distinction is artificial because values always affect cognitive decisions. Here is the nub of their argument. They believe that value biases play a crucial role in the determination of all questions. For example, it would be impossible to investigate the consequences of an affirmative-action program without the investigator's moral attitudes toward affirmative action influencing the outcome of the investigation. Some critics would go so far as to claim that the male chauvinist and the women's liberationist, investigating questions about affirmative action, would inevitably obtain results confirming their own value biases. While not all would go that far, they all share the conviction that values are an integral part of deciding cognitive questions, and hence that there is little point to distinguishing between value judgments and factual statements.

We agree with our critics that values enter into cognitive decision making, but we do not agree that the results are inevitably value biased, nor that the distinction between facts and values is artificial and irrelevant.

Let us analyze *value bias*. What people usually mean by value bias is (1) that the *content* of the values held by the decision maker affects the content of the resulting decisions and (2) that the *content* of values affects the content of the factual statements the decision maker employs in the process. The first statement is true, inevitable, and not really appropriately called bias once the role of values in action decisions is recognized. This conclusion follows from the discussion presented in the previous section of this chapter.

The second statement, however, is not universally true, not inevitable, and, when it does occur, is properly called value bias. The way in which one avoids having the content of value judgments influence the content of factual statements is to develop procedures for evaluating factual statements that are totally independent of the content of the factual statements. Rules of logic and rules of evidence represent attempts to develop procedures for judging factual statements that do not depend upon the content of the factual statements. For example, logic tells us that the pair of statements, "All *A* are *B*" and "Some *A* are not *B*," are inconsistent. That judgment is based on formal properties alone and does not depend upon the content attached to *A* and *B*. There are other similar bodies of rules. Statistical analysis, for example, provides rules for making decisions that do not depend upon the content of the question being decided. These content-less rules do not solve all problems. It is possible to use the most rigorous statistical procedures in a value-biased way; but, insofar as there are instances in which these content-less decision rules work, value bias is not inevitable. In short, value bias exists as a problem to be solved through vigilance and diligent efforts to develop more and more procedures that do not depend upon the content of the question being evaluated. Their failure to distinguish evaluations of the content of statements from content-free judgments leads our critics to believe in the inevitability of value bias.

Our critics' position suffers from a second difficulty: a failure to recognize all the values that impinge upon cognitive decision making in science. In researching a question like affirmative action, more of the investigator's values (other than simply moral attitudes toward affirmative action) enter the process. Deciding a cognitive question according to rules of logic and rules of evidence depends upon a value on truth and a value on striving for objectivity, both of which may be socially and culturally determined. If, however, an investigator is committed to find out whether the statement "An affirmative-action program improves the morale of women" is true or false, that value commitment differs from the commitment of both the male chauvinist and the militant feminist. (One may *desire* the truth of the statement, and the other may *desire* its falsity.)

As a collective social enterprise, science has an organized set of rules that contains values—the values of science, if you will—and scientists must resolve value conflicts in their own work. Such a system of rules, some of which are explicit while others are implicit, may be called a *normative system,* and a scientist operates within the normative system of science. This normative system is only one set of rules that affect the scientist's behavior. The scientist is also a citizen who is affected by the legal rules of his country and the norms of his society. Consider, for example, the dilemma of the medical researcher investigating a potential cure that may have harmful side effects. When does he decide to risk testing the cure on human subjects? Without question, that is a value problem; yet the value conflict has nothing to do with the claim that his treatment is or is not a cure, or the claim that his treatment does or does not have side effects. Furthermore, it is a conflict which can only be resolved by weighing the values that the scientist may hold as a scientist, as a human being, and as a member of a particular society. It is by no means a foregone conclusion that the values a researcher holds as a member of a particular society will always win out over the values he holds as a scientist. If this is granted, bias is not inevitable. While bias may be difficult to control and counteract, our argument thus far demonstrates that value bias is not an inherent feature of resolving cognitive issues.

of the literature. Instead of showing all the relevant information and understanding, a journal article has to focus on the main results of a research project, since the available space is often very limited. In this regard, a book is different. Space is usually not a big problem, so you can be as thorough, thoughtful, complete and detailed as you want. Nevertheless, as a more ambitious project, the book requires not only a more professional and authoritative look but also a more substantial contribution to knowledge advancement. On the part of the author, the book is a more risky investment since it means much greater a loss if it fails in the intellectual market.

For an honors or master's thesis project, it would be considered a remarkable success if some article(s) can be derived out of it and published in academic and/or professional journal(s). For a doctoral candidate, probably publishing the dissertation in the form of a book is the dream for everyone. Yet the chance to succeed is often slim. It is more likely that after the completion of the dissertation project, the investigator will continue to derive one or a series of journal articles if she can manage to accomplish that. While the articles are getting published, new projects will be initiated that may eventually lead to the publication of some book(s), which may or may not be directly related to the dissertation research project.

A general advice for all the students who have done some substantial research work of scientific merit is: Don't wait. Time is of essence. It is true not only in the sense that science needs to move ahead quickly, but that you personally cannot afford missing the opportunity to fully publish a meritorious work. The new graduates are particularly vulnerable in this regard. On one hand, they are facing a tremendous life transition replete of stresses in job hunting, graduation, relocation, and adjusting to a new work environment. When they get settled and feel ready to look back at their theses or dissertations, the work as well as the chance to revise and improve it somehow become remote. On the other hand, however, the graduates are often in desperate need of recognition of their contributions in support of their career development, including but not limited to, getting the jobs they want. It is crucial, therefore, for them to start the publication process as soon as they arrive at some significant results of their research. This is equally important to those who intend to stay in academia and those who want to go into practice. If the students can get some articles or even a book manuscript out before they graduate, chances are that they will finally make it in publishing their work.

A note regarding co-authorship is in order. After a few years working with

A second argument advanced by our critics is that the distinction between factual statements and value judgments is artificial because it does not describe the way people behave.[1] This criticism raises a "straw man." No one would claim that distinguishing facts from values is a normal or natural feature of everyday life. Nor would one claim that people in the ordinary course of events immediately recognize factual questions and value questions and decide each properly. No one believes that we are born with the ability to distinguish between facts and values, as we are born with the ability to distinguish between light and darkness. But a great many things that people learn to do are not normal, natural, or inborn. If people can be trained to distinguish factual issues from value questions, then the distinction becomes an important and useful intellectual tool.

To view science as a normative system is to argue that the distinction between facts and values must be institutionalized as a norm and that scientists must be socialized to behave according to this norm. The distinction, far from being natural, represents one of the great intellectual achievements of science.

Some critics are less sweeping in their charge that the distinction between facts and values is artificial; they exempt the natural sciences from their attack. They argue that the distinction between facts and values may be applicable to the natural sciences, where the subject matter is value-neutral. On the other hand, they argue that in the social sciences one cannot separate factual questions from value judgments because the subject matter inherently involves values. The critics who concede the usefulness of the distinction between facts and values in the natural sciences have virtually lost the argument. Their position rests on the assumption that the nature of the subject matter is the cause of value bias. Such an assumption has little to support it. From the perspective of science as a normative system, such a causal view becomes untenable.

The history of science provides little support for considering the nature of the subject matter as the causal agent. It is a fact that value biases played a key role in natural science for most of the history of natural science. For example, value-bias problems dominated physics and astronomy until relatively recently; the argument about whether or not the earth is the center of the universe took place largely in terms of religious values (i.e., whether or not it was good to question the literal truth of the Bible). It is also a fact that much of natural science has succeeded in rising above these value biases. Natural science surmounted value biases through institutionalizing a set of norms and by socializing generations of scientists to these norms.

Careful study of the history of science would bring out the value problems that once dominated science and the ways in which these were conquered. Such study would support the view that human desire and human effort are much more important causal agents than the nature of the subject matter. If we grant that the

---

1. The question of how close a concept must come to describing reality—in this case, how people behave—is a major methodological question. We reserve consideration of that question for our discussion of concepts in chapter 7.

distinction is useful to the natural sciences and we understand how the distinction became institutionalized in the natural sciences, then it does not require too much faith to assume that a useful distinction between factual statements and value judgments can be institutionalized in the social sciences.

To conclude this section, we should point out that those who assert the artificiality of the distinction between facts and values while also asserting that sociology is inherently value-biased unwittingly get caught in a paradoxical situation. To claim that sociology is inherently value-biased is to make a factual statement; but deciding that claim depends upon being able to distinguish factual questions from value questions. However, if we accept these critics' position, we cannot distinguish factual questions from value questions, and we cannot decide for or against them on the basis of logic or evidence. If we accept their position, to be consistent we must regard the claim that sociology is inherently value-biased as a consequence of their value position. But suppose we do not share their values. By their own argument, a different set of values could well lead to rejecting the view that sociology is inherently value-biased. Not sharing their values, we do not have to accept an argument that they claim is a consequence of a value position. These critics cannot have it both ways: either the statement that sociology is inherently value-biased can be decided by logic and evidence (i.e., it is a factual statement) or there is no need to take it seriously if one approaches the position from a different social and cultural perspective. If our critics make logical and evidential arguments to show that sociology is inherently value-biased, then they implicitly separate factual questions from value questions and attempt to decide independently of the content of their values. Implicitly, then, critics who so argue present a case for our point of view.

CAN SOCIOLOGY BE AS OBJECTIVE AS NATURAL SCIENCE? SHOULD SOCIOLOGY BE AS OBJECTIVE AS NATURAL SCIENCE?

From the discussion in the last section, it follows that the question of the possibility of an objective sociology and the question of the desirability of an objective sociology should not be confused. (By *objective* we mean objective in the same sense as the natural sciences.) Each question can be answered independently. One can agree that an objective sociology is possible but regard it as undesirable; and one can consider an objective sociology impossible yet not be committed to its undesirability. These two questions are at the center of a raging debate in contemporary sociology.

Before we deal with the issues, we must get rid of a terminological confusion that clouds the current debate. Unfortunately, contemporary literature and contemporary discussion center around the possibility or desirability of a value-free sociology. Those who talk about science as value-free, and mean that no value judgments at all enter into the process of deciding the truth or falsity of factual statements, mislead both themselves and their opponents. If the term value-free is used in this completely broad sense, then nothing is value-free. As the previous section argued, values do enter into cognitive decision making, but that need not be a fatal flaw.

The term value-free not only conceals the real issues but has an unfortunate emotional tone. It has become a propaganda symbol, used to paint a false picture of scientists as emotionless automatons. There could not be a more distorted image. Scientists are passionate, are committed to values, and confront value choices in every aspect of their work. Yet the passions of the scientist need not create a biased science. Since the real issues revolve around the possibility and desirability of a sociology that is objective in the same sense that the natural sciences are objective, the analysis of these issues would be better served if we abandoned the term value-free altogether. The reader who encounters the term, however, should be aware that the debate addresses many of the same issues we are considering.

Can sociology be as objective as the natural sciences? Critics, skeptics, and supporters of objectivity all agree that much of contemporary sociology violates even minimal standards of objectivity. Even a superficial glance at the sociological literature reveals many cases where the sociologist, consciously or unconsciously, has allowed values to intrude into judgments of factual questions. On the other hand, closer scrutiny of the literature reveals cases where sociologists have scrupulously adhered to the highest standards of objectivity. Hence, observing the present state of sociology answers our question with both no and yes. However, the present state of affairs does not foreclose future possibilities. As we will show later on, it is erroneous to infer from present facts to what will be in the future. If the present situation does not allow us to characterize the whole discipline of sociology as objective or as biased, the fact that we can cite research which can be described as objective and whose outcome goes against the value biases of the investigator demonstrates that objectivity is not impossible.

Furthermore, a number of logical arguments can be advanced that demonstrate that objectivity is possible in principle. Whether objectivity of sociology as a whole can be achieved depends upon how the majority of sociologists decide the question of the desirability of objectivity.

There are three principal arguments to support the claim that sociology can be objective. The first argument considers the problem of the biases of individual scientists. The second argument looks at science as a collective. Finally, the third argument examines the impact of values at various points in the process of scientific investigation, comparing sociology with the natural sciences.

Those who deny the possibility of an objective sociology make a key assumption about the makeup and behavior of the individual sociologist. They assume that an individual is made up of a constellation of motives and values which result from his socialization in a particular culture and which influence every aspect of his behavior. They claim that his every perception and every judgment are colored by his cultural position and that it is impossible to escape from these biases. Cultural bias operates in the most mundane areas of perception as well as in the most subtle. In support of their position, they cite numerous perception experiments, including one in which children from a culture of poverty judged coins to be larger than did children from a culture of affluence (Bruner and Goodman, 1947), and they interpret that result as an instance of greater desire for money affecting the perception of its size. We can bring to bear much additional documentation of the effects of motives on perception and also of the effects of cultural difference on cognitive judgments. Without question, cultural biases exist and pose a serious problem for those who would develop a science based on observations, evidence, and cognitive judgment free of cultural bias.

What can be questioned, however, is the sweeping nature of the bias claim and the belief that we are unable to counteract these cultural biases. First of all, there is both variability *within* a culture and constancy *across* cultures. Even if we accept the strong deterministic flavor of our critics' view of socialization, we must recognize variability within a culture; otherwise, how would we account for American culture's producing both Marxist and anti-Marxist sociologists? As for constancy across cultures, we have never heard of a member of any culture trying to walk through a wall. Secondly, our Marxist example may have another interpretation; it may indicate the ability of an individual to escape from his cultural biases, to rise above his socialization, and adopt a perspective different from others who have undergone the same socialization experiences. If biases are "trained in," why can't biases be "trained out"? We all have experienced situations in which we can look at something from a point of view that is not our own. To be sure, it takes effort and is not the natural or normal way we behave, but it can be done. It may require great self-consciousness and discipline and, above all, motivation, but these can be products of socialization as well.

The second argument rests on the assumption that sociology as a collective can have properties that individual sociologists do not have. Suppose we grant that an individual can never completely escape from his or her value biases. Nothing in the nature of an objective sociology requires completely objective individuals. Science is a collective enterprise, and perhaps an objective discipline can be achieved by maximizing the range of individual biases within the discipline. Perhaps compensating biases can cancel one another. Nagel (1961) summarizes the argument well, examining value biases:

> They are usually overcome, often only gradually, through the self-corrective mechanisms of science as a social enterprise. For modern science encourages the invention, the mutual exchange, and the free but responsible criticism of ideas; it welcomes competition in the quest for knowledge between independent investigators, even when their intellectual orientations are different; and it progressively diminishes the effects of bias by retaining only those proposed conclusions of its inquiries that survive critical examination by an indefinitely large community of students, whatever be their value preferences or doctrinal commitments. (Pp.489-90)

Nagel states the ideal case. In practice, no science works perfectly. But, clearly, the experience of the natural sciences suggests that corrective mechanisms can work to produce an approximation to objectivity.

Our third argument asserts that, while value judgments enter into scientific activity, value judgments must be excluded from some aspects of science and must be recognized as value judgments in those aspects where they are inevitable. To appreciate this, one must understand the various points in the process of scientific investigation where values may operate and the impact of such values. For the present, let us roughly distinguish five stages of the research process:

1. **Choice of problem and question to be investigated**
2. **Planning the study to answer the question**
3. **Collection and analysis of data**
4. **Interpretation of findings**
5. **Deciding what to do with the results**

What role do value judgments play at each of these stages? Let us look at each stage briefly.

Unquestionably, the values of the investigator play a significant role in the choice of the problem and the questions to be asked. Even when the investigator is motivated solely by curiosity, there is the implicit judgment that curiosity is a good thing. An investigator may ask a particular question because the question is relevant to a pressing social problem; in so doing, he expresses the value judgment that the social problem ought to be solved. Even when an investigator accepts a commission from a client, a series of values must be weighed. Does working for this client compromise one's values? Is the value on scientific independence compatible with the client's vested interest? Value judgments play a role in the choice of problem in the natural sciences as well as the social sciences. That value judgments play such a role at this stage is not only appropriate but inevitable.

Value judgments also enter the planning and design stage of the study, as our earlier example of the medical researcher, deciding whether or not to test his new drug on human subjects, illustrates. No scientist can ever design the perfect study. Compromises must always be made, and such compromises always involve the sacrifice of some values for the sake of other values. On the other hand, at this stage of the research process, some value judgments can produce bias. In our example of planning a study to investigate the effects of an affirmative-action program on men's morale, our investigator planned the study because of valuing an affirmative-action program as a good thing for women. Such a design decision could exemplify the problem about which our critics worry. Usually such decisions are made in the interests of economy and efficiency, with a rationalization such as "We don't have to spend the money to study women—we already know the answer." To be sure, sometimes the answer is known, but more often the expression of certainty merely acts as a cover for implicit values. Without question, this is a problem area, and it is a problem from which the natural sciences are not immune either.

The stage of collecting and analyzing data represents the one area where most people agree that value judgments must be excluded if we are to even approach an ideal of objectivity. Although keeping value judgments out is not trivial, people have been worrying about the possibilities of bias here for a long time; techniques and procedures have been developed to deal with eliminating value judgments from the collection and analysis of data. The idea of taking a random sample of a population, for example, serves to prevent the biased investigator from choosing to talk to people who will give only the answers he or she wants to hear. In part, procedures for collecting data have the objective of minimizing potential bias, both from values and other sources. Similarly, the ramification of analytic techniques for handling data constitutes an elaboration of safeguards against bias. Using well-developed, explicit statistical techniques, for example, minimizes subjective judgments on the part of the investigator, because the rules of the statistical procedures entail a result, whether or not the investigator likes the result. While not all problems of value bias in the collection and analysis of data have been solved, the social sciences have made considerable progress in dealing with bias at this stage of the research process.

When we come to the interpretation of findings, however, value judgments are again inevitable. Part of interpreting the results of a study involves the judgment that some findings are important whereas others are not, or that some findings are more important than others. To question the importance of findings is to ask, "Important for what purpose?" Although deciding on the importance for a given purpose involves many considerations, value judgments enter such decisions. Is the finding that women's morale would be unaffected by an affirmative action program important to the decision to institute such a program? Answering that question depends upon what values such a program attempts to realize. Even if the purpose for which the findings are important is the establishment of a theory, value judgments may operate; and the investigator who values the theory may be more likely to regard a supporting finding as important than the investigator who devalues that theory. Although critics would immediately point to this example as making their case that social science is inherently value-biased, the social sciences are no different from the natural sciences in this regard. The success of the natural sciences allows us to hope that value bias in interpretations is neither inherent nor fatal.

One other aspect of value bias in interpreting results should be mentioned. Some anonymous sociologist has christened this the "fully-only problem." Our male chauvinist and our militant feminist may look at the same data, with the feminist reporting, "Fully 60 percent of the women studied support the institution of an affirmative-action program." The male chauvinist, on the other hand, notes the result as, "Only 60 percent of the women studied support the institution of an affirmative-action program." In our example the bias is as plain as day, but other cases can be more subtle. This example illustrates that the language of reporting and interpreting results can itself be "loaded" (Katzer et al., 1978). Part of the problem of dealing with potentially biased interpretations involves a recognition of such loaded language, a sensitivity on the part of the investigator to avoid loaded language, and alertness on the part of the reader to catch such usage.

The stage of deciding what to do with research results obviously involves value judgments when the objective of the research is to aid in making an action decision. But even when research has as its objective adding to knowledge, value judgments are part of this stage. Consider, for example, results which have controversial consequences, such as a finding that an affirmative action program could decrease the morale of female workers. Since the findings of any study are always tentative and subject to error, the investigator confronts a value dilemma: Is it worse to publish such results, which may be reversed in future research, or to withhold publication until further investigations have been conducted? In controversial areas, the publication of research results is an act that has consequences for a host of values, and the decision to publish must be weighed with those consequences in mind. When research results challenge a well-established theory, this is a particularly touchy question, even though the theory may be esoteric and not involved in any public controversy. The value of scientific caution often conflicts with the value of establishing scientific priority, and that conflict characterizes both natural and social sciences.

Nothing in this analysis discriminates sociology from the natural sciences. While our critics would argue that the asking of value-biased questions does not produce an objective science, the natural sciences are not immune from asking one-sided questions. For example, a scientist studying the effect of adding lead to gasoline on the efficiency of an automobile engine, without studying the effect of lead pollutants in the atmosphere, is asking a one-sided question. The ability to recognize one-sidedness of questions is itself a potential corrective measure for promoting objectivity. The problem is serious only when any scientific study or any scientific theory is regarded as the whole story. Recognizing the partial nature of scientific answers as an inherent limitation of science can promote an appreciation of the sense in which science is objective and of the limitations of that objectivity.

With respect to the later stages in the research process, the example of the natural sciences is instructive. If we consider science as a social institution, then the normative structure of that institution molds the behavior of individuals who are part of the institution. Examination of the natural sciences reveals a clear, explicit normative system, widespread commitment to the norms, sanctions (both rewards and punishments) enforcing the norms, and well-established socialization practices for inculcating the norms into new recruits. To be sure, not every individual scientist conforms fully to the norms, but that should not surprise a sociologist for whom the existence of norms implies some deviant behavior. The point, however, is that natural science has developed a collective, institutional solution to the problem of objectivity.

In the social sciences, on the other hand, one finds no such general consensus on normative standards. Disagreement over standards prevails, and agreement occurs implicitly rather than explicitly; rewards and punishments accrue to individuals, seemingly without regard to any standards; consequently, socialization practices vary widely in accordance with the beliefs of particular socializers. Furthermore, some influential spokesmen in the social sciences make a virtue of normlessness as evidence of the intellectual freedom prevailing in the field. Intellectual freedom is unquestionably a good thing (indeed it is a necessary condition for scientific activity), and it is quite compatible with a clear, explicit normative structure. The danger arises when intellectual freedom is confused with anarchy— and normlessness is akin to anarchy.

The principal necessary conditions for objectivity in a discipline include a clear, explicit normative system, widespread commitment to the norms, mechanisms for enforcing the norms, and effective socialization techniques. The norms must emphasize the importance of the individual striving to recognize and overcome his own biases while also requiring every individual to be alert to the biases of others. The system must enshrine the critical attitude, both for self-criticism and collective criticism. It must encourage the individual to seek out criticism, to be receptive to that criticism, and to be ready always to give criticism.

For criticism to promote objectivity, it must be responsible criticism. As we all know, it is very easy to be critical, but superficial criticism will not accomplish the goal. To be responsible, criticism must be based on penetrating analysis and evidence, not simply on the likes and dislikes of the critics. Anyone can look at a study and dismiss it as, say, middle-class bias, but the critic has the obligation to ask himself how he or she can demonstrate that bias, what results of the study show the bias, and what results are free from bias. Such demonstrations, however, depend on the application of explicit standards, and such standards must be developed, formulated, and transmitted to scientists and laymen. Fortunately, the standards for detecting value bias are closely related to the standards for evaluating scientific products in general. Hence, examining standards for evaluating sociological theory and method should assist us in arriving at standards for searching out value biases.

POSITIVISM, THEORY-LADENNESS, AND THE POSSIBILITY OF OBJECTIVE SCIENCE

Since we have defined objectivity for sociology to mean "objective in the same sense that natural science is objective," we must recognize that there are some who seriously question the possibility of any objective science, natural or social. Many who hold this position believe that they are attacking Positivism and consider themselves as either Anti-Positivists or even Post-Positivists, although the latter position is usually occupied by those who would blunt the attack by modifying what they regard as difficulties with Positivism.

Attempts to grapple with the controversies surrounding Positivism face major difficulties. First of all, the term itself has a number of different usages ranging from an unanalyzed curse word to a name for a specifically defined philosophic position. Even when used to refer to a philosophic position, the label often conceals serious disagreements about the tenets of the position. Halfpenny, in his book *Positivism and Sociology* (1982), analyzes twelve different uses of the term among sociologists; if he had extended his analysis to other disciplines, he probably would have uncovered many more. Secondly, varieties of usage involve more than semantics; writers who consider themselves, or are considered by others, as Positivists differ markedly in their assumptions and their claims. The differences among thinkers like Hume, Comte, Bertrand Russell, and B.F. Skinner are at least as important as the similarities. Furthermore, only at a very abstract, and almost contentless, level can one distill a central core of ideas which these thinkers all share.

These controversies involve a number of different issues, and adopting a position on one issue does not predetermine positions on others. With the variety of different issues and possible positions on these issues, one can be a Positivist, an Anti-Positivist and a Post-Positivist simultaneously. If Positivism means a commitment to using evidence, then this author is a Positivist; if it means that nonobservable entities are inadmissible, then the present writer is an Anti-Positivist. If Post-Positivism represents a concern with the theoretical relevance of observables, then this analyst is a Post-Positivist; and so on. We cannot even survey the range of these issues, but we must discuss one set that is extremely relevant to the preceding discussion in this chapter.

Threats to objectivity arise from other sources of bias in addition to personal values, and one of these sources is the theoretical orientation of the researcher. We agree that this is a source of bias. If the concern with the theory-ladenness of observation were simply an attempt to call attention to this source, we would have no quarrel with those who raise the issue. But those who promote the theory-ladenness critique intentionally or unintentionally attack the fundamental nature of science and deny the possibility of scientific study of social phenomena. Anti-Positivists, as Crews points out,

> think of themselves as staving off a persistent threat of positivist incursion upon human studies. But there must be something else going on here, since by now one might have to repair to the graveyard to find an authentic positivist to kick around. What "antipositivism" really comes down to is a feeling of nonobligation toward [empirical evaluation]—that is, toward the community that expects theory to stay at least somewhat responsive to demonstrable findings....To be a good contemporary antipositivist, then, is to resist the encroachment of science on human studies—to deny...that the natural sciences offer [us] an adequate or even a relevant model....Or, more radically, one can *deny that science itself is really empirical.*
> (Crews, 1987, p. 37; emphasis in original)

Not all Anti-Positivists take such an extreme position, and while this author does subscribe to some elements of Anti-Positivist critiques, he believes that, at a sufficient level of abstraction, the natural sciences do provide relevant principles for the human sciences. Hence, it is incumbent to respond to the position Crews describes, extreme as it may be. We cannot examine all the issues on which this author differs from these "anti-positivists," but one claim that is key to their attack must be analyzed. The claim simply put is: Since observations are "theory-laden," they cannot be used to evaluate theory and as a consequence, evidence does not provide an objective evaluation of theory even in the natural sciences. What does "theory-laden" mean and what are the implications of this claim? Does it mean that a theoretical position determines the nature of the observations? Does it mean that observations are uniquely relevant to a single theory, theoretical orientation, paradigm or metaphysical stance? What does agreeing with the claim require? Does acceptance commit us to a thoroughgoing relativism in which evidence is only for the believers, in which there is no falsity and therefore no truth and ultimately everyone's opinion on every matter is as good as everyone else's? Such a relativism, of course, generates another paradox: It is an absolute relativism.

If those who assert that observations are theory-laden mean simply that observations are filtered through one's presuppositions, theoretical or otherwise, then the idea is neither new nor difficult to accept. What one observes, how one observes, the relevance and significance of what one observes, and so on, are not given in nature. The notion that the human observer functions as a passive receiver of sensations from the phenomenal universe, as a *tabula rasa,* does not command many adherents in today's world. And, while sometimes researchers behave as if observations are *a priori*, or in some other sense privileged, only a few writers are willing to defend the primacy of observations.

In 1934, Cohen and Nagel wrote:

> Even apparently random observation requires the use of hypothesis to interpret what it is that we are sensing. We can claim, indeed, that we "see" the fixed stars, the earth eclipsing the moon, bees gathering nectar for honey, or a storm approaching. But we shall be less ready to maintain that we simply and literally *see* these things, unaided by any theory, if we remember how comparatively recently in human history are these explanations of *what* it is we see. Unless we identify observation with an immediate, ineffable experience, we must employ hypotheses even in observation. For the objects of our seeing, hearing, and so on, acquire meaning for us only when we link up what is directly given in experience with what is not. (Pp. 215–16)

That such an eminently reasonable view can be used as the basis for a subjectivism that represents, in some cases unwittingly, a fundamental attack on science comes as a great shock to many of us. Extending this view to the assertion that all observation is inevitably biased by the theoretical presuppositions of the observer precludes the falsity of any observation—nothing is wrong; it is just a different point of view! Ideas, then, are no longer corrigible by observational evidence, and science as we know it is impossible. Such an extension is unwarranted; nothing in the Cohen and Nagel quotation implies that the content of observations are inevitably biased by the presuppositions of the observers. To be sure, biased observations do occur—and one needs criteria outside of the suppositions of the observers even to identify bias—but biased observations are not inevitable and are regarded as serious problems to be corrected.

Consider reading a speedometer in one's car as an example of a theory-laden observation. Without the theory of classical physics, it would be impossible to construct a speedometer; without that theory, there would be no meaning to the

your mentors, both you and your "bosses" may have a strong desire to publish something out of the research project together. It is ideal and to your best interest, indeed, if you could coauthor some articles and get them published with your advisor and committee members. Things sometimes get complicated, however. As a student you need to publish your work as soon as possible and cannot afford any significant amount of time to be lost. On the other hand, as senior faculty your professor may desire the work to be polished as perfectly as possible before getting it out. They can afford to wait, and in fact they are not so focused on this single project as you are. Therefore, you may find yourself in a dilemma if you think you are ready to publish your work. Co-authorship often involves a prolonged process of communication, discussion, and achieving consensus. It may mean half a year of shelving your book manuscript if you want to get a co-authored article out first. Given your situation in that stressful transition period, you should not hesitate to send your book manuscript right out when it is ready. And your mentors will hopefully understand that and continue to give you their valuable support.

Before starting the first draft of your original work, be it a thesis, dissertation, or some other research project, you need to "organize for writing" (Mauch & Birch, 1993). That might range from getting yourself a computer to having a table or shelf so that you can arrange your notes and data for easy access and lay out your books for reference. In preparing for rewriting for the purpose of publication, you will need to reorganize everything. Especially, you have to recheck your computer files as well as library materials to make fuller use of your resources. You should also obtain the guidelines from the prospective publishers and/or journal editors. Usually publishers have their own author's guide available upon request. Guidelines for contributors to a journal may be found in each issue of the journal. It makes everything easy if you can obtain copies of some books recently published by the prospective publishers for publishing your book and/or recent issues of the prospective journals for publishing your articles. Following the sample formats and reading the guidelines should leave you no problem in preparing the manuscripts (called "typescripts" nowadays, i.e., they must be typed rather than handwritten).

Just as you need an overall plan to start writing your first draft, in rewriting for publication you need to think your general organization. A systematic sequence can now be easily teased out on the basis of your thesis, dissertation, or other kind of research report. The table of contents can be used again as an outline. You need to spend time revising the outline to allow for a smooth flow

sense experience of perceiving a pointer at a given place on a dial or perceiving a number in a digital readout. One may even argue that the particular set of numbers used in the speedometer scale is arbitrary—after all, we have speedometers calibrated in miles and in kilometers per hour. But, and this is the crucial point, it would be absurd to argue that the theory and the observer determine the meter reading *independently* of the motion of the car. Radar traffic surveillance makes even more sophisticated use of physical theory; if a driver is stopped for speeding, however, it would not do any good to argue that the highway patrol is operating in a different theoretical paradigm.

To be sure, using observations to evaluate theories involves much more complexity than deciding whether or not a car is exceeding the speed limit. And the same complex issues arise whether we are evaluating physical or sociological theory. To argue that theory-ladenness necessarily determines the truth value of the observations is to assert the impossibility of using evidence to evaluate theories. The history of science offers many examples of the successful use of evidence to overthrow theories; hence the theory-ladenness argument must be rejected. And, with respect to the natural sciences, most people are willing to reject such an extreme view of the consequences of theory-ladenness. In the social sciences, however, the extreme view has many adherents. For ideological reasons, some people do not want ideas to be constrained by evidence. Others believe that the complexity of social phenomena, the fact that actors and observers have values with respect to the phenomena involved, and the fact that objects of study are free-willed choice makers all combine to bias the truth values of any and all sets of observations.

While a theory may determine the relevance of a set of observations and the procedures for collecting and analyzing the observations, it does not determine the factual truth of all the relevant observations. A theory may even determine the truth of *some* observations, and when this occurs we have a problem of bias similar to value bias and to be dealt with in the ways discussed earlier in this chapter. However, as long as there are some observations whose truth value is not determined by the theory, then there are some observations that can provide an evaluation of the theory independent of the wishes and desires of either the proponents or opponents of the theory, and that after all is one meaning of objectivity.

SUMMARY

Our analysis indicates that sociology can be objective in the same sense that natural science is objective, and that is a powerful sense indeed. To be sure, complete objectivity is an ideal which can never be fully realized. That is true in the natural sciences as well as in the social sciences. Achieving objectivity does not come automatically, but depends upon collective striving to be more objective. While we have shown that in principle sociology can be objective, realizing objectivity in practice depends upon a sensitivity to human limitations, a commitment to a collective enterprise, an awareness of the points at which human limitations may have an impact on the process of scientific investigation, and the collective commitment to develop normative rules to deal with each bias problem as it is recognized.

The only way sociology will strive to improve its objectivity will be to decide that objectivity is desirable. Resolving the "can" question does not answer the "should" question; the desirability issue will continue to be debated. Proponents of "advocacy social science," concerned about the state of the world and impressed by the social problems that confront the world, argue that sociology should not try to be objective but should be committed to moral and ethical ends. They believe that even if sociology could in time be as objective as the natural sciences, time is running out. They worry that unless we use sociology to advocate particular solutions to the social, political, and moral dilemmas of today, we will not have a tomorrow in which to be scientists. This is a hard position to challenge, since it and its challenges boil down to matters of faith.

This author's faith is that we do have the time to develop an objective sociology, and that it is possible to separate one's concerns as a citizen of the world from one's concerns as a sociologist. As citizens, we can advocate value positions, but there is nothing which makes a sociologist's values better than a citizen's values. Social criticism, advocacy of social programs, and calls to action have been around for a long time. On the other hand, the attempt to bring to bear rigorous sociological theory, supported by empirical evidence, on the solution of society's problems is a relatively new endeavor. It is an endeavor in which sociologists may have unique abilities and may make unique contributions to society. But such rigorous, empirically supported theory can only come about through striving for an objective sociology. For these reasons this author believes that sociology should strive to be as objective as the natural sciences.

We have argued that an explicit normative system defining objectivity as a goal and providing standards for criticism represents a crucial requirement. But it remains to be seen whether such norms will gain widespread consensus in sociology. We cannot establish norms by proclamation. While this book aims at developing general standards for the evaluation of theory and methodology in sociology, it will seek acceptance for these standards by presenting a series of problems that theory and method must confront, by promoting an understanding of these problems, and by offering standards as partial solution to these problems. (We emphasize *partial*, for science presumes that any problem solution can be improved upon through the collective application of critical reason.) One could consider the proposals of this book as utopian and as based on an unrealistic faith in human reason, a faith that norms emerge from common attacks and successful solutions to a set of problems. If the readers of this book, however, can develop some consensus, we will be on the way to creating the necessary normative structure.

SUGGESTED READINGS

Gillispie, Charles C. *The Edge of Objectivity*. Princeton, N.J.: Princeton University Press, 1960.

This essay in the history of scientific ideas is both difficult and beautifully written. Gillispie points out that problems of value-bias are part of the history of science, which is a history of striving toward the edge of objectivity.

Goode, William J. "The Place of Values in Social Analysis." In *Explorations in Social Theory*. New York: Oxford University Press, 1973. Pp. 33-63.

Goode's orientation is similar to that presented in this book. He attempts, however, to account for the controversy that exists in sociology over the place of values.

Gouldner, Alvin. "Anti-Minotaur; the Myth of Value-Free Sociology." *In The New Sociology*, ed. Irving L. Horowitz. New York: Oxford University Press, 1966. Pp. 196-217.

Gouldner presents a position diametrically opposed to the position taken in this chapter.

Nagel, Ernest. *The Structure of Science*. New York: Harcourt Brace Jovanovich, 1961. Ch. 13, pp. 473-502.

Nagel presents a rigorous and compelling analysis of the problems of values and subjectivity in a science. He concludes that there are no differences in principle between the natural and social sciences with respect to value bias.

THREE

Two Important Norms of Science

The last chapter argued that objectivity was not a property of the subject matter of a science, but a collective commitment by a group of scientists in their approach to their subject matter. Because a group of people wants to be objective doesn't guarantee that they will succeed. On the other hand, without the collective desire for objectivity there is no chance, whatever the subject matter, that it will be approached objectively.

Looking at science in this way (i.e., as a collective social activity in which the participants share aims and ways of achieving those aims) can dispel some of the mythology that surrounds science. Organized social activities are governed by standards of behavior for the participants. To a greater or lesser degree, these standards are shared, adherence to the standards are rewarded, and violations of the standards are punished by the social group. Sociologists call such standards *norms*; when these are well developed and govern the group's activities, we say that the norms are *institutionalized*. Science is an organized activity and has institutionalized norms. It is these norms that have produced the degree of objectivity that science has achieved. For sociology to be objective, the institutionalization of norms that will promote objectivity is required. In this chapter we will look at the experience of the successful sciences, and we will focus on two important norms of science. We argue that those people who want to "do" scientific sociology must adhere to these norms and those laymen who want to promote scientific sociology must support adherence to these norms.

It is a useful analogy to regard science as a game and the norms of science as the rules of the game. In speaking of a game, we have no intention of playing down the importance or seriousness of science; rather, we use the analogy as a heuristic device—that is, a device to aid us in our search for ideas. Those of us who want a scientific sociology must learn to play the scientific game. When we want to learn to play a new game, the first thing we do is sit down and read the rules, and if the game is well designed, the rule book will be detailed, explicit, and understandable. A well-designed game will also have a set of rules that covers every situation a player could encounter in playing it. Simple games like bingo or checkers can operate with a single page of rules, while more complicated games may require a whole book of rules. When the player gets beyond the rules and into strategies for playing a complicated game, he may have to read several books before he becomes a competent player. Think of how many books have been written about chess or bridge. But the bridge player has an advantage over the player of science: Bridge rules are written down in one place, and bridge experts do not disagree about the rules themselves. (Although bridge experts may argue about the strategies of play, they do not dispute the rules of the game.) Someone who wants to learn the scientific game has greater difficulties. He will not find any single authoritative book of rules, and he will be struck by the controversies that persist over rules of the game. Furthermore, he will have considerable difficulty disentangling arguments about strategy from arguments about the rules themselves. In fact, here our analogy breaks down; it is almost as if we were playing a game with the object of winning, but the game itself consisted of making up rules that maximize the player's chances of winning.

Science cannot separate strategies of play from the rules of the game. If the object of the scientific game is to gain knowledge, then both the rules and the strategies aim to facilitate achieving knowledge and minimizing error. Virtually any rule must also be evaluated on strategic grounds; a norm of science must be judged according to whether it enhances the search for knowledge and whether it increases or decreases the possibilities of error. Since we do not know all of the things that contribute to the search for knowledge or all of the things that produce error, there are bound to be disputes over scientific norms.

One further difficulty confronts the student who wants to learn the rules of the scientific game. The apprentice scientist does not learn rules as rules; he learns implicitly by doing. From the very beginning, he is actively involved in the practice of science, reading and hearing the facts of his field, and watching established scientists formulating new theories or developing new experiments. The apprentice does not memorize a book of rules. Like other norms, the rules of science receive explicit attention only when there are violations. Yet the infrequency of violations testifies to the success of the socialization that takes place. Apparently, there is little need to underline the norms by saying to the student, "This is a rule!"

The power of implicit norms is well illustrated by a recent scandal in the biology laboratory of a major medical school. An investigator reported experimental results which ostensibly demonstrated that a particular chemical applied to the skin of an animal caused tumors. It turned out that no one else could reproduce this experimental result; and on investigation, the results of the original experiment were found to have been faked. When the experimenter admitted that he never ran the experiment but manufactured the results, his scientific career was ended. One could look long and hard and not find any written law against the faking of experimental results; it is one of the rules that goes without saying. Although it is not explicit, it is a powerful control of a scientist's behavior. When a violation of the rule is discovered, the punishment is severe. It is just such rules that constitute the norms that govern science.

Implicit socialization is effective when norms are firmly institutionalized. When norms are in doubt, greater self-consciousness is required. In this chapter we will examine two fundamental norms of science: one, in effect, provides a socially defined purpose for scientific activity; the other is a general rule prescribing how scientific activity should be conducted. The next section discusses the normative orientation of science, and the section following it examines the norm of inter-subjective testability—that is, the requirement that scientific ideas be collectively evaluated using agreed-upon procedures.[1]

THE ORIENTATION OF SCIENCE

One of our purposes in looking at science in general is to develop guidelines for scientific sociology, in the hope that these guidelines will eventually be institutionalized as norms. While there are parts of sociology that are governed by these norms, one cannot say that the norms are institutionalized in the discipline as a whole. Sociologists do many things and play many roles. Sometimes they are scientists, attempting to generate knowledge; sometimes they are applied scientists, attempting to use scientific knowledge; sometimes they are social critics; sometimes they are advocates of a particular cause; and sometimes they are involved in solving practical problems.

Suppose a sociologist writes, ''Political institutions of a society are designed to maintain the economic advantages of the ruling class.'' Which role is that sociologist playing? Is he writing the statement as a scientist, intending it to be evaluated on scientific criteria? Is he writing it as a social critic expressing a particular value position? Is the sociologist championing a cause and writing this in an attempt to rally supporters? Often it is difficult to know which role the sociologist is playing and what criteria are appropriate to evaluate the products of the particular role. Of course, a sociologist has every right to play any of these roles, and nothing in this analysis questions the legitimacy of these roles. What the analysis does suggest, however, is the necessity to distinguish clearly one role from another, in order for the public to apply the appropriate criteria for the particular role. We would evaluate the claim that ''political institutions are designed to preserve the economic interests of the ruling class'' much more stringently if it were intended to be a scientific claim than we would if it were intended to be a battle cry.

As our example points out, the aims of these roles are very different, and our judgments depend upon the aims. Although these roles overlap, on analysis one can see that each role has its own rules and obligations. That norms are not institutionalized in sociology may be the result of failing to distinguish these roles. This certainly poses a problem for the layman in looking at sociology—particularly when

1. In the discussion that follows we will draw heavily on the analyses developed by both sociologists of science and philosophers of science. The list of suggested readings at the end of the chapter will serve as a general reference to the material presented.

reading a newspaper article reporting on some sociological work, since the media rarely indicate which role the sociologist is playing. Promoting scientific sociology, however, requires a clear, self-conscious effort to distinguish the scientific role, and the rules that govern it, from the other roles that a sociologist may play.

In the previous chapter we emphasized the importance of striving for objectivity in scientific sociology, and it is clear now that such striving for objectivity is not compatible with some roles that sociologists may play. For example, objectivity may be quite irrelevant to the role of championing a cause. While the champion may use the results of work done in the scientific role, no one would claim that such use was objective; but this is quite consistent with our discussion of the relationship between knowledge and action—no *use* of knowledge is objective. Hence, it is important to distinguish a role designed to produce knowledge (where striving for objectivity is of primary importance) from roles which use knowledge. In the user role, once the knowledge is accepted, its use is in the service of specific value positions, which by definition are not objective.

There are distinct producer roles and user roles in all of the sciences. Although laymen tend to lump all activities of scientists together as ''science'' and blur distinctions between scientists and engineers, in the well-developed sciences there are clear distinctions. To some extent these distinctions are institutionalized: there are separate university departments of mechanical engineering and physics, for example, and in some places there are even different departments for physics and applied physics. We believe that some activities of sociologists are analogous to the activities of a physicist, some are analogous to the activities of an applied physicist, and some are analogous to the activities of a mechanical engineer. In sociology, however, these different activities are not distinguished by different titles. If we analyze the basis for differentiating between scientists and engineers and between basic and applied scientists, we may gain some insight on how to better delineate roles for sociologists.

If we look at basic science, applied science, and engineering as social activities in general, we can see that the aims of these activities are different; these aims are institutionalized norms that govern the activities. A central norm for each of these activities can be formulated as follows:

Basic science is oriented to the production and evaluation of knowledge claims.

Applied science is oriented to the discovery of new uses for knowledge claims that have been previously evaluated and tentatively accepted.

Engineering is oriented to the solution of technical problems where the problem to be solved is regarded as given.

We will discuss each of these norms individually. But, first, there are some general comments that are applicable to all three. The term *knowledge claim* needs to be defined. We will want to limit what we mean by a scientific knowledge claim, since science and engineering activities do not concern everything that can be called human knowledge. We will consider some limits shortly. For the present, however, let us interpret the term knowledge claim broadly to mean any statement about phenomena that can be accepted or rejected on the basis of some criterion of truth.

By treating these as norms, we are not saying that every basic scientist, applied scientist, or engineer has one of these orientations. In fact, we are saying little about the motivation and commitment of individuals. At the risk of obviousness, we can point to many varied motivations among individual scientists: some are oriented to money, others to fame, others to prestige, and so on. Nevertheless, for individuals to achieve their aims through the roles of scientist or engineer, they must commit themselves to institutional goals at least to some degree. This necessary commitment represents one significant aspect of the distinction between institutional norms and individual motives—it allows a collective outcome to emerge from a variety of individual purposes, even when the purposes may be competing with one another.

We must emphasize that our analysis of norms does not describe individual motivations, and also that the distinctions we make are analytic distinctions. An individual may not share the normative orientation of his discipline, or he may hold several orientations simultaneously. It is not unusual, for example, for a member of a mechanical engineering department to be involved in basic scientific research in physics; and a physicist may at different times do basic physics, applied physics, or engineering. Nevertheless, the analytic distinction is fundamental to understanding scientific activities. The activities are distinct, even though individuals may, and frequently do, cross over from one activity to another. Failure to distinguish among basic science, applied science, and engineering creates confusion among scientists and laymen alike, and confusion leads to unrealistic expectations for, and unreasonable demands upon, science. It is unreasonable, for example, to expect a basic scientist to create knowledge on demand because society feels a problem is pressing. It is unrealistic to say that a sociologist who is researching crime must be able to discover the causes of crime during his three-year research contract. Knowledge must be built on a secure foundation, and just because society regards a problem as important does not mean that the foundation exists on which to build scientific knowledge of the problem.

Let us examine some of the special features of each of these norms (after which we will present a general conclusion of the analysis). The position that basic science is oriented to the production and evaluation of knowledge claims has three distinct features. In the first place, knowledge claims are the focus of concern for the basic scientist. The activities appropriate to the role lead to new knowledge claims or to new evaluations of existing knowledge claims. Second, the standards that the basic scientist uses concern the criteria by which one decides that a statement is, or is not, a knowledge claim, or that it is an acceptable knowledge claim. Third, this norm excludes activities from the domain of basic science: the issues concerning the application of knowledge are outside of the basic scientist's role—

although, as a citizen, the basic scientist may have interests in such questions.

Although the activity of basic science is popularly characterized as problem solving, it is problem solving in a very special sense. The problems of basic science are purely intellectual problems. Regardless of the individual's motives or the society's needs, basic science is successful to the extent that it generates collectively shared understanding of phenomena. Its success or failure can be evaluated according to its own internal standards, and its standards may have little or nothing to do with lay appreciation of its accomplishments.

The criteria for evaluating the productions of basic science are cognitive criteria. To put it crudely, the principal concern of basic science is whether or not a knowledge claim is true; this is an oversimplification that will have to be modified later on, but it is nevertheless helpful in our present discussion. Deciding whether or not a knowledge claim is true in a particular basic science depends upon the development of tools and procedures and specific criteria which are internal to that science. These specific criteria often become very technical—such as, for example, a criterion concerning how many times an experiment must be repeated before there is collective confidence in its results. It is unreasonable to expect that these technical criteria will generally be accessible to laymen; but it is important for the layman to know that such criteria exist, that they are collectively shared, and that they are applied to any claim that is advanced. In short, deciding the truth of a knowledge claim is a responsibility of competent peers who share technical intellectual standards.

The normative statement of the aims of basic science clearly excludes activities from the domain of basic science. The application of knowledge and the solution of practical problems, according to our formulation, are responsibilities of roles other than that of the basic scientist. From this perspective, applied science and engineering are separate and distinct from basic science. Our analytic distinction serves to differentiate responsibilities for resolving the typical question that the lay public addresses to science. To basic science, the appropriate questions are: Is it true? How do we know? Before we can ask whether or not a knowledge claim is useful, it seems appropriate to require that the knowledge to be applied exists—which in our terms means that the knowledge claim has been positively evaluated by the relevant public. Only in this sense is there a priority for basic science, and the reader should remember that it is possible to solve many practical problems without scientific knowledge. After all, we built bridges long before there was a science of mechanics.

One implication of our formulation is of particular significance to laymen. It is an unfair criticism to suggest that the basic scientist has come up with useless knowledge. It is impossible to know what knowledge may be useful in the future. In some sense, the production of knowledge represents a capital investment, the storing away of ideas, some of which may turn out to be useful. While society may legitimately ask how much of its resources it can afford to invest in gambling on a future payoff, to apply standards of usefulness to the activities of every single basic scientist is not appropriate.

To appreciate how long a time perspective is often involved in the capital investment in scientific work, one has only to look at the history of science. Take, for example, the work of Daniel Bernoulli, who in the eighteenth century formulated an impractical principle about the behavior of ideal fluids. At the time, no one could have dreamed of the immense practical payoff that would result from this abstract, impractical investigation. Nothing less than the modern airplane depends upon Bernoulli's Principle; without it, the airplane as we know it today would not be possible. This should give pause to those who believe that the only reason sociology has not solved major social problems is the lack of will. The point is, before Bernoulli's Principle could become practical, much hard work in science and engineering had to take place, and the practical payoff was not realized until nearly one hundred fifty years later. If many current critics of basic science had lived in Bernoulli's time and had had their way, modern society would be very different from what it is.

Consider our norm for applied science. When we say that applied science aims to find uses for knowledge claims that have been tentatively accepted, we want to call attention to the fact that the scientific evaluation process never ends; applied science does not deal with eternal truths. A knowledge claim can always be supplanted by one that comes closer to meeting the standards of collective evaluation. Recognizing this, the layman can appreciate the interplay between evaluating a knowledge claim and using it. The attempt to use a knowledge claim can contribute important information relevant to its evaluation; thus there is a necessary interchange between applied science and basic science. Nevertheless, the primary objective of the applied scientist is the discovery of uses for knowledge, where the knowledge is regarded as given. Whatever the applied scientist contributes to the evaluation of that knowledge claim occurs as a by-product of the principal mission. On the one hand, it would be pointless to try to find uses for something you did not believe to be true. On the other hand, the failure to find a use for a knowledge claim would not question the truth of that claim. If an attempt to increase the productivity of a factory by providing workers with consistent evaluations did not succeed, this would be no reason to question the truth of the principle that workers who receive consistent evaluations are more productive. The principle may or may not be true; the steps necessary to put the principle into practice go far beyond the claim itself, and each step could represent a reason for the failure of application that had nothing to do with the principle itself. Suppose, for example, these evaluations were given by a foreman who was distrusted by the workers and not believed. He could be behaving exactly in accordance with the principle, but without any effect, or even with boomerang effects. We emphasize this distinction because there is a strong tendency among many laymen and practitioners to hastily conclude that the principle itself is false. Instead, the perspective that we are suggesting would raise very different questions and, indeed, would question the truth of the principle only as a very last resort.

Although many writers emphasize the continuity between basic science and applied science—indeed, some go so far as to deny any difference between them—in

of a main theme. After the revision has been made, a copy of the file of the table of contents can be used as a "dummy" (Mauch & Birch, 1993) for the content of the original report to be chosen, revised, and fit in. During the process of rewriting, it is often necessary to go back and rethink the table of contents as the outline again and again. The revised outline, then, will be applied to guide the rewriting further step by step.

The honors or master's thesis usually has limited scope and is suited for deriving some journal articles. Unless it is extraordinarily ambitious in the first place, rewriting it into a book would require enormous amount of work, which might be better considered a new project in order to supply the substance rich enough for a book. For a doctoral dissertation, on the other hand, rewriting it as a book might be much easier than condensing its content into a single journal article, or splitting it into several articles. If you have already planned your dissertation as a book and managed to have it carefully written that way, you may only need to rewrite the introductory and concluding chapters, add a preface, compile an index, and etc.

To publish a book you must find a publisher who would like to undertake a joint venture with you. Simply look at the reference books used for your project, and go to the library for a more thorough search to sort out those publishers whose emphases are closely related to your topic. Pick those whose interests in the subject have remained until recently, and send them an exploration letter with a table of contents for your book. If some of the publishers contacted find interest in your project, they will send you a letter requesting more information, usually a few sample chapters of the manuscript, for review. They may also send you some forms to fill out. You need to provide information with regard to the classification of your subject, numbers of pages, tables, and figures, suggested title, intended audience, relevant professional and academic organizations, contents, unique features and merits, recommendation from experts of the field if any, anticipated completion date, and information about the author. Usually you also need to provide a copy of your curriculum vita, since your capability will be reflected in your qualifications and experiences. In addition to the sample chapters, some publishers may require you to write a formal proposal following some detailed guidelines. All the material, then, will be sent for review of academic and professional content (mostly by persons in the same field), and also subject to market analysis of commercial feasibility. This could be a trying process to you. The outcome could be a deadly blow to your project, although the message is usually sent in a "Dear John" manner. Don't take it personally,

this work we emphasize the distinction. As we have said, individuals may move back and forth between basic and applied roles. The roles do involve different strategies, different emphases, and different givens. While attempting to apply knowledge may lead to discoveries of basic knowledge, such results are by-products and, as by-products, may even detract from the main objective of applying a scientific principle. It is our contention that the search for new knowledge and the search for applications of existing knowledge are different tasks; confusing the two objectives can only suboptimize both discovery and application. A classic example of this occurred when a large corporation asked a social science research team to find out why their product failed to sell. The team accepted the research contract, believing that they could both produce new knowledge and use existing knowledge to answer the client's question. After many years of research, giving questionnaires to a large sample of respondents, analyzing and reanalyzing the data, the team had difficulty pointing to any knowledge that had resulted. On the other hand, management at the client firm felt that the survey, which made use of the most sophisticated available knowledge of consumer preferences, did not tell them anything they did not already know about the failure of their product. In general, the same research techniques are not suited both to generating new knowledge and to answering a practical question. Attempting to maximize one of these objectives involves sacrificing the other. While there might be exceptional cases where the same research can both generate and apply knowledge, such circumstances are indeed exceptional.

It is even more important to distinguish engineering from both applied science and basic science. When we say that engineering is oriented to the solution of technical problems where the problem to be solved is regarded as given, we want to emphasize that solving practical problems involves more than scientific knowledge. Calling engineering problems "technical" in no way belittles them; rather, it limits the problems of concern to those for which a battery of highly developed intellectual tools is required. What our norm emphasizes is that the primary concern of the engineer is with the problem, and the engineer's success depends on whether or not a solution is found and how optimal that solution is. A scientific principle in and of itself almost never provides the optimal solution to a problem. Usually much more is required than scientific knowledge (sometimes this is given the name "clinical insight" or "artistry"). The important point is that when the problem is the main focus of concern, its solution nearly always involves extrapolating from what is known.

In solving a problem, the engineer uses scientific knowledge, technical tools of analysis, historical facts, educated guesses, and even hunches. If an engineer wants to build a bridge over a gorge, he or she requires not only laws of stresses and strains but historical information about the winds in the gorge and the history of avalanches along the sides of the gorge. In short, the engineer requires much more detailed information about the particular problem situation than a scientist investigating an abstract principle would ever need. The engineer combines these details with abstract scientific knowledge, his or her own insights, and a battery of analytic tools to make educated decisions in solving the problem. Furthermore, the engineer usually incorporates safety factors to compensate for the unknown. It should be evident that whatever use an engineer makes of scientific knowledge, he or she must extrapolate from that knowledge rather than simply apply it in a routine, mechanical fashion. And bridges do occasionally fall down.

In a way, it is unfortunate that there is no distinct activity known as *social engineering;* the term scares people, because it implies people being manipulated like robots. Yet sociologists do engage in a great deal of problem solving that is, from our perspective, engineering. Their concern may be making an organization work more effectively, reducing the crime rate in a particular city, contributing to the design of a system of higher education for a developing country, testifying before a court on the effects of racial segregation, or planning for new institutions necessitated by an aging population. In all these activities the problem is given, and the sociologist who chooses to work in the particular area does not have the option of changing the problem to meet the state of available knowledge. For example, a sociologist was once asked to be part of a team to estimate mass transit needs in the United States in the twenty-first century. The sociologist's contribution was intended to be a forecast of the composition of the population, its distribution in space, and its future transportation habits. The sociologist might have wanted to say that sociology had no knowledge relative to future transportation habits, but did have knowledge about why people move—and therefore, instead of forecasting transportation habits, he would prefer to design a program which encouraged people to move back to central cities. However, such a response would have been totally out of bounds as far as the client for the project was concerned. The sociologist had only the option of working on the stated problem or not working on it.

Once the problem is treated as a given, then the engineering focus does not insist that the sociologist work only in those areas where there is sound scientific knowledge. Rather, such a focus demands that the engineer use whatever wisdom, insight, and artistry can be brought to bear. A recognition of this "social engineering" would mean that the sociologist, as an engineer, could lose some inhibition about sticking to "hard knowledge" and could be a good clinician; the client would recognize that he was employing the social engineer for much more than hard knowledge—namely, artistry and insight going beyond what is known.

There are additional reasons for emphasizing the distinct orientations of basic science, applied science, and engineering. Consider, for example, the critics of social science who complain about the atomistic approach of these disciplines. They attack sociology for ignoring the total society and psychology for not treating the whole man. But a science of the whole "thing" is neither possible nor necessary. To be sure, a clinician treating an individual must pay attention to more details of the whole man than can ever be captured in a body of scientific psychological knowledge, and the social engineer must be aware of more features of a social situation than can ever be captured in a body of scientific sociological knowledge. But, as has already been pointed out, the clinician and social engineer operate beyond the limits of what is known, and no body of knowledge will ever eliminate the necessity for artistry and clinical insight. Furthermore, judgment represents the main ingredient in this artistry, and that includes the ability to disregard scientific knowledge as irrelevant to the particular case. If we grant that any action requires extrapolation from scientific knowledge, then it follows that a science of the whole thing is not necessary for action. To demonstrate that a science of the whole thing is impossible requires further development of the properties of a scientific knowledge claim, and so we will return to this issue later on.

The final reason for distinguishing the different norms is that our distinctions point to different foci and different constraints in the process of acquiring and using knowledge. What is assumed, what is problematic, and what can be ignored vary with the different orientations. Consider basic science. Before you know something, you cannot possibly know its uses or whether these uses will be beneficial or harmful; hence, you must assume that the process of evaluation of a knowledge claim is valuable for its own sake and that it does not depend on unknown future possibilities of use or misuse. If science as an institution were to worry about the range of uses of knowledge, the inhibitions would be severe, because the possibilities are infinite and not knowable in advance. However, and this must be emphatic, the individual scientist has obligations to control harmful applications of scientific knowledge. These obligations are part of other roles the scientist plays, not part of the obligations of his scientific role, but nothing gives the scientist the right to abdicate his role as a citizen.

While the scientist has obligations as a citizen, he also has obligations as a scientist, and it is important to keep these obligations separate. For one thing, scientists are not supermen who can solve any problem or who know what is best for the world. Occasionally, overadmiring segments of the public seduce some scientists into believing they can solve problems through sheer brilliance without regard for the relevance or irrelevance of their particular knowledge or competence. On the other hand, some scientists are too easily swayed by critics who condemn them for their irrelevance, and we witness mad scrambles to be relevant. As we have formulated the norm of basic science, it does not require that science work toward solving mankind's problems or even toward developing useful knowledge. Other institutions, or scientists themselves in other roles, may be oriented to the uses of knowledge; and certainly, the use of knowledge is a basic concern of society at large. Although society may support basic science, counting on the knowledge produced to be useful, the basic scientist who focuses only on what is presently thought to be useful will probably not produce any general knowledge, and in the long run will not produce anything new that has a chance in the future to be useful. In sum, the basic scientist must assume that the use of knowledge is the province of others and that it can safely be ignored until knowledge has been produced and evaluated; for basic science, only the cognitive status of knowledge claims is problematic.

The applied scientist and the engineer, however, cannot treat the scientific knowledge they use as problematic. They must assume that it has been scrutinized and evaluated according to the proper canons of science. What is problematic is the ability to use knowledge to achieve a result that will be considered a desired result; but desired by whom? For an applied scientist, demonstrating that he can use scientific knowledge to achieve some practical outcome may be valuable in and of itself. The engineer, on the other hand, often takes his client's definition of the desirable result as a given, so the evaluation of outcomes and the value judgments on which they depend are not problematic.

The value issues in applied science and in engineering, however, cannot be dealt with so easily. Action in the real world is an institutional commitment for applied science and engineering; hence, it seems that such a commitment entails the responsibility to evaluate the value consequences of the action from a broader perspective than that of the particular applied scientist or engineer. Those who criticize the narrowness of engineers, for example, point in part to the frequent failure of engineers to look beyond their client's value position. Consider, for example, the highway engineers whose clients want a high-speed freeway designed to minimize construction costs. Until very recently, how many highway engineers considered ecological damage in their calculation of costs? How this responsibility can be handled institutionally without crippling applied science or engineering poses serious problems, but we have passed the time when society will automatically applaud any technical result.

Some will object that we are biased toward basic science, since we have allowed it a freedom that we do not allow to engineering and applied science; that is, we have claimed that basic scientists are not obligated to consider the value gains and losses that might result from the use of their knowledge. One might object that, even if we restrict basic science to thought rather than action, thoughts have consequences for social values. Our defense to the objection is two-fold. First, the actions of applied science and engineering are relatively focused, so it is practically possible to consider value gains and costs of these actions; but the possible uses of abstract knowledge are not focused, making it necessary to consider the total range of human values for every knowledge claim considered, and that is clearly impractical. Secondly, imposing the obligation to consider value issues not only when acting but also when thinking constitutes a form of thought control. Even if desirable, effective thought control is not feasible. Society can tell a person what to do and what not to do, and it may be fairly effective, but it cannot similarly tell a person what to think.

Issues of social control of science have received considerable attention in recent years. The objective of this discussion has been to bring out some of the central features of social control, where it is located, and how it is exercised. Consumers have a vital interest in the problem of social control, so it is unfortunate that we cannot pursue the discussion; but a full treatment would require a book in itself. Before leaving this topic, however, we should draw two implications of our analysis of norms. First, social control occurs both within the institution and from other institutions of society, but internal and external control have different purposes. The position here, for example, argues that control over the uses of basic scientific knowledge belongs outside of basic science, but that applied science and engineering share with other institutions of society the responsibility for controlling the uses of knowledge. Secondly, an understanding of the normative structure of science should enable scientists and laymen to exercise more effectively their appropriate controls. Conversely, if society required science to exercise society's responsibilities, or if science abdicated its own responsibilities to society, the results would be counterproductive. Science is not competent to make the value decisions involved in the use of knowledge. For societal pressure to force science to get involved in these value issues will not protect society from the misuse of knowledge, and it may prevent science from doing what it is most competent to do—produce and evaluate knowledge claims.

Society has every right to demand that basic science exercise stringent controls over the production and evaluation of knowledge claims. Furthermore, exercising these controls is not a trivial matter. There are disputes within science over what constitutes a knowledge claim, what kinds of knowledge claims are properly within the realm of science, what standards are used to evaluate knowledge claims, and how knowledge claims are produced. These problems, as they relate to sociology, constitute our concern for the remainder of this book. While the issues seem elementary, the reader should recognize that they are not simple (*elementary* also means *fundamental*, and often elementary issues are the most profound).

Before turning to the specific problems of producing and evaluating knowledge claims, we must examine one additional general norm. This norm, while it operates to some extent in applied science and engineering, is crucial to basic science.

INTERSUBJECTIVE TESTABILITY AS A NORM

The second norm for examination is:

Scientific knowledge claims are intersubjectively testable.

Previous discussion introduced the ideas that are incorporated in this norm. The requirement of intersubjective testability emphasizes the collective responsibility for the evaluation of knowledge claims and excludes from science those claims which cannot be subjected to collective evaluation.

Consider the following example. Social scientists have been interested in community power structures for many years. A large number of studies have focused on how power is organized in cities and towns. Methods for studying community power structures have been developed and utilized in the attempt to evaluate something like the following knowledge claim: "In most local communities the power structure is pyramidal." By pyramidal we mean that power is monopolistically held by a single cohesive leadership group.

Two principal methods have been put forward to evaluate this kind of knowledge claim. One is known as the *reputational method,* in which informants are used and are asked to identify the most influential people in the community. In competition with the reputational method, the *decision-making method* involves historical reconstructions of community decisions, using available documents and defining as leaders those who are active participants in the decisions. A less formal method that has also been used to study community power involves a sophisticated observer doing a *case study* of the community. Approximately half of the studies using the reputational method find that community power is pyramidal, no studies using the decision-making method find that power is pyramidal, and about a third of the studies using the case study approach find that power is pyramidal (Walton, 1966).

What, then, do we say about the proposition that most community power structures are pyramidal? All three methods seem to support the conclusion that the proposition is not true, but the agreement ends there. What do we substitute for this knowledge claim? If we look at the research using the decision-making method, we might formulate the knowledge claim that no community structures are pyramidal; if we look at the results using either of the other methods, we might put forth the knowledge claim that some local community power structures are pyramidal. Since these two knowledge claims are contradictory, we are in an intolerable situation.

What makes the problem worse is that the evaluation of knowledge claims not only differs for differing methods of observation, but also differs for the disciplinary affiliation of the researcher. Sociologists tend to use the reputational method, while political scientists and anthropologists use the decision-making and case-study methods. In his analysis of fifty-five such studies of community power structure, Walton argues that the disciplines have different ideological perspectives and hence choose different methods and arrive at different conclusions. Not only does he point to a problem due to the possible bias of a method, but he suggests an ideological bias as well.

If such biases were inevitable, collective evaluation of knowledge claims about community power would be impossible. One's evaluation would depend upon the method one used and the ideological bias that this method reflected. If this were inevitable, scientists would regard knowledge claims about community power structure as outside the realm of science. Fortunately, there is no need to regard the present situation as inevitable. Even with this example there is the minimum accomplishment of collectively rejecting the knowledge claim that most community power structures are pyramidal. With more adequate formulation of knowledge claims and greater refinement of methods of study, it is possible to greatly enhance the possibilities for collective agreement in the evaluation of knowledge claims. Indeed, Walton concludes that ''studies of local power structure will benefit from use of a combination of research methods as protection against this source of bias''[2] (1966, p. 689).

---

2. Walton's proposal may not solve the problem, especially if different methods provide contradictory information about a particular community. In fact, we will suggest in the next chapter that such unqualified claims as ''In most local communities the power structure is pyramidal'' are not appropriate scientific knowledge claims. We will argue that a better way to formulate the problem (and one that is solvable) is to ask the question, Under what conditions is a local community power structure pyramidal?

The norm of intersubjective testability requires that scientists be on constant alert for sources of bias and that they make every effort to develop procedures for the elimination of bias once it is identified. This norm constitutes a reformulation of the older notion of scientific knowledge as objective. Few scientists today would claim that their knowledge is objective—in the sense that it corresponds to reality and is independent of the knower. They recognize that one cannot determine objectivity in this sense; any determination involves human beings and thus depends on the knower. Furthermore, they understand that scientific knowledge claims do not capture reality, but are abstractions from reality. As abstractions, scientific knowledge claims leave out much of reality, sharpen some aspects of reality, and rearrange the elements of reality. Science does not hold up a mirror to reality, and scientific knowledge claims are not photographs of reality.

Objectivity remains an ideal of science, although there is the realization that it can never be achieved. Striving for this unrealizable goal forms much of the character of modern science. But we must strip away one connotation of objectivity: the belief that, once attained, scientific knowledge would be true for all time. The past hundred years have dramatically altered this view. Laws of nature—the very name implies eternal truth—have been repealed in a succession of scientific revolutions. The fact that scientific laws, like other human constructions, are modifiable has made older views of objectivity untenable.

Requiring collective evaluation raises the question of what constitutes the collective. A scientific knowledge claim is not subject to the evaluation of everyone and anyone, since not everyone is competent to evaluate it. Although the norm necessitates public evaluation, the jury for this evaluation is narrowly limited. An institutionally defined ''relevant public'' exercises the right of evaluation, and admission to the relevant public requires certification of competence, symbolized today by the doctorate and a list of publications in the field. Although some critics compare relevant publics to priestly cults as far as natural science is concerned, most of us accept the necessity of delegating these evaluations to the relevant public just as we accept the necessity of delegating judgments of guilt or innocence to a particular criminal trial jury. The sociologist certainly does not feel competent to evaluate the knowledge claims of a physicist.

It is interesting to observe that, while laymen are quite willing to delegate evaluation of natural science knowledge claims, they do not have the same willingness to delegate the evaluation of social science knowledge claims. Laymen, particularly well-educated laymen, do not hesitate to judge social science. They seem to feel that being human automatically guarantees competence to judge claims about human behavior. This is symptomatic of the lack of institutionalization of social science, and it is detrimental to scientific development of social science. Complexities associated with the testability feature of the norm of intersubjective testability should demonstrate conclusively that most of these judgments require technical competence and provide one set of guidelines for occasions when consumers and laymen should reserve judgment.

Scientific knowledge claims are testable in two senses: they are subject to the evaluation of critical reason ad they are subject to evaluation by evidence. Scientific norms represent a unique combination of rational and empirical standards. Although most discussions concentrate on the empirical aspect of science, the rational apect cannot be neglected. Questions such as what constitutes an empirical test can only be resolved through reasoned argument and analysis as well as collective acceptance of the implications of the argument.

The norm of intersubjective testability requires collective agreement on what constitutes a test of a knowledge claim and what is the outcome of the test. The example of research on community power structure illustrates the problem of obtaining agreement on the outcome of a test—that is, bringing to bear observations to evaluate a knowledge claim. On the other hand, it could be argued that one or all of the methods used to study community power structure are inappropriate tests—that is, they are irrelevant to the knowledge claim. Deciding what constitutes a test (i.e., what observations are relevant to a knowledge claim) and what is the outcome of a test (i.e., whether the observations confirm or do not confirm the knowledge claim) involves exceedingly difficult and complex issues.

Consider the following example: Several years ago a sociologist put forth a claim about social change in the United States; another sociologist decided to test that claim empirically through a content analysis of the themes used in magazine advertising. The second investigator argued that, because advertising in women's magazines should be especially sensitive to social change, an analysis of a sample of these magazines over a period of years should provide evidence relevant to the claim. The first sociologist was enthusiastic about the project until it turned out that the content analysis did not support his claim. Apparently, the outcome of the test diminished its relevance as a test. Of course, there are serious questions about whether or not magazine ads provide relevant evidence for a claim about social change in the United States; arguing for relevance depends on a number of assumptions about the United States and about magazine ads. For example, we can argue for or against the assumption that magazine ads are sensitive reflectors of cultural standards, and the relevance of the study depends on our decision about that and other assumptions. Besides substantive assumptions about social change, the United States, and magazine ads, other considerations such as the adequacy of the sampling procedure enter into judging whether the empirical study constitutes a test of the first sociologist's claim. What should not enter into this judgment, however, is our like or dislike of the outcome of the study. Agreement on what constitutes a test cannot depend on whether or not the test supports our desires for the knowledge claim to be true or to be false.

Agreement on what constitutes a test and agreement on the outcome of a test pose difficult technical issues that cannot be settled by a snap of the fingers. The reader may be surprised that obtaining agreement on the outcome of a test is not more clear-cut; most people believe that a test clearly shows a knowledge claim to be either true or false. But unfortunately, as we shall see later on, empirical truth is very difficult to determine. For the sake of discussion, let us assume that the magazine study constitutes a test. If the magazine study were consistent with the claimed social change, would we say the claim was true? Suppose another study was inconsistent with the claim. What would we say then? What if the magazine study was inconsistent with the claim? Would we regard the claim as false? Clearly, the results of one study are not enough. But then, how many studies are required? Is this even an issue that can be settled by establishing a standard number of required tests? The outcome of one test or of even a large number of tests is not a simple decision about whether the knowledge claim is true or false. The problem with the community power studies, mentioned earlier, remained after fifty-five different studies were analyzed.

Obtaining agreement on the observational outcome of a single empirical study poses no insurmountable problems, but it does require the formulation of explicit standards, the acceptance of these standards, and the technical competence necessary to apply them. Sociologists, particularly methodologists, have paid a great deal of attention to these problems; so we know at least where to begin. The more difficult problems of obtaining intersubjective agreement on the interpretation of observations and on the relevance of observations to knowledge claims have received little attention. But these are important, albeit difficult, problems, and we cannot duck them; they involve nothing less than the empirical character of science: Science as an institution is firmly committed to the position that the empirical world has the capacity to modify scientific knowledge claims.

The strategy for tackling the problem of agreement on interpretation and relevance of observations also involves formulating explicit standards, promoting the acceptance of those standards, and developing the technical competence to apply them. But this strategy requires leaving the empirical realm and the realm of what is traditionally called methodology. Sociological interpretation of observations and decisions about the relevance of observations to sociological knowledge claims are issues of sociological theory. Not only are theory and method intimately related, but the solutions to problems of theory construction in sociology will also contribute to solving some knotty methodological problems.

In keeping with the view of science as a normative system, we can look at problems of theory and theory construction as requiring the formulation of subsidiary norms for the implementation of the norm of intersubjective testability. Defining what we mean by sociological theory and developing standards for evaluating sociological theory are both norm-setting activities. Clearly, value judgments are involved, but the cognitive content of these standards allows them to be judged on grounds apart from the values or tastes of the judge. Consider, for example, the norm requiring scientific knowledge claims to be precise, which can also be expressed, "Precision is desirable for scientific knowledge claims." This makes the value aspects explicit. But we can look at this standard in another way; namely, that the standard of precision is instrumental to achieving the standard of intersubjective testability. This latter statement no longer depends on value judgments but is a purely factual claim that can be evaluated on logical and evidential grounds.

As the last comment implies, we will approach theory and method in an effort to develop standards that are instrumental to achieving conformity to the norm of intersubjective testability. It is our belief that scientific sociology requires institutional and individual commitments to the two fundamental norms presented in this chapter: (1) sociology as a basic science must be primarily oriented to the production and evaluation of knowledge claims and (2) sociological knowledge claims must be intersubjectively testable. These commitments can only be made on value grounds. To those who share our commitment, we would point out that once committed to these two norms, we can leave the arena of value judgments and consider the proposals put forward in terms of whether they are instrumental to the norms. The reader who accepts these basic normative commitments, then, has a cognitive criterion against which to evaluate the standards for theory and method that we develop here.

SUGGESTED READINGS

Feigl, Herbert. ''The Scientific Outlook: Naturalism and Humanism.'' In *Readings in the Philosophy of Science,* ed. Herbert Feigl and May Broadbeck. Englewood Cliffs, N.J.: Prentice-Hall, 1953. Pp. 8-18.

Feigl presents five pricipal criteria of the scientific method. We believe his criteria represent norms of science and note that he gives a prominent place to intersubjective testablility.

Kuhn, Thomas S. *The Structure of Scientific Revolutions,* 2d ed. Chicago, Ill.: University of Chicago Press, 1970.

Kuhn provides an interesting and controversial account of the way science changes.

Merton, Robert K. ''The Normative Structure of Science.'' In the *Sociology of Science,* ed. Norman W. Storer. Chicago, Ill.: University of Chicago Press, 1973. Pp. 267-78.

This is an early and pioneering work in analyzing scientific norms; it was originally written in 1942.

Popper, Karl R. *The Logic of Scientific Discovery.* New York: Basic Books, 1959.

In this fundamental work, Popper tackles the central problem of what distinguishes scientific knowledge from other types of human knowledge.

and keep trying, as long as you are confident about the value of your work. Eventually, and hopefully, you will be thrilled by the offer of a publishing contract, which usually comes with an author's guide on how to prepare the manuscript. Some publishers offer editorial assistance; others may require the author to provide a "camera-ready copy" (CRC) of the typescript (ready for the printer). In any case, you must be able to produce copies of your manuscript that meet the publisher's specific requirements. Nowadays, this usually means computer-generated files and laser-printed fine documents. If you are required to provide camera-ready copy, the burden is on you for all the editing and typesetting tasks (you certainly can have someone else do them if you trust her and you can afford to pay). In addition, you must follow a specified citation format (this is easier than reading a whole publication manual!). You must also obtain written permissions for all copyrighted materials used in your work. Here you should pay close attention to that matter since copyright provisions are usually more generous to theses/dissertations than to formal publications. Therefore, something you took for granted in your first writing may result in violation of law if you do not change it when rewriting for publication. Generally speaking, the "golden rule" is to cite only when it is necessary. This is especially true for long, direct citations. Note, however, your experience in writing the thesis or dissertation may be different. Some of your readers (e.g., advisor and/or committee members) would take the amount of cited work as a criterion for evaluation. Some would not even allow you to miss any major work in the field. Some journal reviewers also make their judgments of the manuscripts partly based on this factor. The best you can do, therefore, is always to try to use your own words but provide full citations of all the major related works.

Writing journal articles is similar to writing books in certain ways. To publish an article you must find a suitable journal. Look at the articles used for your project, and go to the library to sort out those journals whose emphases are closely related to your topic. Contributing to a journal is nevertheless simpler than submitting a book proposal. You just follow the guidelines of the journal and send your article directly to the editor (be careful: the address of the publisher and the address of the journal editor are often different). The article will usually be reviewed by anonymous peer reviewers, which is a time-consuming process, and thus is a real test of your patience. The outcome could be that your manuscript gets rejected, with the reviewers' comments being sent to you. You should keep trying by turning to another journal, as long as you are

FOUR

Ideas,
Observations, and
Knowledge Claims

In the last chapter we learned that basic science was oriented to the production and evaluation of knowledge claims and that these knowledge claims had to be intersubjectively testable. In this chapter we will look closely at scientific knowledge claims to determine what they are, how they are arrived at, and how they are used in science. From now on, when we use the word *science,* we will mean basic science.

Constructing, evaluating, and using scientific knowledge claims takes us into almost every aspect of the research process. Broadly speaking, one gets an idea and considers where that idea applies, what observations are relevant to the idea, what limitations and qualifications are necessary, what constitutes a test of the idea, and when an idea passes the test. This broad-brush picture of the process summarizes the way a great many scientists operate. Here we will take up the requirements for such ideas, how they are limited and qualified, and how they are made relevant to observations. Our purpose is to provide a much more careful and rigorous formulation of this broad-brush picture.

In order to present a careful formulation of the research process, we must introduce several fundamental concepts. Most of these concepts will be explicitly defined. The key concepts and their relationships are illustrated in figure 4.1. The arrows in the chart may be interpreted as meaning "lead to"; an idea leads to both a scientific knowledge claim and an observation statement. The boxes indicate part-to-whole relationships where the parts are numbered; thus, scope conditions are a part of both a simple knowledge structure and a theory. The chart illustrates the major concepts that will guide our analysis of the research process for the remainder of the book. These concepts are new and are not easy to understand, and so we will devote this chapter to defining, developing, and illustrating them. That task, in turn, will require introducing additional concepts that will be used to develop the central ideas. These include theoretical knowledge, historical knowledge, abstract statement, universal statement, and conditional statement. It will be necessary to explicitly define these as well.

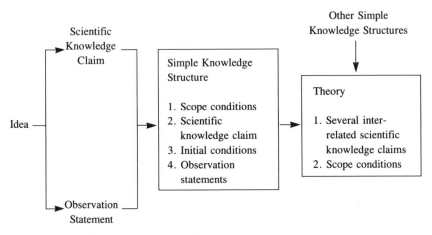

Figure 4.1 Key Concepts and Their Relationships

Some terms will not be defined precisely; for these we must assume that the reader's intuitive notions correspond to our intended meaning. The principal example is the term *idea*. While it is possible to give examples of ideas—"Prestige in society corresponds to status" or "Red tape is a property of Washington bureaucracy"—it is virtually impossible to rigorously define the concept *idea*. A quick glance at a dictionary will confirm this. As we will see in a later chapter, it is always necessary when developing a system of definitions to begin with some undefined terms which depend upon intuitive understanding.

We should say something about our choice of terminology. The terminology used in this chapter is new. Some of the terms do not correspond to existing usage either in sociology or in science in general. There are two reasons for introducing a new terminology. First, existing terms have a variety of different meanings, only some of which correspond to the meanings we want to convey. A good example of this problem is the usage of the term *theory* in sociology. The term is used very broadly, so that almost any hunch or idea is labeled a theory. In natural science the term theory has a much more restricted meaning, and the definition we will present shortly will be faithful to that usage. It is important for the reader to keep in mind that the use of the term theory in this book differs from its usage in sociology in general; thus, it is inappropriate to read all meanings sociologists attach to that term into our usage. The second reason for introducing a new terminology is that there is often a number of different terms for essentially similar ideas. Thus, one finds such terms as *proposition, hypothesis, empirical generalization, assumption, axiom, principle, postulate, theorem,* and *corollary* used by some writers interchangeably and by other writers to represent distinctly different ideas. The concepts to which these terms refer are similar *but not identical;* one cannot really fault either set of writers. Since our purpose is to convey specific meanings for our concepts, however, we run a great risk of confusion if we rely on existing terminology. There are some times when we borrow existing terms—for example, *theory* and *initial conditions*—but we hope that in those instances we are being faithful to usage in science. We run a risk with the term theory, because our meaning is much more restricted than that in sociology, but no better term is available to us.

We recognize that the introduction of new terminology puts a burden on the reader, but this is a trade-off for guaranteeing that the reader shares our intended meanings. The issues to be considered are controversial, and the controversies can only be understood by cutting through the confusion that results from varying terminology. Our hope is that precise formulation of our concepts will at least serve to clarify the issues.

DEBUNKING A MYTH

There is a widespread myth that any science begins with observations of nature. Most laymen encounter this view at a very early age and generally believe in it. To be sure, it is an ancient and honored belief, having its roots in Francis Bacon's writings over three hundred years ago and having champions in every age. Despite its veneration, however, it is still a myth.

A simple exercise suffices to demonstrate the falsity of the view that science starts with observations. Suppose we send students out to observe an ongoing social institution. They are sent to a courtroom and told to observe the social structure of the courtroom and to prepare a set of field notes that contains anything and everything that may be relevant to observing social structure. Even though we do not define social structure for the students before they go into the field, simply telling them to "look at" social structure already removes the exercise one step from pure observation. Nevertheless, what happens with even this small degree of structuring demonstrates clearly that observations themselves depend upon preexisting ideas in the minds of the observers.

This exercise generates a great deal of frustration in the students. Imagine how high the frustration would be if we sent a group of students out into the field with the instruction, "Go and look." Students quickly realize that they cannot observe everything. Even a setting as restricted and organized as a courtroom provides infinite possibilities for an observer to note. The first thing students learn is that they must be selective, and their frustration stems from the lack of clear instructions concerning what to select from the mass of sensations that strike them in the field setting.

When students return from the field and compare their field notes, they discover two important features of observation. First of all, they find some things which nearly all observers notice, so that some selectivity at least is shared. For example, all students note that there is a judge in the courtroom who sits higher than everybody else and wears a distinctive costume. Secondly, each observer realizes that he has noted some things which no other observer recorded. One student, for example, will concentrate on the physical arrangement of the courtroom, going into copious detail in describing the physical features of the setting, even going so far as to count the number of chairs in the room. Another student will focus on the case being tried and will present an elaborate description of the testimony of the witnesses. A third observer becomes fascinated with the clothing of all the actors in the courtroom drama, and this student's field notes resemble a description of a fashion show.

What the students readily grasp from the comparison and analysis of their field notes is that their shared observations to a large extent represent a shared cultural frame of reference which provides categories to order their perceptions. This shared culture prior to the process of observing generates agreement among observers as to what is perceived and how it is organized. When students from another culture are part of the group of observers, their presence emphasizes the importance of a shared cultural frame of reference. Often, features of the courtroom which American students take for granted are totally strange to students from another culture. Students from societies which do not have jury systems, for example, usually give much less weight to observations of jurors in their field notes. Obtaining agreed-upon observations of any phenomenon, then, depends on much more than the physical reality to which the observers expose themselves. It depends upon preexisting ideas that direct the observers and provide categories for ordering the experience.

Even when an observer has a unique basis for selecting and recording his observations, the observations can be traced to preexisting ideas. Our example of an observer's interest in clothing generating an attention to details of fashion certainly supports this claim. In short, the phenomenon does not speak for itself; observations are not determined by the thing observed. At the very beginning of the process—the decision about what to observe—ideas play a crucial role, even though the ideas may be only partly conscious and not at all articulated. Moreover, it follows that shared observations depend upon shared ideas.

One more problem deserves mention. Observation is an active process: physical sensations from reality do not record themselves; every observation involves inference on the part of the observer. We do not, for example, observe anger. We observe facial expressions, tones of voice, things that are said, and we infer that a person is angry. As we attempt to observe instances of more abstract concepts (for example, when attempting to observe social roles in action), the process of inference increases in complexity. The fact that the observer must make inferences in order to observe poses problems that we must confront and solve. If we are concerned with shared observations, and if these observations depend upon inferences of individual observers, then for the observations to be shared the inferences must be made in the same way by the observers. In short, observation depends upon shared collective modes of inference.

As the preceding discussion illustrates, we cannot accept the view that the starting point of science is observation. While it is clear that observations represent a crucial element in science, raw observation, uncolored by mental activities of the observer, does not exist. The philosopher John Locke's view that the human observer is a blank tablet—a *tabula rasa*— is simply untenable. Particularly when we require collective agreement on the nature of observation, something clearly must precede the act of observing. We have already seen examples of how a common culture provides implicit ideas that nevertheless offer a basis for shared observations. In science the development of ideas and the training in applying these ideas provide the common frame of reference necessary for agreement on observations.

In fact, one major purpose of theory in science is to provide a common frame of reference for collecting and analyzing observations. If we provisionally define a scientific theory (we will discuss theory in chapters 10, 11, and 12), we can illustrate its role in providing a common frame of reference.

Definition:
A scientific theory is a set of interrelated statements, some of which are definitions and some of which are relationships assumed to be true, together with a set of rules for the manipulation of these statements to arrive at new statements.

The statements assumed to be true are usually called the *assertions* of the theory, and the rules are called the *syntax* or *logic* of the theory. If, for example, we have a theory about status, the theory would contain a definition of *status,* which would represent a collection of properties that one would look for when attempting to observe status. If one property attributed to status in the theory was deferential behavior (for example, a person of lower status defers to a person of higher status), then all observers using the theory would be alerted to look for instances of such deference. As we have defined the theory, it would contain many more things than simply the property of deference, so observers would have a number of elements that they shared in observing status. Researchers using the theory would spell out explicitly what to observe and how to draw inferences from what is observed.

Even in many everyday observations, we make use of scientific theory, though we may not be aware of it. As we noted in chapter 2, to observe the speed of an automobile, using a speedometer, involves us in a complex chain of scientific theory which, when we read the needle, we may not recognize but which is a vital part of our observation.

A scientific theory provides one basis for the process of observation. But to claim that science begins with theory is misleading. As we have defined the term theory, it involves the explicit statement of many ideas and the interrelationships among these ideas. It is too much to expect that one can start out with such a well-developed formulation. In short, scientific theory does not spring full-blown from the head of Zeus. If scientists had to wait for the kind of well-developed formulation that scientific theory entails before they did anything, they would be paralyzed.

Furthermore, scientific theories would never develop without observations; so observation cannot wait until a theory exists. Hence, just as we must reject the myth that observations are the starting point of science, we must also reject the view that science begins with theories as we have defined them.

IDEAS—THE RAW MATERIAL OF SCIENCE

Science starts with ideas, and an idea is not a theory. An initial idea may be somewhat vague, may have many implicit elements, and may not be connected to many other ideas. An idea is the raw material from which theories emerge, and the process that leads to the emergence of theories is a long and difficult one. A theory emerges from the development and collective evaluation of many ideas. One may have an initial idea that higher status people have more power in society than lower status people, but that is a long way from having a theory of the consequences of social status. The problem is how to get from the crude idea to the collectively accepted theory. If we focus on the idea, the problem is really three-fold: *I must be able to evaluate my own idea; others must be able to evaluate it; and "reality" must be able to affect the evaluation of ideas.* Through the process of evaluation, an idea is extended, modified, sometimes rejected, almost always superseded by a better statement of the idea, and always related to other ideas. The one point which must be emphasized again and again is that the evaluation of ideas is a process, one successful culmination of which is a theory.

We must recognize however that, while ideas which successfully undergo this process develop into scientific theories, not all ideas become scientific theories. Some ideas are inherently outside of science; they are not subject to collective evaluation. My preference for chocolate ice cream is not a collective issue. The triviality of this example should not, however, obscure the fundamental importance of the point. While everyone would agree that a preference for chocolate ice cream is a matter of individual taste, if I had said that my preference is for democratic institutions, we probably would have more argument. Nevertheless, these two examples deal with the same issue. Things that are matters of individual taste are not subject to collective evaluation and are therefore outside the realm of science.

Some ideas are not related to "reality." They are nonempirical—by which we mean that there is no way to relate the ideas to sense experience. Thomas Hobbes characterized the initial state of society as "a war of all against all." As his statement stands, the idea is nonempirical despite its obvious appeal to the pessimists among us. It is impossible to return to the initial state of society (the famous "state of nature" which has a central place in much classical political theory) and impossible to obtain any information about this state of nature. We therefore cannot collect observations that will enable us to decide whether the initial state of society was a paradise or a "war of all against all." Because we are unable to observe the initial state of society, there is no way that any reality can change the opinion of someone who holds the Hobbesian belief.

One way to view the process of development of scientific ideas is to consider it in part as a transformation of ideas into knowledge claims. Those ideas which cannot be transformed are then outside of science. Such things as value judgments, preferences, and nonempirical ideas, while they may influence the course of science, are not the material from which scientific theories emerge. Only those ideas which can be collectively evaluated through observations from reality are the ideas with which science deals.

The perspective that science begins with ideas and transforms these ideas into knowledge claims has two very important implications. First is the implication that ideas are not reality; they are abstractions from reality, and necessarily selective abstractions. While reality may be exceedingly complex, scientific ideas always involve simplification. The distinction between ideas and reality cannot be over-emphasized. A great danger arises from forgetting this distinction and confusing ideas with reality. Consider, for example, the definition of an apple. In a dictionary we will find something like "a round, firm, fleshy, edible fruit with a green, yellow, or red skin and small seeds." It is very clear that any apple that one selects at a fruit stand has many more properties than that definition captures. Its taste can be sweet, sour, sharp, or delicate; its size can range from a diameter of one inch to a diameter of six inches; thus any idea of an apple omits many properties that could be used to describe a particular apple. The idea of an apple is an abstract idea selecting a limited number of properties from a much larger collection embodied in any real apple. While the idea of an apple is helpful, it is not the same thing as a real apple. Indeed, sticking strictly to the dictionary definition, we might call things "apples" that are commonly defined as other fruit; some pears would fit the definition. What is important is to recognize the possibility of error in using the idea. If we confuse the idea with the reality, we will frequently make mistakes and be totally unaware of them. Confusing the idea with the reality is known as committing the *error of reification*.

THE ERROR OF REIFICATION

The error of reification has occurred quite frequently in the social sciences. Perhaps the best example is the concept of "group mind." As an abstract idea standing for certain properties of group activity, the idea can be useful. When the group mind is reified, however, we attribute "thinking" to groups in a totally inappropriate way. This reification has generated many pointless controversies over whether or not groups actually think. If we are sensitive to the problem of reification and recognize "group mind" as an abstract idea, we will avoid such controversies.

The problem of reification becomes particularly acute in the use of scientific ideas. While scientific ideas may apply to a particular situation—for example, we can analyze a particular hospital as a bureaucracy—the application always introduces a certain amount of distortion of the situation. Even if we were applying a fully developed theory of bureaucracy to analyze Metropolitan General Hospital, we must remember that our analysis ignores many things going on in the hospital that are nonbureaucratic or are even antibureaucratic; it should come as no surprise that the friendship between Dr. Jones and Nurse Smith should result in violating the bureaucratic rules which may govern the hospital. Describing the hospital as a bureaucracy is perfectly legitimate, as long as we recognize that the bureaucratic analysis is a *partial* description, not a total description. Partial descriptions are indeed very useful, as long as we do not confuse the description with the thing being described, that is, as long as we do not commit the error of reification.

THE IMPRACTICALITY OF HOLISM

The second implication of this perspective is that the approach to scientific understanding of phenomena is analytic rather than holistic. In the last chapter we asserted that a science of the whole thing was impossible. Now we can say that, since any description of reality is partial, attempts at total description must suffer from the error of reification. Beginning researchers often fall into the trap as, for example, when one wanted to do a study of a classroom describing everything that went on.

No ideas or collections of ideas can capture all aspects of a single human being or a total society. Of necessity, the formulation of an idea about the bureaucratic aspects of a general hospital leaves things out, overemphasizes some aspects of the phenomenon of the general hospital, underemphasizes other aspects, and is thus not a holistic description. The very act of formulating an idea—that is, abstracting from the reality—is an analytic activity. Unfortunately, many of the proponents of holistic approaches fail to recognize this fundamental fact. What they really mean when they say that psychology should consider the whole person is not that we shouldn't analyze elements of the person, but that our ideas should represent a different analysis than is traditionally done. We should form ideas, for example, that relate emotions to learning, rather than simply look at learning as responses to stimuli. Of necessity, the proposal is no less analytic than the ideas of a stimulus-response psychologist who looks at how rewards and punishments affect responses to stimuli. Similarly, the holist who argues that one cannot understand a group from examining characteristics of group members is proposing a different set of ideas to analyze the phenomenon of a group, but he or she is still arriving at ideas analytically. Any formulation of ideas about phenomena must depend upon an analytic approach. As Phillips (1976) put it, "The analytic...method...is such a moderate and reasonable position that no scientist, not even a holist, can avoid putting it into practice. By contrast, holism—taken seriously—is an eminently unworkable doctrine."

In short, science is not an attempt to photograph reality, to capture the totality of the situation, or to know all variables. If one recognizes that ideas are abstractions from reality and that, of necessity, they omit significant aspects of reality, one cannot be a holist. To be sure, it is a difficult issue to decide between partial descriptions that are useful (there may be a better way to analyze Metropolitan General Hospital than looking at it as a bureaucracy) and partial descriptions that are inadequate. As we develop criteria for evaluating knowledge claims, we should be in a better position to evaluate partial descriptions. For the present, however, we should be suspicious of descriptions that claim to be complete.

SCIENTIFIC KNOWLEDGE CLAIMS: STATEMENTS ABOUT EVENTS

Reality comes in many disguises. We often hear that science describes phenomena, or that science explains events, or that science predicts future occurrences. In that way it is claimed that events or phenomena are the subject matter of science. If such claims are a shorthand way of speaking, we are all right, provided that we remember that the shorthand stands for ideas about phenomena or ideas about events. If we forget that, there is the danger of reification because total phenomena or whole events are taken as the object of science. Prediction and explanation in science are also partial. They are predictions or explanations of an *aspect* of an event or an *aspect* of a phenomenon. Ideas are not phenomena or events; ideas are statements. They are verbal expressions about aspects of phenomena or aspects of events.

The argument can be summarized with a few important assertions:

Science is concerned with making statements about events or phenomena. These statements have particular properties.

One of the principal properties of scientific statements is that the statements can be collectively evaluated by reason and evidence.

Any limitations on the kind of statement we should consider are limitations designed to facilitate collective evaluation by reason and evidence.

When ideas are transformed into statements that can be collectively evaluated by reason and evidence, they become scientific knowledge claims. We are now ready for a more precise definition of a *scientific knowledge claim:*

A scientific knowledge claim is an abstract, universal, conditional statement in the form of a declarative sentence that has a subject and a predicate (i.e., is a grammatical sentence) where the predicate asserts something about the subject of the sentence.

As we will use the term scientific knowledge claim, the sentence "American society in the 1980s taboos incest" is *not* a scientific knowledge claim, whereas the sentence "For societies which have independent nuclear families, all such societies will taboo incest among members of the nuclear family" is a scientific knowledge

confident about the value of your work. Eventually, you may be thrilled by an acceptance letter, which usually comes with the reviewers' comments for improving the manuscript. Unlike a book project, the submission of an article to a journal for publication does not involve complicated procedure. The rejection rate is high, however. After a back-and-forth process of rejection and resubmission, it would take a considerable period of time (months or over a year) for an article to finally get published.

It's worth the effort: Where there is a will, there is a way

Translating a thesis, a dissertation, or a research report into a journal article or even a book poses a real challenge to the beginning researcher. For a new college graduate, there is only a fleeting opportunity while she is swamped in the major life transition following graduation. This opportunity, however, is unique for any researcher since there is hardly any other comparable one that is so good for starting publication. Here are the reasons. First, if writing for publication is a risky investment on the part of the researcher, the risk has been reduced to the minium since the thesis, dissertation, or funded research report has been written anyway. The additional investment of time and energy, though maybe still considerable, is minor compared to creating a brand new product. Second, the thoroughness of the work is unique since in few, if any, other research occasions the researcher would be so serious about and invest so much time and energy in the project. Thirdly, the author will probably never be aided in such a way again by experienced faculty who guide the author through each stage. Indeed, the thesis or dissertation is a milestone in one's career, and often reaches a depth that has profound impact on one's future work. The thesis or dissertation, of course, may involve many mistakes or appear immature in many ways. But its gathering of information and its contribution to understanding may also be significant. A sapient reader will gain more from a work containing both mistakes and contributions than one that contains neither. In fact, few works are without mistakes. It is not uncommon that a publication that looked perfect in the first place would turn out to be terribly wrong at least in some aspects. A significant product should not wait too long for publication for fear of minor mistakes, or the researcher may miss the publication opportunity at all. As the author becomes more mature, his product will be getting gradually improved.

Those fresh researchers who want to rewrite their work into journal articles

208

claim. The reason the first sentence is not a scientific knowledge claim, whereas the second is, will become clear after we have discussed the properties of abstractness, universality, and conditional nature.

From now on, when we use the term *knowledge claim* we will mean scientific knowledge claim. We do not mean by this usage that the only knowledge is scientific knowledge. Knowledge of a fact, for example, may be well accepted as knowledge but not as scientific knowledge. In this discussion we are using the term knowledge claim to refer to a technical concept which we have defined above, and we are abbreviating its label simply to avoid tedious repetition.

Knowledge Claims Are Abstract

All statements are abstract. When one says, "This is a table," the statement is the result of complex mental processes that are not identical with the act of perceiving an object. A whole range of experience is summarized in "This is a table"—from the physical experience of light rays impinging on the retina, to the past learning of a social convention that objects with four legs and a flat surface are called tables. Statements, however, range from the more concrete to the more abstract. Although we cannot precisely quantify a statement's degree of abstractness, we can still agree that the statement "This is a role" is more abstract than the statement "This is a table." Furthermore, we are all aware that the statements of higher mathematics are probably the most abstract we will ever deal with. Consider for example the mathematical idea of a *set*, which is usually defined as "a collection of elements." That idea is considerably more abstract than the statement "This is a table."

The dictionary defines the term *abstract* as "considered apart from any application to a particular object." In that sense, "This is a table" when used to refer to a particular table would not be an abstract idea, whereas the statement "Tables generally have four legs" would be a statement considered apart from any particular table.

Our notion that knowledge claims are abstract, then, requires that knowledge claims be considered apart from application to any particular object. The first implication that follows is that one cannot have knowledge claims about Stanford University or about the United States where Stanford University and the United States are considered particular objects. Statements about Stanford University or about the United States do play an important role in the evaluation of knowledge claims, but they are not themselves knowledge claims. Another way to put this is that what people generally regard as facts are not scientific knowledge claims, because facts usually apply to particular objects. (We will see later on that facts are also not universal because they refer to particular times and/or places.) Thus, the fact that the estimated population of the United States in 1986 was 239 million refers to a particular object, the United States in 1986, and is not sufficiently abstract to be regarded as a knowledge claim.

We all have difficulty dealing with abstract ideas. Not the least of these difficulties occurs in trying to communicate abstract ideas to others so that others think about them the same way we do. We all prefer to think in concrete imagery. To talk about the details of a particular murder—the use of a .38 caliber gun which was fired at close range into the head of a male victim—is easier than to characterize murder abstractly. For example, in formulating an abstract idea of murder, we cannot restrict ourselves to talking about guns, but must allow a wide range of weapons. One can readily see how much easier it is to describe to a friend a particular murder in a TV show than it is to talk with the friend about an abstract idea of murder; yet murder isn't even a very abstract idea. If we tried to talk about deviance, we would have even more problems of thinking and communicating what we have thought.

The more abstract an idea is, the more it is removed from an individual's experience. Hence, difficulties in communication arise because there are no points of reference in shared experience on which to hang the ideas. When I use the idea of deviance, readers may feel unsure about precisely what I intend to convey. From this we must recognize that the demands of communicating abstract ideas so that they are understood impose very severe constraints on how abstractions should be used. I cannot, for example, assume that when I use the word deviance my audience will understand only what I intend to communicate, nothing more or nothing less. Furthermore, a common danger of totally abstract discussions is that such discussions can be empty of content. To solve these problems when we are dealing with abstract ideas, we must have very explicit rules governing our usage of abstractions. If we go back to mathematics for a moment, we find that the idea of sets only becomes a useful abstraction when one thinks about sets in terms of explicit rules that make up what is known as set theory. One can look at mathematics generally as systems of rules for manipulating abstract ideas. From mathematics we learn that the possibility of thinking precisely about abstract ideas depends upon having systems of explicit rules to govern our thought.

Knowledge claims are not as abstract as mathematical ideas. While the abstractions in mathematics do not have to be tied to concrete objects (in geometry one can talk about points and lines as two-dimensional—even though such ideas can have no actual realization, because any point drawn has a third dimension), scientific abstractions eventually must be linked to concrete objects. Without such linkage, science could not have an empirical character, and the world of experience could never operate to modify scientific knowledge claims.

Since we have emphasized the difficulty of dealing with abstract ideas, someone may ask, Why can't knowledge claims be concrete? There are many reasons for requiring knowledge claims to be abstract, but the simplest reason is that concrete statements are not very useful. The more a description captures a particular concrete object, the more tied to that particular object the description will be and the less likely that that description would fit anything else. We could spend our lives refining statements about a particular grain of sand, since every grain of sand is unique, and still never know anything about the beach.

Since we require knowledge claims to be abstract, the process of abstraction becomes a crucial element in the development of knowledge claims. One major issue, for example, concerns how one evaluates alternative modes of abstraction. We will return to this issue when we consider problems of definition.

Knowledge Claims Are Universal

We are now ready for a definition of a universal statement:

A universal statement is a statement whose truth is independent of time, space, or historical circumstance.

"Power increases with status" would be an example of a universal statement, whereas "Lawyers have more power than doctors in industrial societies in 1980" represents not a universal statement but a singular statement. While we might argue about whether or not the use of a term like *lawyers* restricts the statement to a particular historical context, the fact that the statement has a particular date means that it is not applicable to anything but one period of historical time.

A singular statement is a statement whose truth depends on a particular time, place, or historical circumstance.

Singular statements, for example, usually contain proper nouns or terms whose meaning depends on a particular historical context. Singular statements do play an important role in science, but they are not the focus of scientific concern. Here again, the emphasis is on a distinction between the statements that we are concerned with evaluating and those which are used in the evaluation process.

What we have presented thus far has an important implication for empirical research in science. Of necessity, empirical studies take place in particular times at particular places; hence, statements which report research results are necessarily singular statements. Implicitly, they contain time and place specifications; that is, the time and place in which the study was done qualifies the description of the empirical results. From what we have said it follows that empirical research represents the process of using "reality" to evaluate knowledge claims. While empirical research is essential to basic science, it is not an end in itself nor even the main focus of scientific concern.

Requiring universality for knowledge claims actually formulates a norm of science—the norm that science concerns itself with general knowledge, not historical knowledge.[1] One consequence of this norm is that there is not one set

---

1. The reader may wonder, What about the geological reconstruction of the earth's history? Is that not science? From the present formulation, such activities would not be considered basic science. Insofar as historical geology involves scientific principles from physics, chemistry, or geology, it has aspects of an applied science. Nevertheless, the present point of view emphasizes a difference between general scientific knowledge and historical knowledge. While the distinction is fundamental and has consequences for how one approaches one's subject matter, the present formulation in no sense implies that scientific knowledge is better or worse than historical knowledge. They have different purposes and different uses, and one type of knowledge may be well suited for one purpose and totally unsuited for another.

of principles for the twentieth century and another set for the sixteenth century, and there is not one set of principles for the United States and another for the bushmen of Australia.

The reader is probably aware that there is much dispute over whether anything but historical knowledge is possible in the social sciences. A respectable body of opinion claims that all knowledge of social phenomena is relative to particular historical periods. (Incidentally, the argument is similar to the argument about cultural bias, considered earlier.) These critics are clearly correct if knowledge of social phenomena is restricted to the results of empirical studies, which are historically limited. If, however, we mean something else by knowledge of social phenomena—and we do—then we can show that nonhistorical knowledge claims can be generated and can be evaluated. Such knowledge claims must be sufficiently abstract so that what they assert is not restricted to a particular collection of spatial, temporal objects. To be a knowledge claim, the statement "Higher status people have more power in society" must apply to more than people in the United States in the twentieth century.

We do have a problem. On the one hand, knowledge claims must be universal and abstract, hence not tied to particular objects in space and time; on the other hand, empirical research generates only singular statements directly tied to spatial, temporal objects—and therefore not knowledge claims. Yet empirical research must have an impact on knowledge claims; otherwise knowledge claims are not modifiable by experience, and we do not have empirical science.

The resolution of the problem involves linking knowledge claims to singular statements; that is, knowledge claims are not themselves singular, but must be tied to singular statements in such a way that the singular statements can affect the evaluation of the knowledge claims. This leads to the idea of a "simple knowledge structure" in science. We will return to the simple knowledge structure after we consider a third property of knowledge claims.

Knowledge Claims Are Conditional

The conditional requirement represents one of the most difficult and least understood properties of knowledge claims. It is difficult because there are several different senses in which knowledge claims are conditional, and these different usages often blend with one another. Furthermore, the part played by "conditions" has not been sufficiently analyzed by the scientists who use them or by those trying to understand science. To be sure, many unsolved problems remain in connection with the requirement of conditionalization. Nevertheless, this requirement affects scientific thinking in many important ways, so that we must make an effort to understand what conditions are and why they are so important.

Perhaps if we analyze an example it will aid our understanding. At an early age we all learn that water boils at 212°F. When we become a little older and a little more sophisticated, our learning is modified so that we learn, "At sea level water boils at 212°F.'' Then we become still more sophisticated and learn, ''Given air pressure equal to 14.7 pounds per square inch, water boils at 212°F.''

Recognize that our earliest learning, ''Water boils at 212°F,'' represents an unconditional universal statement. The statement has no qualifications and no references to time or place. Furthermore, as any mountain climber can testify, the unconstrained statement has many exceptions. These exceptions force us to regard as false the unconstrained statement—which we will call an *unconstrained universal*. As is the case in our example, unconstrained universals are usually false.

When we qualify the claim with ''at sea level,'' we have a conditional statement that guides us to particular situations in which the claim should be true. The qualification, however, does not provide as much guidance as it might. To be sure, our conditional statement does constitute an improvement over the unconstrained universal; yet situations exist which are not at sea level where water will boil at 212 °F, and situations exist at sea level (although rare) in which water will not boil at 212 °. The qualification ''at sea level'' can be considered a *singular condition* because it refers to a collection of particular places. Singular conditionalization, while both useful and essential, represents only part of the story.

Initial Conditions. Let us call such singular statements of conditions *initial conditions*, and recognize that statements of initial conditions play a central role in linking abstract knowledge claims to singular statements describing empirical research. To illustrate this, let us expand the boiling water example even though it may seem trivial to do so. Spelling out the example is not only necessary for understanding conditionalization, but is consistent with the view that collective evaluation of statements requires explicitly dotting all the i's and crossing all the t's. Only by making every statement explicit can we expose the entire structure of an argument to critical collective evaluation.

Suppose we wanted to test the claim that at sea level water boils at 212 °F. What makes this trivial is the fact that we believe in the truth of the statement so strongly that no one would ever undertake such an exercise; yet that should not obscure the point behind the example. Suppose we place a pot of water on our stove, put a cooking thermometer in the water, and watch. As the result of our watching, we note that when the silver column of mercury rests at the line representing 212 ° we see steam rising from our pot of water. Having conducted this terribly inefficient experiment, we now have two problems: first, we must formulate a statement which describes our observations; second, we must connect that statement in some way to the claim we want to evaluate. Neither of these is a trivial problem that has a simple mechanical solution; on the contrary, solving such problems requires imagination and creativity.

In trying to formulate a statement describing our observations, we immediately confront the fact that many different statements will all describe what we observed, a fact which is a major reason for calling our experiment terribly inefficient. In other words, the experiment leaves itself open to many different descriptive statements. For the moment, we can only recognize the problem and note that we require some guidelines to help us formulate such statements, which we shall call *observation statements*. Since observation statements describe the results of observing at a particular time and at a particular place, observation statements are of necessity, singular; and, even for our trivial experiment, the class of possible observation statements is very large. For the purpose of our example, however, we can arbitrarily choose one of these possibilities. Suppose we formulate the following statement: "On January 3, 1985, at 7:00 A.M., steam rose from a pot of water on my stove in Stanford, California, when the thermometer in that pot registered 212°F." Spelling this out so explicitly should indicate something of the range of possibilities for constructing observation statements. We could, for example, have described what actually happened, in place of the phrase, "when the thermometer registered 212°F." We have, of course, spelled out things much more explicitly than is usually the case, for many of the elements in our statement are taken for granted. Sometimes, taking something for granted creates no problems. For example, we have no need to spell out in detail what was involved in arriving at the phrase, "when the thermometer registered 212°F." In areas which are more problematic, however, as in much of social science, what can be taken for granted is often not clear. The point need not be pursued further as long as we recognize that constructing observation statements poses serious problems which must be solved in the research process and, further, as long as we recognize the dramatic illustration of a point made earlier—namely, that observations do not speak for themselves.

The next problem concerns using the observation statement we have constructed to evaluate the claim that at sea level water boils at 212°F. Somehow, we must connect our observation statement to the claim. We accomplish this by linking terms in the observation statement to terms in the claim by means of additional statements of initial conditions. We have to assert, first of all, that the pot of water on my stove on January 3, 1985 at 7:00 A.M., in Stanford, California, represented an instance of "water at sea level." Secondly, we must assert that my observation of steam rising from this pot is an instance of "boiling." We could extend the example further, but these two assertions should be sufficient to illustrate the structure of this kind of argument. We thus have four statements, as follows:

1. At sea level, water boils at 212°F.
2. A pot of water on my stove on January 3, 1985, at 7:00 A.M. in Stanford, California, is an instance of "water at sea level."
3. The observation of steam rising from the pot of water on my stove on January 3, 1985, at 7:00 A.M., in Stanford, California, is an instance of "boiling."
4. There was steam rising from the pot of water, which occurred when the thermometer registered 212°F.

With this chain of statements we have established a linkage between the claim and the results of our experiment. We can investigate our statements to see if indeed there is a clear, logically rigorous linkage, but we will save that for another time. What is important in the present context is the necessity for such linkage before observation statements can have any bearing on the evaluation of claims.

What we have done so far exposes an important feature of the argument. Suppose, for example, we deny the truth of assertion 2 by showing that my stove in Stanford, California, was actually 800 feet above sea level. The falsity of the initial condition that water on my stove represented an instance at sea level quickly leads us to decide that my experiment was totally irrelevant to the claim about what happens to water at sea level. The other initial conditions that we could state operate in a similar fashion. What we have done in this example has been to develop a chain of singular statements of initial conditions which refer back to the terms in our original claim—that is, in statement 1. Some of these initial conditions link abstract concepts like "boiling" to particular instances in the observation statement. Other singular conditions establish a chain referring back to the original initial condition of "at sea level." As we shall see, these chains of reasoning constitute crucial features in the development of scientific theory and in the use of empirical research to evaluate theory.

Before introducing another type of conditionalization, one more feature of our experiment should be emphasized. Suppose we are willing to accept the truth of our initial conditions. Does our observation statement prove our claim? Clearly, it does not. It is always possible, and indeed even likely, that someone who repeated our experiment, using different initial conditions which he or she regarded as true, would fail to observe steam rising when the thermometer registered 212°F. No matter how many times, under different initial conditions, we observed the steam rising, the possibility always would remain that someone else would fail to observe it. The most important point to be gained from this discussion is that observation statements never prove that claims are true. The best that we can hope for from observation statements is that they are *consistent with* the claims they are used to evaluate. This may disappoint those who believe that science proves the empirical truth of its claims but, as we shall show later, it is impossible to prove empirical truth. We should point out, however, that the ability to demonstrate consistency or inconsistency between observation statements and knowledge claims represents a very powerful intellectual and practical tool.

Scope conditions. We have said that initial conditions do not represent the whole story. Our most sophisticated learning—as represented by the statement, "Given air pressure equal to 14.7 pounds per square inch, water boils at 212°F"—is still a different type of condition; it is more helpful, for the condition guides us to many more situations than the condition "at sea level." The qualification is stated in terms of an abstract idea—pressure—and it is universal, that is, not constrained to particular times and places. The qualification tells us that in order to apply the claim, "Water boils at 212°F," we don't have to worry where we are; we only have to measure pressure. In other words, it gives us a general rule for working with knowledge claims.

Statements of conditions that are abstract and universal we will call *scope conditions*. More precisely, we offer the following definition:

Scope conditions are a set of universal statements that define the class of circumstances in which a knowledge claim (or a set of knowledge claims) is applicable.

In our analysis we shall use initial conditions and scope conditions extensively. It may be helpful to reexamine the structure of argument that we developed above in statements 1 through 4 with the addition of scope conditions. We can simply add a statement numbered zero (0):

0. The principle "water boils at 212°F" applies to situations when air pressure is 14.7 pounds per square inch.

Statement 1 would then be rewritten dropping out the phrase "at sea level." Then we would insert another initial condition, statement 1a, as follows:

1a. Sea level is an instance of air pressure equaling 14.7 pounds per square inch.

With these revisions, we could use the remainder of the argument as presented earlier in statements 2 through 4—although, of course, using such a scope condition as statement zero (0) would enable us to simplify other parts of the argument as well. For our present purpose, that is not necessary. The revised argument should demonstrate, however, why we regard "at sea level" as an initial condition, since it very clearly plays that role in the revised argument where the scope condition allows us to have many instances not at sea level substitutable for statement 1a.

Finally, the most important feature of the idea of scope conditions is that it allows us to preserve the universal character of knowledge claims and, at the same time, to recognize that *universality does not mean "without exception."* In other words, to argue that some historical phenomena are exceptions to a universal principle does not require the belief that only historical knowledge is possible. For example, some sociologists claim that a valid sociological principle must apply cross-culturally. Indeed, some would go so far as to say that it must apply to all cultures. The idea of scope conditions allows us to reject that position. It is possible to have valid sociological principles as long as they are valid for those cultures which satisfy the scope conditions. This does not require that all cultures, or indeed any culture, satisfy the scope conditions. The significance of the idea of scope conditions is simply that *wherever scope conditions are met, a knowledge claim is applicable.* Nothing in either the scope statement or the knowledge claim guarantees that any particular situation will meet scope conditions, and this is the fundamental import of the requirement that knowledge claims be conditional.

For a knowledge claim to be conditional, then, requires that its scope of application be explicitly stated and that it be linked to the empirical world through the formulation of statements of initial conditions.

Antecedent conditions. There is one more sense in which it is desirable for a knowledge claim to be conditional. Knowledge claims formulate statements of relationship between what are called *antecedent conditions* and *consequent conditions*. For example, in the statement, "If a group is formally organized, then it has a status hierarchy," the *if*-clause represents the antecedent conditions and the *then*-clause represents the consequent conditions. Roughly speaking, knowledge claims "if. . .then" construction fulfill this standard. By and large, this sense of conditional is a criterion concerning the form of the statement that makes up the knowledge claim. By putting statements in this form, we are able to use the tools of deductive logic in order to deduce new statements from our knowledge claim and statements of initial conditions. This sense of the requirement that knowledge claims be conditional is well understood and widely used in the sociological literature. While such usage is necessary, we contend that it is insufficient. For us, in order to be conditional a knowledge claim should meet three criteria:

1. **The knowledge claim should be stated in the antecedent-consequent form.**
2. **There should be at least one explicitly stated scope condition which can be joined with the knowledge claim.**
3. **There should be at least one statement of initial conditions that can be formulated containing terms used in the knowledge claim and the scope statement.**

The reason for criterion 1 is that it allows us to use powerful tools to manipulate statements to arrive at new statements, and it emphasizes the relational nature of knowledge claims. According to this criterion, the statement "Crime exists" is not a knowledge claim, whereas the statement "If poverty increases, then crime increases," has the proper form for a knowledge claim. If, for example, we had the following two statements— "If industrial productivity decreases, then poverty increases" and "If poverty increases, then crime increases"—we can use elementary logic to derive a new statement: "If industrial productivity decreases, then crime increases." As has already been said, this sense of *conditional* is well understood and widely employed.

The reason for criterion 2 is that it provides guidelines to those situations where we expect the knowledge claim to be true, or at least to those situations where we are willing to submit the knowledge claim to a test. This criterion allows us to have laws, even when there are exceptions. Unconstrained universal laws do not occur in any science. Every law of physics has exceptions. We even know the conditions when the law "Everything that goes up must come down" will fail. The requirement that we explicitly state scope conditions allows us to deal with exceptions by defining or redefining the circumstances of applicability of a knowledge claim.

Finally, criterion 3 enables us to link abstract universal statements to singular observation statements. It is this ability to establish such a linkage that allows us to use the empirical world to evaluate abstract universal statements. Recall our use of sea level as an instance of air pressure at 14.7 pounds per square inch. What we have said so far illustrates the path along which an idea must travel in order to be scientifically usable. Simply to make a statement of the idea is not enough; at least two other explicit statements are required. But that forces us to think about where the idea applies and how to translate the idea into observations, and it forces us to do so by making explicit statements so that others can think along the same lines.

Our discussion of the boiling of water illustrated the statements that were involved in expressing a simple physical idea as a knowledge claim. Statement zero (0) is a scope statement; a rewritten statement 1 is our knowledge claim; statements 1a, 2, and 3 are statements of initial conditions.[2] The statements concerning boiling water, however, included one additional statement, number 4, which is an observation statement. Once we have formulated scope statements, a knowledge claim, and statements of initial conditions, we can use logic to deduce one or more observation statements. Having done so, we have an explicit, logical argument which we call a *simple knowledge structure.*

SIMPLE KNOWLEDGE STRUCTURES

Consider the following example. Suppose we start with the knowledge claim, "If *P* is the person who speaks most frequently, then *P* will be the person other group members speak to most frequently." Then suppose we assert the following scope statement: "The relationship between frequency of speaking and frequency of being spoken to applies to task-oriented groups." To this we add two statements of initial conditions:

1. Seminars in sociology at Stanford in 1987 are task-oriented groups.
2. An observer's ranking of who speaks most frequently and is spoken to most frequently in sociology seminars at Stanford in 1987 are instances of "person who speaks most frequently" and "person other group members speak to most frequently."

With these statements we can then deduce the observation statement, "In sociology seminars at Stanford in 1987 an observer's ranking of the person speaking most frequently will correspond to that observer's ranking of the person other group members speak to most frequently." Here we have a simple knowledge structure

---

2. The alert reader will have noticed that the statement "Water boils at 212°F" is not in the proper form according to criterion 1 above. That is easily remedied as follows: "If X is water, then X boils at 212°F." All the statements could be rewritten in proper form now that we have introduced the criterion concerning the form of statements.

and/or books, however, may be discouraged by publishers, journal editors, or even their own supervisors. Imagine, for instance, over ten publishers or journal editors have turned down your manuscript. Or you have to make major revisions since a reader of your thesis or dissertation has rated your work as "substandard." In such a case, should you still try to publish your work or not?

The current peer and supervisor review system adopted by academic and professional communities certainly has its merit. Unlike the notorious and controversial censorship system in other aspects of social life, few people, if any, have questioned the academic and professional gate-keeping practice and scrutinized the downside of its consequences. A one-sided view, however, may let us overlook some potential problems, which may have been posing an unnecessary hurdle against the full-wing advancement of our intellectual undertaking. The key issue is that there is no perfect judge, especially a judge of scientific work, in the world. On the surface, peers in the same field would be able to provide the best judgment. This is true oftentimes, but not always. The history of natural science provides useful lessons, not to mention the very political social sciences. Not all great scientists received fair treatment from the peers of their times. Indeed, a monkey could be killed by its peers if it dared to be too different. Humans tend to judge by evidence of mistake rather than of merit. Even more problematic, a fellow who believes in an orthodox model of the research process may not tolerate the writing of a researcher who sees the potential problem of that particular model and thus does not follow it. Certainly, human beings cannot afford making too many mistakes, yet may be even more vulnerable to missing just a single contribution such as Galileo or Einstein's. Overly critical is dangerous, especially when critics function as gatekeepers. Natural scientists may be a little more fortunate since fewer prejudices or prejudgments would be involved. For the behavioral and social sciences, however, a more tolerant and lenient attitude should be advocated.

In real terms, there are practical problems associated with any academic and professional review system. For example, review by peers "in the same field" is usually a rough or arbitrary statement. Even if the reviewer's interest is close enough to the researcher's topic, the reviewer may not have spent as much time and energy as the researcher in that particular aspect. Because of her superior position as an academic or professional judge, however, she may not have the time to fully appreciate the author's intent and focus, or try to understand the author's bias just as that of her own. Indeed, considering the burden of reviewing others' manuscripts, it is sometimes unrealistic to expect the reviewer to come

which spells out the entire logical argument and provides a basis for critical evaluation. We can, for example, reject initial condition 2, in which case the observation statement would not be relevant to the knowledge claim. But if we accept the premises of the argument, then we cannot dispute the relevance of the observation statement to the evaluation of the knowledge claim.

Finally, in this example we have a knowledge claim that is abstract, universal, and conditional, and yet does not refer to every group that has existed or will exist in the future. What this statement asserts is that the claim relating speaking and being spoken to applies only to those groups that are task oriented. Such an assertion means that the claim is irrelevant to groups that are not task oriented, such as cocktail parties; it therefore provides a guideline that tells us not to use cocktail parties to investigate this knowledge claim.

If ideas are the raw material of science, they, like other raw materials, are not really usable until refined. Initially an idea is quite crude, and in this crude form almost impossible to evaluate. A new idea is often vague, not usually universal, and not linked to empirical phenomena. On the other hand, a new idea may be very concrete; then, of course, it is specific and it is linked to empirical phenomena. As we said earlier, however, concrete ideas have little scientific utility; they do not promote thinking or generate research. The fact that a new idea is initially not very useful is not a problem—or even something to criticize. It becomes a problem if the idea remains in its initial crude form.

The scientific research process involves reworking and rethinking an initial idea to transform it into a knowledge claim and to make it subject to collective public evaluation. Turning an idea into a knowledge claim requires formulating it in abstract and universal terms. Making the idea subject to collective public evaluation by reason and evidence requires embedding the knowledge claim in a structure. This structure, to which we have given the name *simple knowledge structure,* accomplishes the linkage between a knowledge claim and empirical phenomena; the structure indicates those situations in which the knowledge claim applies; and, finally, the structure turns unconstrained universals into conditional universals.

A simple knowledge structure is made up of four different kinds of statements: (1) It contains an abstract universal knowledge claim; (2) it has an abstract universal statement of scope conditions; (3) it includes at least one singular observation statement; and (4) it involves at least one statement linking the knowledge claim and the scope conditions to the observation statement or statements—that is, there is at least one singular statement of initial conditions. We have presented two examples of simple knowledge structures, one built around the knowledge claim that water boils at 212°F and the other built around the claim relating frequency of speaking to frequency of being spoken to. In the next chapter we will examine in detail the research implications of another sociological example of a simple knowledge structure.

Constructing, using, and evaluating simple knowledge structures takes us into almost every aspect of the research process. The purpose of formulating the idea of a simple knowledge structure is to help make the features of this process explicit and open to analysis, since very often these stages occur implicitly in the mind of the researcher. By making these aspects of the process explicit, we will see that the process involves criteria and standards to guide both the researcher and the relevant public who evaluate the work. Formulating an idea, for example, requires consideration of concepts and conceptualization, problems of definition and abstraction. While conceptualization and definition are creative acts, they do not occur in a vacuum. Both the creative process and the evaluation of its product occur in accordance with normative standards. The objective of turning an idea into a knowledge claim and embedding it in a simple knowledge structure imposes constraints on the formulation of concepts, the definition of terms, and the relationship between an idea and its empirical representation. We will look more closely at these issues in considering the problems of definition and conceptualization.

The idea of a simple knowledge structure provides a framework for much of the analysis that follows. One important caution must be emphasized: simple scientific knowledge structures are not theories. Although the development of a simple knowledge structure can be viewed as a theoretical activity—it involves thinking, analysis, and the use of critical reason—it does not deal with one central problem. The problem we have not discussed thus far concerns the relation of one abstract universal idea to another. Later on, when we discuss scientific theory, we will require that a theory deal explicitly with the interrelationships of a set of knowledge claims. Hence, theories are more complex than simple knowledge structures. (The reader will be forgiven if he or she questions the use of the term *simple* in connection with a structure involving four different kinds of statements.) Although theories will be more complex, they will involve the same basic ingredients as simple knowledge structures. Hence, at this point, we can say that simple knowledge structures are the elements out of which scientists construct theories.

SUGGESTED READINGS

Cohen, Bernard P. "The Conditional Nature of Scientific Knowledge." In *Theoretical Methods in Sociology: Seven Essays*, ed. Lee Freese. Pittsburgh, Penn.: University of Pittsburgh Press, 1980.

This essay goes more deeply into the issues presented in this chapter and is suitable for the reader who wishes to pursue a more technical treatment than is presented here.

Wallace, Walter, *The Logic of Science in Sociology*. Chicago, Ill.: Aldine, 1971. Pp. 31-60.

This introductory book, although it shares a similar orientation to that presented in this chapter, provides an introduction to the same set of issues from a quite different approach.

Willer, David, and Judith Willer, *Systematic Empiricism: Critique of a Pseudo-Science*. Englewood Cliffs, N.J.: Prentice-Hall, 1973.

The Willers' polemical attack on empiricism in sociology points to the need for a more theoretical orientation. Although the attack is overly strong, it does contain some telling criticisms.

FIVE

Simple Knowledge Structures

Chapter 3 asserted that basic science is concerned with the production and collective evaluation of knowledge claims. Chapter 4 defined a knowledge claim and described its important properties—namely, that a knowledge claim is abstract, universal, and conditional. We also asserted that the collective evaluation of a knowledge claim, using reason and evidence, requires that the knowledge claim be embedded in a structure of statements which we have called a *simple knowledge structure*.

The research literature and the literature on methodology do not discuss simple knowledge structures. The simple knowledge structure is a model for analyzing what sociologists do and what they should be doing. As a model, it is not in general use, but it is presented here because it provides a useful tool for analyzing sociological research. Up to now, sociological studies have not been guided by explicitly formulated simple knowledge structures. Some research, however, has been based on what we might call an implicit simple knowledge structure. By analyzing such studies we can bring out and make explicit the statements that have guided research. More frequently, however, using the simple knowledge structure as a model to analyze a study will identify missing features and bring out problems of study. If the idea of a simple knowledge structure helps us to understand why a particular study is or is not a contribution to sociological knowledge, then the model is clearly justified as an analytic tool.

Perhaps the most important virtue of the idea of simple knowledge structure is that the model helps to relate apparently unrelated studies. The emphasis in sociological research, and in methodological discussions as well, has been on the isolated single study. There are both historical and practical reasons for this emphasis. From the point of view of developing sociology as a science and generating a body of sociological knowledge that meets the tests of collective evaluation, however, the emphasis on the isolated single study may be counterproductive. At the very least, the emphasis on the single study to the exclusion of the methodology of a cumulative series of studies represents a serious obstacle to the growth of knowledge. The training of sociologists, for example, deals with all of the requirements for doing a "good" study, but it almost never discusses the problems of relating one study to another. The implicit assumption is that, to produce scientific knowledge, the researcher need only to make sure that one single piece of research is done as well as the state of the art permits. Without denigrating the importance of doing a good study, a single isolated, good study is only the beginning of the process of producing and evaluating knowledge. Perhaps the most important consequence of using the simple knowledge structure as a model lies in the demonstration that a single study, of necessity, must have gaps and must raise as many questions as it answers. Such a result points to the need for a cumulative series of related studies before we can have any confidence that a knowledge claim has been submitted to serious and searching collective evaluation. In order to document this claim, we will analyze a research example in fine detail.

The objectives of this chapter then are:

1. To provide further illustration of the ideas introduced in Chapter 4
2. To show how a simple knowledge structure guides the planning and execution of empirical research
3. To illustrate the kinds of questions that arise in planning and carrying out a study
4. To demonstrate the need for a series of studies in order to seriously evaluate a knowledge claim because of the inherent limitations of any single study

A STUDY BASED ON A SIMPLE KNOWLEDGE STRUCTURE

What would a study guided by a simple knowledge structure look like? While there are some examples in the literature which approximate research guided by an explicit simple knowledge structure, such examples are exceedingly complex and would raise issues that are irrelevant to our present purposes, obscuring the points we want to make. In fact, any piece of actual research involves a whole range of considerations which, from the point of view of our analysis, would be distractions. So it is necessary to sacrifice realism and construct a hypothetical example. Consider the following knowledge claim:

High-status people participate more than low-status people in political activities.

Here we have a statement that is a declarative sentence asserting something that can be either true or false. The statement is abstract and it is universal. It remains to be conditionalized. Contrast this formulation with the following two statements:

a. People with more education vote more frequently in presidential elections in the United States.
b. People with more education vote more frequently.

Statement *a* is obviously singular, since it refers to the United States as a particular place, but statement *b* is also singular, though less obviously so. The idea of voting, to be meaningful, depends upon a particular historical circumstance and a particular kind of political system.[1] The idea of political activities is also more general than the idea of voting; it includes such things as joining organizations, making contributions, and circulating petitions. Formulating this more general idea suggests that these various things are interrelated kinds of behavior—a suggestion that is lost if we focus solely on voting. Furthermore, our knowledge claim as it is formulated includes much more than simply voting in national or local elections. Our knowledge claim is not restricted to the analysis of societies. It could, for example, also refer to political behavior within an organization, from a small local club to a body like the United Nations. That there are similarities between the behavior of United Nations members and the way citizens vote in a national election in the United States remains to be determined; the formulation of our knowledge claim suggests looking for such similarities.

As our knowledge claim is stated, there are clearly examples which would falsify it. For example, in the days of the urban political machine in places like Boston or Chicago, it was probably not true that high-status people voted more than low-status people, since the constituency of the machine was made up largely of people of lower socioeconomic background, and the purpose of the machine was to maintain power by mobilizing this constituency to participate. Furthermore, some of the activist movements of recent years have been mass movements of lower-status people, designed to increase their participation so that it exceeds that of high-status people. Hence our knowledge claim as an unconstrained universal does not apply to any situation without exception.

As guidelines for the choice of appropriate empirical situations in which to evaluate the knowledge claim, we need to formulate scope conditions. Without scope conditions the researcher could go out and choose any election in any place in any organization at any time. If the researcher had chosen to test the knowledge claim in Mayor Daley's Chicago, he might have come up with negative evidence; if the researcher had chosen to evaluate the claim in a local election in suburbia, he might have come up with supporting evidence—and we would have a situation familiar to readers of the social science literature: "Previous studies have led to contradictory and inconsistent findings." The attempt to conceptualize where a knowledge claim applies should go a long way in reducing collections of study results that are inconsistent.

---

1. Some people would like to introduce the idea of degrees of universality, and hence would regard statement *b* as "more universal" than statement *a* but "less universal" than our original knowledge claim. This may be a useful way to modify the formulation of a knowledge claim. However, it is premature to introduce that idea, since it has many problems that remain to be worked out. The biggest danger with statements of "middling universality" is that we tend to think in concrete historical terms; so, a proposition about voting limits our thinking to elections in democratic political systems, or even to American presidential elections. Regarding universality as an all-or-nothing property forces our thinking to break with concrete historical circumstances.

For the sake of simplicity, we will formulate only two scope conditions for our knowledge claim. These are:

1. The knowledge claim applies to those political activities where participation is totally voluntary.
2. The knowledge claim applies to political activities where there are no organized efforts to bring out participation of one type of person to the exclusion of other types of people.

The second condition rules out those circumstances where effective political machines operate; the first scope condition eliminates dictatorial societies where participation is forced and where we would expect little difference between the participation of high- and low-status people, since both are subject to severe penalty. The scope of applicability of our knowledge claim is still pretty broad, and in actual practice we might want to limit it further. Nevertheless, we have already indicated large areas where we would not expect our knowledge claim to be true.

As the knowledge claim stands, it is still too vague to guide research. The term "political activities," for example, could include almost anything. Furthermore, terms like "participation" and "status" also need to be specified. In order to make these vague ideas usable in research, we must formulate statements of initial conditions. At the outset, the researcher has a wide range of choices, and one study could not incorporate even a small fraction of the possibilities. But compare, for example, a study which limited itself to a statement such as "People with more education vote more frequently for president of the United States" or "Members of old families belong to more civic organizations than newcomers." These two ideas would be the focus of two very distinct studies, and focusing on one would be unlikely to generate an interest in the other assertion. That people with higher education and members of old families are both instances of 'high-status people," and that voting in an election and joining civic organizations are instances of "political activities," means that all of these ideas could be included in statements of initial conditions. Again, this way of formulating our ideas calls attention to a broad range of relationships. In other words, there are a number of directions in which we can go at an early stage of investigation of our ideas, it is often not clear which is the most fruitful one to pursue. To develop our example, however, let us formulate a few statements of initial conditions:

In Place *P* at Time *T*—

1. Participation in political activities is totally voluntary (i.e. Place *P* at Time *T* is an instance where scope condition 1 is true).
2. There are no organized efforts to bring out participation of one type of person.
3. Level of education is an instance of level of status.
4. Level of income is an instance of level of status.
5. Sex is an instance of status, with male being higher than female.

6. Voting in a local community election is an instance of participation in political activities.
7. Voting in a U.S. presidential election is an instance of participation in political activities.
8. Discussing candidates for political office with friends and neighbors is an instance of participation in political activities.

At this point it is already possible to argue about the merits of what these initial conditions assert. Surely there are readers who will take strong exception to the initial condition which asserts that males are higher status than females. But the task of our research would be in part to evaluate whether these initial conditions are met; that is, to validate that these assertions are true in the situation we choose to study. Let us therefore postpone such arguments until we examine what an empirical study would do to establish our initial conditions.

Thus far we have developed our simple knowledge structure to the point where collective evaluation on rational criteria is possible. But we have not yet allowed for the empirical evaluation of our knowledge claim. We must take one more step in this chain of reasoning in order to make evidence from observations relevant to evaluating our knowledge claim. That is, we must formulate observation statements which can be checked against what we actually observe in an empirical study. From what we have formulated so far, we can derive the following observation statements:

In Place P at Time T—

1. Highly educated people vote more in a local community election than people with less education.
2. Higher income people vote more in a local community election than lower income people.
3. Males vote more in local community elections than females.
4. Highly educated people vote more in a presidential election than people with less education.
5. Higher income people vote more in a presidential election than lower income people.
6. Males vote more in a presidential election than females.
7. Highly educated people discuss candidates for political office with friends and neighbors more than people with less education.
8. Higher income people discuss candidates for political office with friends and neighbors more than lower income people.
9. Males discuss candidates for political office with friends and neighbors more than females.

Strictly speaking, these are not observation statements. Except for sex, the terms in statements 1 through 9 are not directly observable. For example, we do not directly observe a person's education; we infer it from a measure that we construct, either a response to a question on a questionnaire or from information in a document. This may seem like hairsplitting, but the distinction is important to keep in mind, since inferences are always subject to error. If we use questionnaire responses to measure people's education level, we must assume that these responses give an accurate reflection of the educational level of the people studied. From sad experience we know that this assumption is not always justified—people exaggerate their education because having more education is valued, particularly in our society. If anyone doubts this, one only has to look at all the schools in the United States that call themselves colleges especially the so-called colleges that advertise on the inside of matchbook covers. We will leave these observation statements in their present abstract form, emphasizing the need to keep our caution in mind.

Our knowledge claim, two scope statements, eight statements of initial conditions, and nine observation statements together constitute our simple knowledge structure. By making things explicit we have already developed a complex apparatus which may seem to be misnamed "simple," but it *is* simple in the way it functions. Now we are ready to see how the objective of evaluating this simple knowledge structure can guide empirical research. Some sociologists might object that empirical research is really not done this way. To be sure, it is not always necessary to spell things out. However, in order to understand and analyze what goes on, or what should go on, in empirical research, we need to be explicit.

A SIMPLE KNOWLEDGE STRUCTURE AND WHEN, WHERE, AND HOW TO DO A STUDY

The first step in conducting a study is to choose a place, P, and time, T, where we can assume that our first two initial conditions hold true. Here, historical knowledge, hunches, and educated guesses are all important in deciding where and when to do the study—that is, where and when initial conditions 1 and 2 are true. (In certain types of studies we can build in checks to test whether these conditions hold for the time and place we have chosen.) In most elections held in the United States, for example, it is a good bet that the first initial condition is true, since there are no penalties for nonvoting. Some other countries, however, assess fines for those people who do not vote; our formulation of initial condition 1 guides us away from choosing one of those countries. But U.S. elections sometimes involve "get-out-the-vote" drives which are directed towards specific segments of the population. Often, for example, labor unions conduct intensive voter turnout campaigns among their members. Our second statement of initial conditions tells us *not* to study such elections. Furthermore, if we want to investigate all our observation statements in one study, the fact that three of them deal with presidential elections also limits our choice of time and place. For example, we could study French elections, but British elections would be clearly inappropriate since they do not have a presidential system.

In choosing times and places for our study, it is possible to think about the best way to do the study. But here considerations of cost, effort, and available resources play a major role in our decision. While it might be desirable to do a comparative survey in several countries, all of which meet our initial conditions, the cost would be prohibitive. Practical considerations always necessitate compromising with the ideal way to evaluate knowledge claims. What should be clear thus far from use of our simple knowledge structure is that one cannot set out to do a study of any election in any place at any time. The desire to evaluate ideas constrains our choices, and practicality constrains them even further.

A quick glance at observation statements 1 through 9 indicates that we require observations on voting in presidential elections, voting in local elections, discussing candidates with friends and neighbors, educational level, income level, and sex. Let us consider observation of "discussing candidate." To obtain direct observations of how frequently people discuss candidates for a large number of people would be practically impossible. Since people do not discuss politics all the time, directly observing how often one person discusses candidates with friends and neighbors would involve an observer following that person around for days. Since our study cannot be based on only one person, we would have to multiply that effort many times, and it quickly becomes clear why direct observation is not feasible. Hence we must use indirect observation, such as asking people to report how often they discuss candidates with their friends and neighbors. But self-reporting techniques also have problems that we must recognize. For example, people are often asked to report things that they do not remember. If discussing politics is not important to people, then the chances of their not remembering discussions of politics are high. Nevertheless, this is the best we can do, and we must be cautious in interpreting the results of our study. The difficulties with self-report of discussion of candidates could lead to falsely positive or falsely negative results for observation statements 7, 8, and 9.

The concerns reflected in our observation statements have a direct consequence in choosing the type of study that we would do. The set of nine observation statements more or less limits us to doing some type of face-to-face survey—a poll. If we were interested only in the first six observation statements, and if we chose to study an American city, we could use available records. For example, the U.S. Census Bureau compiles information on small geographic units called census tracts. They characterize census tracts by education, income, and proportion of males and females. Cities in turn keep voting-turnout statistics by geographic units. Thus we could compare the proportion of people voting in geographical units populated by predominantly high-status people with the proportion voting in geographical units populated predominantly by low-status people.

We would have to be cautious in interpreting these results. It is hard to find geographic units that are "pure," that is, either all high-status or all low-status people, and it is especially hard to find geographic units with real differences in the number of males and females. If there were any differences in geographic units in voting turnout, we would not know whether these differences were due to males

afford the time for a lot of such references, I would suggest you focus more on research itself following the instructions of your program guide when you write. You can also make reference to the successful examples of the graduates of your school by looking at the format of their theses/dissertations recently filed at the university library.

to the same level of understanding as that of the researcher who has spent months or years on the project. Surely, since no one has the same mind, and since the reviewer is chosen from a group of academic/professional elites who supposedly have extensive training and experience, he will always have something to offer that is new and/or suggestive to the author. Therefore, either in terms of principle or in terms of practical benefit, the researcher should always take review comments positively to help improve the research work. In terms of publication opportunity, however, this does not mean that the reviewer's opinion will necessarily reflect the total value of the work. Compare the comments from different reviewers, and you will see how different views may be involved with regard to the same work. It is not impossible that someone who has never written anything significant on the topic would conclude that your work is "not of much interest." In a sense, this is often a political matter, specifically a matter of achieving consensus. A tension between the authors/researchers and the reviewers/publishers can easily be observed in the scientific community.

Yet surprisingly few, if any, studies have been conducted on this enormously disturbing subject. On the part of the researcher, she may well consider this as actually a matter of a power relationship. The potential problem is that those who are responsible for judging what is right and what is wrong may only be "politically correct." Therefore, if you trust the value of your work, it is really your responsibility to get it published. After all the painstaking labor, you have become a uniquely qualified expert on the particular subject. Even if it is just a master's thesis that has been regarded as too lengthy by your faculty advisor, you may still have all the reasons to rewrite it as a book. This is your product anyway, which is different from the thesis/dissertation stage when your project must be approved by your advisor, thesis/dissertation committee, or agency research supervisory board. It is you, the author and researcher, that has the final say on this matter (of effort).

Probably there are very few, if any, other topics in academic and scientific life like "publication" where so many myths go unchallenged. Surprisingly, relatively few publications about professional or academic publication appear in any field's literature. Research in this regard seems to have helped rather than prevented creating some of the myths. A citationist view, for example, asserts that a large number of citations of a work in the publications of others suggests that it has extraordinary significance. The meaning of such a statement, however, depends on how "significance" is defined. There is critical thinking in

not voting, females not voting, or both sexes not voting in equal proportions. Suppose, for example, we were comparing a district which had seven males and three females with a district that had five males and five females. We would expect from our observation statement that more people would vote in the first district than the second. But since the districts are not pure, that could happen in a way that was directly contradictory to our observation statement, and we would not know it—since we were only dealing with overall statistics characterizing the districts, not the behavior of individuals. Seven people could have voted in the first district, four males and three females, and the five people in the second district, all females. The "male district" had a higher turnout than the "less male district," yet females clearly voted more than males.

The example suggests that we must be very careful in choosing our unit of analysis. When our observation statements talk about individuals as the unit of analysis, we should avoid using aggregates of individuals, such as geographical units, to evaluate our knowledge claim. If we could find pure areas, using aggregates would not be a problem, but pure geographical units are very hard to come by. Here again the formulation of explicit statements, with individual people as the unit of analysis, constrains the choice of type of study.

The observation statements we have formulated, together with practical considerations, rule out both direct observation studies and the use of available records. Hence, it seems that we should do a survey where we ask people to report their education, income, and voting behavior, and their discussions.

Suppose we choose to do such a survey among voters in an American city where there are no organized efforts to bring one particular group of people out to vote. Our simple knowledge structure imposes additional requirements on our choice of city and time. Observation statements 1 through 6 require us to measure voting in both presidential and local community elections, but we cannot choose a city that holds its local elections on the same day as the presidential election. In such places and times, we would not be able to assess observation statements 1 through 3 independently of statements 4 through 6, since coming out to vote in the presidential election is equivalent to coming out to vote in the local election, because both elections would be on the same ballot. In fact, we would not be assessing *any* of our six observation statements, because we would not know whether people were participating primarily in the presidential election and only incidentally in the local election, or vice versa. When the local election is held at the same time as the presidential election, many people probably vote in the local election simply because they are already in the voting booth to vote for president; these people would not bother to come out to vote if it were just a local election. On the other hand, some people whose major political concerns focus on local elections would not bother to vote for president if it required any extra effort. These possibilities raise serious doubts about choosing to study a city that holds its presidential and local elections simultaneously.

On the other hand, the necessity to choose a city that holds its local election at a different time from the presidential election, combined with our choice to do a survey based on self-report, poses a dilemma. We have already noted that self-report has problems, such as asking people to report things they may not remember. If we choose a city where the local election is widely separated in time from the presidential election, our respondents may not remember whether they voted in one or the other election. The bias due to selective "forgetting" is particularly serious when studying voting. Voting is regarded as a good thing; it is the duty of every good citizen to vote. To admit to not voting is equivalent to admitting poor citizenship. Hence a person who does not remember is likely to give the normative answer of good citizenship, thus producing a systematic bias in our observations. Such biases cannot be completely eliminated; even some people who remember that they did not vote will not admit it. We can, however, strive to minimize this kind of bias by the way we write our questions and the way we choose a city in which to do our survey. In short, we must choose a city where presidential and local elections are held at different times, but at times that are not too far apart. While we cannot say how far is too far, it is clear we do not want a city that has its local election two years away from its presidential election.

One other general problem about choosing a particular city at a particular time must be discussed. In choosing a particular city, we run the risk that there is something special about the city we have chosen that will affect the evaluation of our knowledge claim. There may be something we are not aware of and have not captured in our formulation of a simple knowledge structure. Suppose we chose a city so dominated by one ethnic group that other ethnic groups felt alienated and did not vote. If the dominant ethnic group were composed of mostly people with little education, and the alienated ethnic groups composed of people with more education, we might get negative results for observation statements 1, 4, and 7; but such results might be atypical of results for cities generally. While the ethnic composition of the population could be checked, there is always the possibility of other unknown and uncheckable factors that operate in a single study to produce atypical observations.

Consider another example. Suppose we chose an election that was unique in some way, such as a local election in which all the candidates for office were women. This might produce negative results with respect to observation statement 3 that might well be unique to that particular election. While we would probably recognize the uniqueness of an election in which all the candidates were women and avoid it by choosing a different election, there is always the possibility of a uniqueness which we did not recognize and which operates to produce unique results in the evaluation of our observation statements.

The problem of peculiarity, or uniqueness, is a problem that no single study can avoid. Every time and place has unique and peculiar features that might affect the observations taken in a study. This is true whether the study is a survey, a laboratory experiment, an investigation of historical records, or whatever. The only way to eliminate the possibility that the results obtained are unique or peculiar is to do a series of studies in different places at different times and in different contexts, but studies which all meet the scope conditions of our simple knowledge structure. Explicit scope conditions provide the means for choosing times, places, and contexts that generate a cumulative series of comparable studies. Finally, this possibility of uniqueness of results is a major reason for our assertion that we cannot have much confidence in the evaluation of knowledge claim based on a single empirical study.

Using a Simple Knowledge Structure in Design and Analysis

The analysis thus far has shown how a simple knowledge structure guides the choice of type of study and the choices of time and place in which to do the study. Now let us look at how the simple knowledge structure guides other phases of our study. We will consider three phases:

1. The development of a questionnaire
2. The selection of people to be interviewed
3. The analysis and interpretation of the data from the study

In designing a questionnaire, it is obvious from our observation statements that we must formulate questions to obtain information about the education level, the income level, and the sex of the people we study. It is also obvious that we must elicit information about voting in the presidential election in year Y in our city, voting in the local election in the same year, and also information about whether or not these people discuss candidates with friends and neighbors. The problem is to design questions and plan interviews so that we obtain information in which we have some confidence.

Obtaining information on the educational level of respondents poses a problem. There are a number of different ways to ask a question about respondents' educational levels, and different questions may produce somewhat different answers. For example, look at the following two questions: How many years of schooling have you completed? and, What is the last grade you completed in school? "Years of schooling" may be confusing to many people, particularly when asked orally in an interview. How many of us instantly recognize that high school graduates have twelve years of schooling? On the other hand, "last grade you completed" also is problematic: What is the last grade that a person with a Ph.D. has completed? Furthermore, we have already noted the potential for upward bias in self-reports of educational level. Responses concerning educational level are particularly susceptible to this kind of bias, since people tend to upgrade their education in order to look good in the eyes of the interviewer. The problem of bias is compounded with both of our questions because they are ambiguous and do not constrain the respondent to a specific answer. If educational level were incidental to our study, we might be able to tolerate errors due to ambiguity of the question and biases due to upgrading on the part of the respondent. However, educational level is central to this study: three of our nine observation statements deal with educational level. Because of its importance, we should not rely on a single question to elicit a person's educational level. Multiple questions would allow us to eliminate ambiguity and would provide cross-checks to estimate biases.

We might use a series of questions, such as the series illustrated in figure 5.1. The questions appear in regular type, and the instructions to the interviewer appear in bold type. The interviewer proceeds with the series of questions depending upon the answers the respondent gives. While there are some technical problems with this series of questions, the series is far superior to any single question. The series eliminates a great deal of ambiguity—"grade in elementary school" is much less ambiguous than "grade in school." Furthermore, since it requires a series of specific responses, it reduces the tendency to upgrade.

With responses to the series of questions in figure 5.1 it is possible to classify people into several categories of educational level:

a. Persons with less than eighth-grade level of education
b. Persons with more than eighth-grade but less than high-school-graduate level of education
c. Persons with high-school-graduate level of education
d. Persons with more than high-school-graduate but less than college-graduate level of education
e. Persons with college-graduate level of education
f. Persons with more than college-graduate level of education

The question series might even lend itself to a finer classification with more than these six categories but, because of the way we have formulated observation statements 1, 4, and 7, we have no need for a finer classification. In fact, if we could decide in advance where to draw the line, we would need only two categories— highly educated people and people who are not highly educated. Deciding where to draw the line—that is, deciding what to call "highly educated"—is a technical question that is not trivial. We will have more to say about this later on. We should, however, point out that the formulation of our questions depends very heavily on the type of observation statement we have formulated.

The reader may be surprised that a matter such as ascertaining a person's educational level requires so much thought and effort. Obtaining trustworthy observations is never a simple matter. It depends upon paying careful attention to the ideas we want to evaluate, the technical problems of obtaining observations, and the possible sources of error. Although one might not think so, especially in view of the number of questionnaires that bombard all of us, developing usable questionnaires is a major undertaking requiring both technical skill and considerable artistry. Constructing a questionnaire is not a task for an amateur, even an intelligent, dedicated amateur.

While it may be obvious that we need questions to obtain information relative to our observation statements, a simple knowledge structure guides the construction of a questionnaire in another important way that is not so obvious. The simple

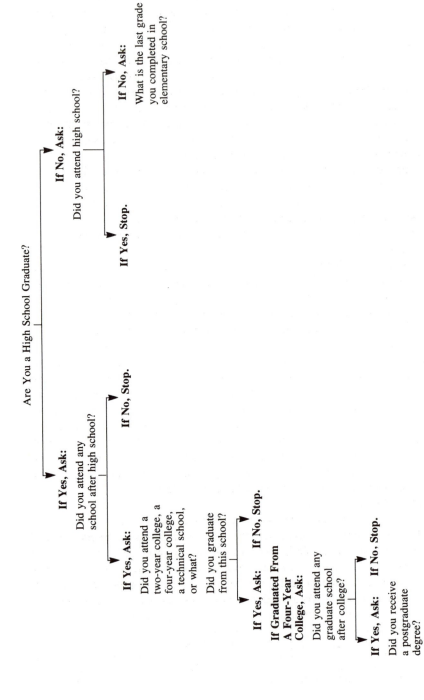

Are You a High School Graduate?

If Yes, Ask:
Did you attend any
school after high school?

If Yes, Ask: **If No, Stop.**
Did you attend a
two-year college, a
four-year college,
a technical school,
or what?

Did you graduate
from this school?

If Yes, Ask: **If No, Stop.**

**If Graduated From
A Four-Year
College, Ask:**
Did you attend any
graduate school
after college?

If Yes, Ask: **If No. Stop.**
Did you receive
a postgraduate
degree?

If No, Ask:
Did you attend high school?

If Yes, Stop.

If No, Ask:
What is the last grade
you completed in
elementary school?

Figure 5.1. Question Series on Education

knowledge structure directs us to obtain information in order to check whether the elections in the city we have chosen meet the initial conditions we have asserted. Consider initial condition 5: Sex is an instance of status, with male being higher than female. Suppose that statement is false for the city we have chosen at the time of the elections we are studying. If that is the case, then observation statements 3, 6, and 9 are all irrelevant to the evaluation of our knowledge claim, "High-status people participate more than low-status people in political activities." Males could vote more than females, or less than females, or in the same proportion as females, and none of these results would tell us anything about the participation in political activities of high- and low-status people, since we have supposed that it is not true that sex is an instance of status.[2]

Formulating questions to check whether initial condition 5 is true for the time and place of our study can test our ingenuity. We cannot ask the bald question, "Are males higher status than females in City *X*?" Many respondents would simply look bewildered, while those who had some notions about the idea of status might very well laugh at the question. We can virtually guarantee that among those who could attach some meaning to the term status we would have a range of different meanings; by allowing each respondent to interpret the term status, in effect we would be asking different questions of different people. What we need is a conception of status that would aid us in formulating questions that were meaningful to our respondents and for which we could assume a uniformity of meaning among our respondents.[3] Such a conception would associate some properties to the idea of status. For example, we could assume that one property of status was the ability to influence other people, that high-status people were more influential than low-status people. With this assumption, we could ask our respondents to name three people whose opinions influence them most, and then count the relative frequency of males and females among the people named. If both male and female respondents gave a preponderance of male names in answer to this question, we could regard that as evidence that initial condition 5 was true for our city. Here again, we would not want to rely on a single question. The construction of additional questions, however, depends upon explicating a concept of status. In order to do that we need some understanding of concept formation, a topic to be discussed in chapter 7.

Before leaving this example, we must point out that the explicit statement of initial conditions draws our attention to the need to go beyond the "common-sense" idea of status to a more refined idea. An initial study would not start out with a well-developed conception of status; over a series of studies, the idea should become more and more refined. Once the need for going beyond the "common-sense" idea of status is recognized, the opportunity arises to relate the study to

---

2. There are several ways in which this initial condition could be false. For example, sex could still be an instance of status, but females might be of higher status than males.

3. In constructing questions in general, it is a good idea to avoid very abstract terms for which there are ranges of possible meanings, since the meaning a person might respond to could be very different from the meaning the questioner intended.

other studies having nothing whatsoever to do with political participation, and to draw on these studies to aid in the formulation of concept of status.

The next phase of our study involves the selection of people to be interviewed. Let us briefly examine the way in which a simple knowledge structure guides this phase.

We could interview the entire adult population of the city we have chosen, but that is both impractical and unnecessary. We can get the same information from a sample at much less cost. Most readers have some notions about polling and will conclude that we want a representative sample of the adult population of our city. Some readers will be aware that we can obtain a representative sample by taking a random sample. It is true that a random sample of a population will allow us to make statements about that population with known limits of error. A major part of the discipline of statistics, *sampling theory*, is devoted to the technical issues of drawing samples and to the kind of inferences about the population that are legitimately based on sample results.

Conventional wisdom might suggest that our problem is readily solved by employing well-known techniques and drawing a random sample of the adult population of our city. But conventional wisdom is inadequate to our task. If we think about our objective in doing this study, we can see why conventional wisdom fails.

We are interested in evaluating a knowledge claim, not in making statements about the adult population of our city. Since our knowledge claim is a universal statement with restricted scope, we are interested in making statements about all past, present, and future voluntary participation in political activities where no special efforts to induce participation occurs. The population we want to sample is the population of all past, present, and future possible participants. In general, a random sample of the adult population of our city will *not* be a random sample of the population about which we wish to make inferences. Suppose, for example, the city we chose to study was a fashionable suburb. That city would likely have a preponderance of high-status people. A random sample of the adult population would reflect this preponderance, but there is good reason to assume that the population we are interested in would not have a preponderance of high-status people.

There might still be good reasons to use random sampling of our city,[4] but we cannot mechanically apply statistical theory to justify making statements about the population of all voluntary participants in all political activity, past, present, and future. We will have more to say about this when we discuss the problem of drawing inferences from empirical research.

---

4. We might want to use random sampling procedures to guarantee that our interviewers did not bias the results of our study by the way they chose people to interview. For example, one very experienced interviewer advised neophytes that the easiest way to obtain interviews was to choose undertakers as respondents, explaining that undertakers often had a great deal of time to spare and were quite willing to be interviewed. Since there is some reason to believe that undertakers might participate in political activities in a way that differs from the participation of other people, we might be quite unhappy if a sample of undertakers resulted from our interviewing. However, this is very different from using random sampling to justify inferences to a population.

In addition to calling attention to the fact that the population of our city is *not* the population we wish to represent by our knowledge claim, the simple knowledge structure guides the selection of people to be interviewed in another important way. Suppose we chose a fashionable suburban city for our study. Further suppose that 90 percent of the adults in that city were college graduates. If we chose 100 people at random to interview, we would expect 90 people to be college graduates. At best this would produce ten people who could be considered less than highly educated, and we would hardly feel comfortable evaluating our knowledge claim on the basis of so few people. In such a city, focusing on our simple knowledge structure would alert us to the need to oversample people who are less than higly educated. We would then design our sampling procedure to insure that we have a sufficient number of people for the analyses we plan to do. Unfortunately, studies that are not guided by an explicit formulation often turn out to have too few people of a particular type, and are therefore unable to carry out the analysis required to evaluate the ideas of the study.

Using a Simple Knowledge Structure to Plan Data Analysis

One of the main advantages of a simple knowledge structure is that it provides a clear plan for the analysis of the data *before* the study is conducted. In a sense, explicitly setting down our ideas provides a road map leading through what otherwise might be great complexity in dealing with the observations of our study. Having this road map in advance enables us to anticipate problems that might arise in the analysis while it is still possible to do something about them. Survey methodologists typically advise the survey researcher to set up *dummy tables* (tables that represent the relationships to be examined but that contain no numbers or contain fictitious numbers) in advance of collecting data. Constructing dummy tables and thinking about them can suggest the need for additional observations, and the survey questionnaire could then be modified accordingly. Research papers reporting survey results often contain the lament, ''If we had recognized this problem in advance, we would have collected additional information.'' A simple knowledge structure, while it will not anticipate all the problems that might emerge in the data analysis, does provide some foresight to substitute for hindsight. It can reveal potential problems, enabling the researcher to cope with them before they become real problems.

Our simple knowledge structure directs us to two main types of analyses: tests of the observation statements and tests of the statements of initial conditions. It also suggests some additional analyses. Let us look at dummy tables for each of these types.

From our observation statements, we can construct nine tables, of which table 5.1 is typical. The boxes labeled *a, b, c,* and *d* in table 5.1 are called *cells*. Sex is called the *row variable* (there are two rows, corresponding to male and female) and presidential-voting behavior is called the *column variable*. The row totals ($a+b$

Table 5.1. The Relationship of Sex to Presidential Voting Behavior

	Voted	Did Not Vote	Total
Male	(a)	(b)	(a+b)
Female	(c)	(d)	(c+d)
Total	(a+c)	(b+d)	Grand Total

and $c+d$) indicate the total number of males and total number of females in our sample. The column totals ($a+c$ and $b+d$) indicate the number of people who voted in our sample and the number of people who did not vote. The grand total represents the number of people interviewed in our sample. Note that both the row totals and the column totals add to the same amount, the grand total. In cell a we classify respondents who are male *and* report that they voted in the presidential election. Each cell of the table is filled by classifying and counting people according to their joint possession of two attributes. Hence the number of people in cell d is the number of females in our sample who reported that they did not vote in the presidential election.

Observation statement 6 tells us to compare cell a and cell c. But suppose our sample had many more females than males and our table looked like table 5.2. In this dummy table, we have interviewed 150 people, of whom 50 were male and 100 were female. Comparing the numbers in cell a with the numbers in cell c would be misleading, simply because of the accident that our sample contained twice as many females as males. We can take into account the unequal number of males and females by comparing percentages rather than numbers of people and by asking the question, Is the percentage of males who voted greater or less than the percentage of females who voted in the presidential election? Because our observation statement calls for comparing males with females, we calculate percentages across the rows of the table and obtain table 5.3, in which each row adds to 100 percent and column totals are no longer meaningful.[5]

---

5. Many people who approach tables like this for the first time have trouble deciding which way to percentage the table. Should the number in cell a of table 5.2 be expressed as a percentage of the row total, percentage of the column total, or a percentage of the grand total? Which way to percentage depends upon what comparison is required. Our observation statement requires us to compare males and females; hence we want to percentage across the rows, that is, express each number in a cell as a percentage of its row total. Percentaging down the columns of table 5.2 answers a different question: What percentage of voters are male compared to what percentage of nonvoters are male? Readers should convince themselves that this question is different from the question, Do males vote more than females?

Table 5.2. The Relationship of Sex to Presidential Voting Behavior

	Voted	Did Not Vote	Total
Male	(a) 40	(b) 10	50
Female	(c) 60	(d) 40	100
Total	100	50	150

Table 5.3. The Relationship of Sex to Presidential Voting Behavior

	Voted	Did Not Vote	Total
Male	(a) 80%	(b) 20%	100%
Female	(c) 60%	(d) 40%	100%

Table 5.3 shows that a higher percentage of males than females voted. Note that we do not compare cells *b* and *d*, because that comparison is already determined by comparing cells *a* and *c* and by the fact that each row must add to 100 percent. With only two categories, saying that a greater percentage of males than females voted is the same as saying that a greater percentage of females did not vote. Usually, in the percentaging of survey results, tables 5.2 and 5.3 would be combined, with the raw numbers or frequencies in parentheses next to the percentages, so that we have table 5.4.[6]

Table 5.4 is typical of the dummy tables generated by our observation statements when we plug in fictitious numbers. There are nine such tables because our observation statements deal with three row variables (sex, income level, and educational level), each of which is related to three column variables (voting in the presidential election, voting in the local election, and discussing candidates with friends). Consider table 5.5 for educational level and voting in local elections. We have already noted that our question series in figure 5.1 gives rise to six categories. How do we decide which categories represent highly educated people? We could

---

6. Often the frequencies are presented first, with the percentages in parentheses. That does not make a difference as long as it is clear which number is a frequency and which is a percentage.

philosophy and ample evidence in sociology and other studies of science showing that the achievement of academic and professional status is not just a "scientific" matter. It is also a "social" and "behavioral" thing. As the citationist viewpoint is indiscriminately adopted, however, the complicated matter is reduced to a single indicator and the tradition of analytical and critical thinking is lost. The various consequences of positivism in this regard continue to go unnoticed, and it is ironical that sophisticated scientists including social researchers would become so straightforward and absolute in dealing with their own livelihood. As a matter of fact, if you happen to choose a publisher or a journal that has very limited circulation, the difference could be huge in the chance of being cited as compared to the choice of a more popular one. The personal choice, however, is not necessarily related to the quality or significance of your work. This is just one of many factors that possibly influence the rate of citation. Sometimes people cite you because you are well-known, or ignore your work because you are nobody.

The history of science does not establish evidence for a discrimination against books, or a preference for journal articles to book publications. Rather, credit seems to have been given to any significant work no matter it was published in a journal article or a book. The book might have been deemed to contain usually a greater amount of work than that in an article. Recently, however, there seems to be a trend to play down the role of book publications in favor of journal articles in the surge of professionalism and gate-keeping. Explicit and ardent proponents of this trend of course include those who would claim little, if any, success in publishing books. The concern is both based on a belief about knowledge development and associated with professional benefits such as hiring, retention, and promotion. One argument is that new and important findings should be published first in journal articles rather than in books. The reason why there should come today such a requirement is not clear, and there are researchers who still consider books their most important publications. For those who personally only accept journal articles, probably it is not that they forget how they have been brought up and helped by books in their careers. It could be that they consider journals have better gate-keeping systems in terms of peer reviews. As indicated earlier, however, censorship of any kind will not work perfectly to determine the value of a genuine scientific work, although policing the scientific community is needed to deal with such crimes as forging and plagiarism. For a genuine scientific work, it is really just a technical issue whether to appear in a book or a journal article. Yet the research effort could be

**Table 5.4. The Relationship of Sex to
Presidential Voting Behavior**

	Voted	Did Not Vote	Total
Male	(a) 80% (40)	(b) 20% (10)	100% (50)
Female	(c) 60% (60)	(d) 40% (40)	100% (100)
Total	(100)	(50)	(150)

**Table 5.5. The Relationship of Education to
Local Election Voting Behavior**

		Voted	Did Not Vote	Total
Educational Level	Highly Educated	(a)	(b)	
	Less than Highly Educated	(c)	(d)	
	Total			Grand Total

arbitrarily decide that college graduates and persons with more than college-graduate level of education were our highly educated people, or we could wait until we had collected our actual data and then decide to draw the line "highly educated" and "less than highly educated" so that approximately 50 percent of our sample fell into each category. In that case the category "highly educated" would be completely relative to our particular sample. On the other hand, on examining our dummy table we could decide that, rather than some arbitrary drawing of the line, we would prefer to use what the people we are studying *perceive* as "highly educated." In that case, we would need to develop new questions to ascertain how our respondents perceive "education." While this too would be relative to the community we studied, it would not represent an arbitrary decision of the researcher. Thinking about the dummy table in advance of collecting the data allows us to consider alternative strategies for classifying people, some of which might necessitate asking additional questions on our questionnaire.

Table 5.6. The Relationship of Respondent's Sex to Sex of Persons Mentioned as Influential

		No. of Males Mentioned	*No. of Females Mentioned*
Sex of Respondent	Male	(a)	(b)
	Female	(c)	(d)

Earlier, we mentioned the desirability of checking as far as we can whether or not our statements of initial conditions are true for the elections in the city we have chosen. Sometimes we are quite content simply to assume that an initial condition holds true. For example, while there might be bizarre circumstances in which "voting in a presidential election" was not "participation in political activities," by and large we are content to assume that that initial condition holds. On the other hand, we have noted the problematic character of sex as an instance of status in our city. The question we formulated can lead to table 5.6 as a test of whether initial condition 5 is true.

Recall that we assumed that high-status people were more influential than lowstatus people, we asked our respondents to name three people whose opinions influenced them most, and we counted the relative frequency of males and females among the people named. If we assume that each of our respondents mentions exactly three people, no more and no less, and that we can classify these three people according to their sex, we can then fill in table 5.6. Cell *a*, for example, is the total number of males mentioned as influential by male respondents, and the row total (*a*+*b*) should be equal to three times the number of males in our sample. Thinking about this dummy table, however, immediately reveals a problem, which suggests that this might not be an appropriate analysis. In our earlier discussion we had decided that if both male and female respondents gave a preponderance of male names in answer to this question, we could regard that as evidence that initial condition 5 was true for our city. In other words, to accept that initial condition 5 holds, we require that the frequency in cell *a* be greater than the frequency in cell *b* and also that the frequency in cell *c* be greater than the frequency in cell *d*. But suppose our sample had five females who each mentioned three females as influentials, and ten females who each mentioned two males and one female as influentials. The second row of table 5.6 would then read as follows:

Female	(c) 20	(d) 25

The greater frequency in cell *d* would lead us to reject the truth of initial condition 5. But it might be premature to reject initial condition 5: the way we have constructed the example, two-thirds of our female subjects behaved in accordance with initial condition 5 holding true. This dummy table, then, suggests that there are serious ambiguities possible in analyzing the data this way. Setting up a dummy table forces us to rethink how we will deal with this issue. We might be able to come up with other ways to analyze the question we have chosen, or we might want to come up with alternative questions.

The simple knowledge structure also suggests additional analyses that might be done. For example, since we regard voting in the presidential election, voting in the local election, and discussing candidates with friends all as instances of participating in political activities, we might want to look at the interrelationship of these three kinds of behavior. We might expect, for example, a relationship between whether or not one voted in a presidential election and whether or not one voted in a local election. Hence, we could construct three tables such as tables 5.7, 5.8, and 5.9. If there were a relationship among these three types of participation, most of our respondents would fall in the *a* and *d* cells of the three tables. If there were absolutely no relationship, we would expect roughly equal proportions of respondents in each of the cells.

Table 5.7. The Relationship of Presidential and Local Election Voting Behavior

		Local Voting	
		Voted	*Did Not Vote*
Presidential Voting	Voted	(a)	(b)
	Did Not Vote	(c)	(d)

Table 5.8. The Relationship of Presidential Voting to Discussion Behavior

		Discussed	*Did Not Discuss*
Presidential Voting	Voted	(a)	(b)
	Did Not Vote	(c)	(d)

Table 5.9. The Relationship of Local Election
Voting to Discussion Behavior

		Discussed	Did Not Discuss
Local Voting	Voted	(a)	(b)
	Did Not Vote	(c)	(d)

Suppose that the *b* and *c* cells of all three tables were empty. That would mean that the three kinds of participation were perfectly correlated; in effect, we were not observing three different kinds of political behavior, but were observing one kind. That would suggest that our observation statements were redundant and would force us to reformulate them. On the other hand, if the tables showed no relationship between the various kinds of political participation, we might suspect they were *not* all instances of the same, more abstract idea. That too would suggest reformulation. Putting down our ideas as explicitly as we have done in the simple knowledge structure forces us to raise these kinds of questions. It is only because we consider "discussing with friends" and "voting in elections" as instances of the more abstract idea of "participation" that we would be led to do these kinds of analysis.

The dummy tables we have presented represent the simplest kind of analysis of the data. To be sure, the observation statements in their present crude form only require very simple analysis. In actual research, however, much more sophisticated analyses would take place and much more complicated dummy tables would be generated.[7] In the interests of clear presentation of our ideas to readers without any statistical background, we have tried to keep our example free of technicalities. We should point out that even with much more sophisticated statistical analysis, the issues and ideas that we have presented remain the same.

Let us turn to the problem of interpreting data that might be collected in our study. The main question of interpretation concerns whether or not the actual data support the truth of our observation statements; that is, are the actual tables which result from our analysis consistent with the truth of our observation statements? We are not interested in any one of our observation statements for its own sake. Rather, we use the observation statements in order to evaluate our knowledge claim. Since our observation statements are logically derived from the knowledge claim taken together with one or more statements of initial conditions, data which support the truth of our observation statement also support the whole simple knowledge

---

7. More sophisticated analysis (for example, using multivariate techniques) would deal with the relationship of several factors simultaneously, whereas our examples examine the relationship of only two factors at a time. For some of these more advanced techniques, equations or diagrams would substitute for dummy tables. Writing down an equation or drawing a diagram in advance of the collection of data serves the same purpose as our dummy tables.

structure, and in turn support the truth of the knowledge claim that is central to the simple knowledge structure. Logical connections among the statements mean that evidence for one statement is evidence for all the others. This fact allows us to use observation statements that are empirically testable to evaluate a knowledge claim that itself is not directly testable empirically. It is the chain of reasoning embodied in the simple knowledge structure that ties the abstract universal statement to the empirical world.

Deciding whether or not the data support our simple knowledge structure is not always a simple matter. If our analysis yielded a table like table 5.10, we would all probably agree that the data support observation statement 5. Since the observation statement asserts that highly educated people vote more in a presidential election than people with less education, we can regard 80 percent of the highly educated people voting compared to 10 percent of the less-than-highly educated people voting as consistent with the observation statement. We would then regard table 5.10 as supportive of our knowledge claim. Suppose, however, that our analysis yielded a table like table 5.11. While the percentages in table 5.11 are still consistent with our observation statement, the differences are so small that they might have occurred by chance. There is a question of whether it is supportive of our statement. Here we see why sociological researchers need a knowledge of statistics. The discipline of statistics provides a body of techniques to help resolve such questions, and the researcher must understand which techniques are appropriate under which set of circumstances.

When we encounter a table like table 5.11, in addition to statistical tests, our simple knowledge structure comes to a rescue. Since we are not interested in this table for its own sake but rather for its help in evaluating our knowledge claim, the fact that we have eight other observation statements generating tables for us becomes very important. Suppose that all of the tables generated by our observation statements looked like table 5.11, or suppose that the other eight tables were all better than table 5.11 in the sense that they more closely approximated table 5.10. In that case, we would regard the consistent pattern of the tables as indicating general support for our knowledge claim and would regard table 5.11 as

Table 5.10. The Relationship of Education to Presidential Voting Behavior

		Voted	Did Not Vote
Educational Level	Highly Educated	(a) 80%	(b) 20%
	Less than Highly Educated	(c) 10%	(d) 90%

Table 5.11. The Relationship of Education to
Presidential Voting Behavior

		Voted	*Did Not Vote*
Educational Level	Highly Educated	(a) 52%	(b) 48%
	Less than Highly Educated	(c) 48%	(d) 52%

forming part of a consistent picture. The fact that table 5.11 was part of a consistent overall pattern would be more relevant than the fact that it shows small differences in the percentages. In short, the simple knowledge structure cues us to look for consistent overall patterns rather than to concentrate on the specific numbers in any one table.

As long as the tables show a consistent pattern, our problems of interpretation are simplified. Suppose, however, that the tables do not show a consistent pattern. Say we have a table like table 5.10, a table like table 5.11, and a table that shows that higher percentage of females vote in presidential elections than males. In the initial stages of research, this case is far more typical than having all results come out neatly and cleanly supporting or disconfirming our observation statements. If we get these inconsistencies, our task is to try to understand why they occur—to explore the data we have collected in an effort to explain the inconsistencies. The first place to look would be to see whether the initial conditions are true or not for the study we have done. If it turns out that sex does not meet initial condition 5 for our city, then we are free to discount an inconsistency that occurs there. On the other hand, sometimes it happens that all the initial conditions that can be tested hold, yet one still finds inconsistencies in the data. This tests the mettle of the researcher in generating ideas to explain the inconsistencies and then proceeding to do additional research to test those ideas. Although it happens in the published literature, it is really an abdication of responsibility to say that some of our results support our knowledge claim and others do not, without developing an explanation for why this has happened. Once again, this points to the necessity for a series of studies, since examining the data generates new ideas whose evaluation requires new observations and new research designs.

What if all the tables that we generate support all the observation statements? Have we proved that our knowledge claim is true? The answer is an emphatic no. There is never a guarantee that someone repeating our identical study will not come up with results that contradict ours. There is no such thing as empirical proof. We will examine this issue more thoroughly in chapter 13. What must be emphasized, however, is that it is impossible empirically to prove any knowledge claim.

The notion of proof and the term itself apply to logical analysis, not to empirical evidence. For this reason we use terms such as *support, confirm, disconfirm,* and *fails to support* instead of terms such as *prove* and *disprove.* We can never prove an idea; all we can do is show that it is consistent with the available evidence, not foreclosing the possibility that additional evidence might turn up which is inconsistent with our idea. This terminological discussion is not just a quibble. It once again calls attention to the provisional nature of all scientific ideas. No matter how well established an idea is, there is always the possibility that it will be overturned.

SUMMARY

In this chapter we have presented a number of new ideas. Although we have tried to present them as simply as possible, the ideas are not all simple and they merit considerable reflection. As we have said before, elementary considerations are often the most fundamental and profound.

Research is a complicated activity. Our illustrations only scratch the surface of the questions that arise in planning and executing a study. If one takes these questions seriously, then it is important to recognize two principles: First, no study, no matter how well thought out and how well carried out, can answer all the questions that arise. Second, dealing with the issues of research requires a strategy; without a strategy, the researcher will be overwhelmed by the number of issues that confront any study.

Issues of uniqueness, questions concerning initial conditions, considerations of scope limitations, and the impossibility of empirical proof all demonstrate the inherent limitations of a single empirical study. As long as the focus remains on a single study, the unanswered issues overshadow any answers the study provides. There are even reasons to question whether the study provides any answers at all, as we have illustrated in our analysis. Only by embedding a single study in a research program of cumulative comparable studies can we have any confidence that we have dealt with these issues and that we have produced meaningful findings.

As the reader reviews the questions that we have illustrated in approaching an empirical study to evaluate a knowledge claim, the necessity for a plan of attack should be clear. The number of diverse issues that research must attack can only generate confusion in the absence of a coherent strategy. The simple knowledge structure provides one such strategy. It organizes the attack. It enables the researcher to ignore some issues as peripheral to the purpose of the research, and then to concentrate on central questions. It helps the researcher make the assumptions of the research explicit, and it points to those assumptions which can be checked during the course of the study. It also brings out those assumptions which cannot be checked and which the researcher must tentatively accept on faith in order to conduct the study. It is a very rare circumstance in which all the assumptions can be checked out in one single study.

At the most general level, a simple knowledge structure represents an orientation to research. This orientation is first and foremost to the evaluation and production of knowledge claims. In the next chapter we will contrast this orientation, which we call a *generalizing orientation*, with another important orientation to the research process, the *particularizing orientation*.

SUGGESTED READINGS

Katzer, Jeffrey, Kenneth H. Cook, and Wayne W. Crouch, *Evaluating Information: A Guide for Users of Social Science Research.* Reading, Mass.: Addison-Wesley, 1978.

This book, written from a consumer's point of view, provides a range of questions that one can ask about social research.

Riley, Mathilda White, ed. *Sociological Research: A Case Approach.* New York: Harcourt Brace Jovanovich, 1963.

This collection of papers provides an excellent introduction to the various approaches to sociological investigation.

Warwick, Donald P., and Charles A. Lininger, *The Sample Survey: Theory and Practice,* New York: McGraw Hill, 1975. Chs. 2, 3, 6, and 7.

This is an intermediate-level coverage of the issues that arise in the planning and conduct of survey research. It deals not only with the uses but also with the abuses and limitations of the sample survey.

SIX

Particularizing and Generalizing Strategies

The sociological literature represents a range of different orientations to research. Published studies reflect a variety of different research strategies. Some strategies quite explicitly reflect the researcher's orientation; other studies are influenced by a fundamental orientation, but less self-consciously so. Such sociological schools of thought as Structural Functionalism, Ethnomethodology, Symbolic Interactionism, Marxist Sociology, and Causal Modeling provide their adherents with very distinct strategies. These schools disagree quite profoundly in their orientation to research and its objectives and, indeed, to the nature of knowledge itself. We cannot survey here the range of distinctive strategies and the variety of orientations of sociologists, but it is important for us to look at one general contrast that underlies some of the disagreements among sociological schools. This contrast is between a particularizing orientation and a generalizing orientation, or between a particularizing strategy[1] and a generalizing strategy.

The issues involved in these distinctions have a long history. They have been discussed under such rubrics as "idiographic versus nomothetic knowledge," "statistical versus clinical prediction," and "experimental versus case study research." In many previous analyses, the "versus" has been emphasized; writers have been intent in supporting one side while demolishing the other. Distinctions were too sharply drawn and underlying issues frequently were lost in polemics. In this book, we reject the categorical approach which argues that one orientation is good and the other bad. Rather, we insist that each orientation serves different

---

1. In the first edition of this book, we used the terms "historical strategies" and "historical orientation" for what we are now calling "particularizing strategies" and "particularizing orientation." Although our earlier usage was correct according to our definition of historical knowledge as knowledge of particular time–space situations, it also raised many extraneous issues and was erroneously interpreted as an attack on historical sociology. Since we believe that historical sociology can employ generalizing *or* particularizing strategies, we certainly intended no sweeping criticism of this subfield of sociology. We hope that the term "particularizing" will convey a clear sense of our meaning, will avoid side issues that detract from the analysis, and will free us from mistaken attributions of attacks on historical sociology.

hurt by scientific community's emphasis on crime control rather than on science promotion. To some, the ambivalence toward books is probably just a utilitarian one: publishing a book without first deriving a couple of articles is a real waste. This is true when it comes to counting the numbers of publications in making hiring, retention, promotion, and other personnel decisions. The ideal solution would be a gradual approach in publication, that is, first to derive a series of journal articles, and then to present the whole thing in a book. Ideals are often hard to realize, however. For a new Ph.D., for instance, the choice given to her is so limited that she will either publish one or a few articles, or a book, or nothing at all, simply because most graduates do not have the time, energy, and opportunity required to publish both articles and books. Remember all the stresses immediately following their graduation. Ironically, for some of those who have passed this stage, a piece of research that does not make a book can be played up into many journal articles as well as a lot of conference presentations. This kind of publication practice may be justified as a good job in knowledge dissemination as well as personal recognition. Indeed, no researcher should be required to publish only one or a certain amount of articles or books per project. There is a difference, however, between the creation of knowledge and the advertising of it.

The above discussion is not intended to depreciate the role of academic and professional journals and the value of journal articles as a whole. It is only concerned with unjustified depreciation of academic/professional book publications. Further to this concern, it should be noted that there are also many "myths" regarding book publishers. A typical one would consider university publishers, especially those run by topnotch universities, as better than commercial publishers. There is no established evidence, however, that would show higher efficiency in knowledge dissemination or greater interest in new knowledge of university publishers over commercial publishers. The difference seems to lie more with the reputation of the universities that would glorify their publications as well, which may have overshadowed their cost-benefit concerns or some other problems. Research is needed with regard to the contributions of publications produced by different publishers as well as their procedures of production.

The above discussion of the issues in research publication may make less and less sense as information technology rapidly makes its advances. The new means of electronic publication, for example, puts all publications on the same computer screen. We do not know what publication will look like and what it

objectives so that the strategy that might be optimal for one set of purposes is likely to be suboptimal for another.

The particularizing orientation involves a concern with events that take place in a *particular time-space context,* and the knowledge of those events that it generates is historical knowledge. For many purposes, historical knowledge is essential, but it differs from theoretical knowledge because the latter is not limited to particular space-time contexts. The particularizing orientation concentrates on sequences of concrete historical events and attempts to link together as many events as possible. An interest in linking prior events such as the personalities of candidates for public office, the degree of interest in the election, and the strength of campaign efforts to voter turnout would be attempting to link a sequence of prior historical events to a resulting historical event. The time period can be a particular year or a century or many centuries and the place can be a particular laboratory or a particular set of societies like Western countries. As long as the specification of particular times or places are necessary to the truth of the claim, the knowledge is historical. But we must be careful in this connection. The time-place context must be an inherent part of the claim. If, for example, we specify abstract properties that as far as we know have only occurred at one particular time or in one particular place, we are not necessarily dealing with historical knowledge *as long as nothing precludes* a future realization of these abstract properties in a future time or place. Knowledge about Europe, the United States, and Japan in the nineteenth and twentieth centuries is necessarily historical knowledge, but knowledge about industrial societies—although these may completely overlap with Europe, the United States, and Japan—need not be historical knowledge. Nineteenth and twentieth century Europe, the United States, and Japan cannot occur ever again after the year 2000, but industrial societies can exist in the far distant future.

The defining element of the particularizing orientation is its concern with an event or a phenomenon in a particular time and place. The event or phenomenon is of interest for its own sake. Concerns for understanding the American Revolution, or the operation of a particular hospital, or voter turnout in presidential elections, or sex discrimination are historical concerns insofar as any of these concerns are for the object in and of itself. In the particularizing orientation, the goal is to understand the object, and the investigator employs whatever means are instrumental to that goal. The means may include scientific theories, but they need not. If, for example, the goal is to understand voter turnout in the United States, then statements about turnout in previous elections as well as statements about the political participation of high-status people may both be means to that goal. However, if the knowledge claim is not particularly helpful in understanding voter turnout, it is easily abandoned, because the objective is understanding voter turnout.

The generalizing orientation is concerned with developing scientific knowledge. In the first chapter we asserted that scientific knowledge was theoretical knowledge. Now we can say that such theoretical knowledge is abstract, conditional, and universal. It consists of statements, the truth of which do not depend upon time or space and which are considered apart from their application to any particular object.[2] For the generalizing orientation, the goal is to evaluate knowledge claims or theories, and empirical studies are one means instrumental to that goal. An investigator may choose to investigate voter turnout in an election as a means to the goal of evaluating the statement about the political participation of high-status people. However, other time-place situations may just as easily be chosen. Because the concern is with evaluating the statement, the generalizing orientation readily abandons studying voter turnout as a means, in favor of a better situation; but since the statement is the focus of concern, it is not abandoned. In the interest of evaluating knowledge claims, situations are substitutable; that is the essence of the generalizing orientation. For the particularizing orientation, in the interests of understanding a situation, statements are substitutable.

We should emphasize that scientific knowledge is not the only kind of knowledge. In an earlier chapter we noted that scientific knowledge need not be useful knowledge. The other side of that coin is that other kinds of knowledge can be practically useful in ways that scientific knowledge may not be. Attempting to solve a practical problem is at least as dependent on historical knowledge of the problem as it is on scientific theories. The engineer needs both historical and scientific knowledge. In short, scientific knowledge is not better or worse than historical knowledge. We are all familiar with the conflict between "book learning" and "experience." The chief engineer who complains that his newly hired engineering school graduates are useless because "all they have is book learning" is, in effect, complaining that the general principles of book learning may not be relevant to the problems of the job. When our chief engineer bemoans the lack of experience of the new engineers, he is really commenting on their lack of historical knowledge. On the other hand, it is unlikely that our chief engineer would hire anyone without book learning, that is, anyone who had no scientific knowledge to bring to bear.

The particularizing orientation, then, aims at developing historical knowledge, while the generalizing orientation aims at developing scientific knowledge.

The objectives of the two orientations are different. The strategies they lead to are different. Problems arise when the two orientations are confused, as, for example, when one attempts to achieve a scientific objective with a particularizing strategy. In the next section we will show that the particularizing orientation is suboptimal for producing scientific knowledge. Here we should note that the generalizing orientation is suboptimal for solving practical problems. The objective of producing and evaluating universal statements may, and usually does, lead away from the most important features of the historical situation that hold the key to solving the practical problem.

Consider, for example, the person oriented to evaluating the knowledge claim, "Higher status people participate more in political activities." Such a person is quite likely to ignore the effect of weather on voter turnout (unless, of course, weather can be formulated as an instance of some more general sociological idea

---

2. The reader who recalls our definition of *abstract* will recognize that property of theoretical knowledge.

relative to the evaluation of the knowledge claim). From the point of view of a generalizing orientation, it is perfectly legitimate to ignore the role of weather in political participation. On the other hand, the politician, concerned with voter turnout in a very practical way, will not ignore the effect of weather. Whereas weather may represent a distraction if one's objective is to evaluate a universal statement about the relationship between status and participation, the universal statement may also be a distraction to an investigator worried about the major factors affecting voter turnout, and weather might be the most important factor.

As we have pointed out, the two types of strategy differ in what they leave out. That difference essentially represents the two sides of a trade-off. Since historical strategies generate detailed descriptions of the event of concern, they are much less likely to apply to a new situation. The more detailed the description, the more it is likely to be unique. This strategy sacrifices application to new situations for a close fit to the situation of concern. The generalizing strategy achieves applicability to a wide range of situations at the price of a very partial description of any one situation. Understandably, the practitioner with a practical problem and the scientist with a stake in universal knowledge adopt very different positions in this trade-off.

Some researchers believe that a particularizing strategy can produce scientific knowledge by generalizing from a collection of historical cases. As we will show in chapter 13, such a view cannot be justified. Particularizing strategies can be useful in generating theoretical ideas, but it is not an optimal strategy for such an objective because of the very likely possibility of being seduced by the richness of the particular time-space phenomenon. One gets caught up with interest in the situation for its own sake, and making statements about the situation becomes an end in itself. The resulting statements have little connection with one another and this, as we will indicate in chapter 10, is not conducive to formulating theories.

THE ORIENTATION TO PHENOMENA

One form of particularizing orientation is the orientation to phenomena. Let us first examine the features of this orientation and its accompanying strategy, and then consider why it is suboptimal for producing and evaluating scientific knowledge claims.

The orientation to phenomena has as its focus a particular phenomenon. It represents an attempt to understand everything that is practically possible about, say, participation in the political process, or status equality, or sex discrimination. The phenomenon per se is the focus of concern, and any factor that relates to this phenomenon is a legitimate interest of the researcher. Thus one is concerned with all of the important factors that affect, say, political participation. This orientation contrasts sharply with the orientation to evaluating knowledge claims, where one is concerned with a statement relating one general idea to another rather than with a phenomenon for its own sake.

One consequence of an orientation to a phenomenon is that it leads to a strategy of generating lists of statements about that phenomenon. For example, a classic study of participation in political activities (Berelson et al., 1954) lists in its appendix eighteen different statements relating some factor to voter turnout. A few examples from this list would be informative:

1. The higher the political interest, the greater the turnout will be.
2. Men vote more than women.
3. The higher the educational level, the more the turnout will be.
4. Residents of metropolitan areas vote more than residents of towns and cities, who in turn vote more than rural residents.
5. Members of labor unions turn out more than nonmembers.
6. In this period, people with a Republican-vote intention turn out more than people with a Democratic-vote intention.

The first thing to note about this list is that almost the only thing the statements have in common is that they all deal with voter turnout. In contrast, a researcher studying political participation with a generalizing orientation would focus on more limited sets of *interrelated* statements. Secondly, in the list of examples—and throughout the appendix to the Berelson book—each statement has exactly the same importance attached to it as every other statement. There is no distinction among types of statements; all are relationships to be studied for their own sake. In fact, a general feature of this strategy is that any statement of the form, "Higher _____ leads to higher voter turnout," is admissible because it says something about the phenomenon of concern.

Furthermore, we can regard all the statements in the list as singular statements. A statement like number 6 represents no commitment to when or where such a statement might again be true. "In this period" refers to the presidential election of 1948. Possibly this statement applied to that election because of unique and idiosyncratic features of the presidential race that year. (In 1948 President Truman faced Thomas Dewey, and everyone believed Truman had no chance to win. Hence, one could regard statement 6 as applicable to that election on the assumption that Democrats were discouraged about the possibility of their vote having any impact.)

The other statements we regard as singular because they contain the term "voting" or the term "turnout." "Turning out to vote" is a less abstract idea than "political participation" and depends upon a particular historical context for its meaning. What we have in this list of examples is essentially a set of observation statements. In general the orientation to phenomena produces lists of observation statements, although many may be more complicated than the statements of our example.

Moreover, in presenting this list, the authors of the study made no attempt to relate any of these statements to one another. For example, they do not consider any statement in this list as an instance of a more general idea. The authors present these statements for their own sake and provide no basis for ordering the list.

Finally, no statements on the list are conditional statements, and there are no statements of scope conditions. The phrase "in this period" in number 6 may be considered a condition, but it is a singular condition analogous to the phrase "at sea level" discussed in the section on scope conditions in chapter 4. The other statements have no constraints and are treated as if they have universal applicability.

The example we have chosen in many ways typifies results that emerge when an investigator is primarily oriented to a phenomenon.[3] We can regard such results as a collection of facts about a phenomenon; and, of course, if the objective is to deal with that phenomenon, one often needs an extensive list of facts.

To generate scientific knowledge, however, an extensive list of facts may not be necessary and may even be counterproductive. Facts may start a thinking process that results in universal ideas, and facts certainly play an important role in the evaluation of knowledge claims. The longer the list of facts, however, the less likely that either purpose will be served. It is not at all clear that one can think about a range of unrelated factors (such as political interest, membership in labor unions, size of residential area, party affiliation, etc.) without some more general ideas to organize these factors. Yet the orientation to phenomena does not require organizing these factors; it is sufficient that at some time in some place each relates to the phenomenon of interest. Whether they will all relate to voter turnout in the same way at different times and in different places is an open question.

Each fact in the list relates some factor to voter turnout, and that is about all they have in common. Thinking more abstractly about, for example, statements 2 and 3, we might arrive at the idea that educational level and sex have something in common, in that they are both instances of the more general idea of status. But to try to deal with the whole list at once inhibits general thinking, since the list contains too many diverse ideas to handle simultaneously. We cannot ask, for a long list of factors, the question, What does Factor A have in common with Factor B? We might ask the question, What does political interest have in common with residence in a metropolitan area? Adding a third factor, however, such as union membership, increases the difficulty of finding a common element that all share. Once one starts to organize and order the list, many of the statements would drop out as irrelevant to the ordering chosen. But the development of general ideas requires ordering; hence the generalizing strategy requires one to think very selectively about the facts, while the particularizing strategy discourages selectivity.

The fact that studies done with the particularizing orientation make no distinctions among types of statements—scope conditions, knowledge claims, initial conditions, and observation statements—but treat all statements as findings, provides no guidelines to the next researcher who comes along. The second researcher can assume that all findings of the previous study will hold, that no findings of

---

3. The example comes from a study done more than thirty years ago. But many current investigators, although they may use multivariate techniques of analysis, exemplify the same orientation. For example, studies which attempt to "maximize explained variance" of some particular factor, such as voter turnout, are exclusively concerned with the phenomenon to be explained. The objective of such studies is to account for all the variation, within limits of error, or some phenomenon.

the previous study will hold, or that some findings will hold. But which? How can the second researcher use the first study? Are there any guidelines other than the researcher's own interests and tastes? To be sure, the second researcher will always select things from the first study to use; but can that selection be rational? In short, with the orientation to phenomena, it is difficult for one study to build in any significant sense on previous research, except to add to the list of factors that might be related to the phenomenon.

Simply adding to the list of factors related to the phenomenon fails to distinguish between relationships that are unique to a particular study and relationships that are general; between relationships that are important and relationships that are trivial; and between relationships that are stable and relationships that are unstable. Furthermore, this orientation does not recognize that some observed relationships are conditional on other things being true, and fails to consider questions of scope. These features of the particularizing orientation almost guarantee that two studies of the same phenomenon will produce some contradictory findings.

Many sociologists sincerely believe that the accumulation of facts—that is, of lists of factors relating to a given phenomenon—will eventually lead to a theory of that phenomenon. Facts are observation statements, and observation statements are not scientific knowledge. The faith that collecting such lists of observation statements will generate scientific knowledge may be justified in the long run. However, one can reasonably argue that, given an interest in scientific knowledge, focusing directly on knowledge claims to be evaluated constitutes a strategy likely to pay off more quickly.

PARTICULARIZING AND GENERALIZING STRATEGIES AND THE PROBLEM OF PREDICTION

Those who believe that the particularizing strategy is appropriate for generating scientific knowledge base their belief in part on the notion that scientific knowledge is predictive knowledge. They empasize the importance of prediction in scientific activities, and they argue that one predicts a phenomenon by knowing all of the important factors that affect that phenomenon. The desire to predict justifies the orientation to a phenomenon.

The view that science predicts phenomena rests on an inadequate understanding of scientific prediction. To be sure, prediction is important for the generalizing orientation, but it is a special kind of prediction. If we look closely at the role of prediction in science, we will have a better appreciation of its nature and also of the reasons why prediction does not justify a particularizing strategy.

Let us distinguish two types of prediction, *prophecy* and *conditional prediction*. Prophecy is an attempt to unconditionally predict the future. For example, when a public opinion poll attempts to predict the winner of an election by ascertaining, some time before the election, how a sample of voters intends to vote, it is attempting to prophesy the future. Although many people regard such polls as "scientific," it is impossible to scientifically predict the future. Scientific prediction is conditional prediction. A scientist makes a prediction, *given the realization of certain conditions*. The scientist in no way predicts whether or not these conditions will be realized. A scientist would say, for example, "If there are no efforts to turn out voters of any particular type, then high-status people will vote more than lower status people." The "if" condition in the statement is crucial, but even that may not be sufficient for a scientist to venture a prediction. The scientist is not foretelling the future, because the prediction takes no stance on whether or not there will be an effort to turn out voters. If there is such an effort, the prediction does not apply; that is, no prediction at all is made. Conditional prediction, then, applies only if explicit scope conditions are met; the scientist does not predict whether or not these conditions will ever be met by any future event. While conditional prediction is the stock-in-trade of the scientist, the gift of prophecy is no more given to scientists than to other mortals.

If we understand the difference between prophecy and conditional prediction, we will understand why the orientation to phenomena at its best does not generate conditional predictions; at its worst, it represents prophecy. Suppose an investigator oriented to phenomena wants to develop a prediction scheme for predicting voter turnout. If the scheme had wide applicability, some sociologist would be willing to regard it as a scientific knowledge claim, even though it dealt with something as concrete as voter turnout. To create such a scheme, our investigator would want to include all the important factors that affect voter turnout. Suppose our investigator based the predictive scheme on the six statements we listed. As we already noted, "intention to vote Republican" may be important in determining voter turnout in one time-place situation; indeed, in that one situation "intention to vote Republican" may be the most important factor. But it is unlikely to be important in a wide range of situations. Hence, if our investigator built the prediction scheme from studying that situation, the scheme would be likely to fail when applied to new situations. Even if our investigator were confident that all of these factors were of general significance, his predictive scheme would fail because of some idiosyncratic feature of the new situation for which prediction is attempted. A severe storm, for example, could knock the prediction for a loop. In other words, there is no way to develop a foolproof scheme because there is no way to take into account all factors that might conceivably affect the phenomenon to be predicted.

That such schemes are not possible does not rule out the science of sociology. The natural sciences cannot foretell the future either; despite the extensive knowledge of hydrodynamics, no physicist would attempt to predict the behavior of my teakettle on my stove.

The orientation to phenomena, then, with its penchant for generating lists of factors, either represents the attempt to prophesy the future, or it puts the emphasis in the wrong place for generating conditional predictions. No matter how long the list of factors, nothing prevents the emergence of an additional factor in a new situation. Furthermore, the longer the list of factors, the more likely it is that some of those factors were unique to previous historical situations and were therefore not applicable to the new situation. Prophecy sometimes works, just as horoscopes sometimes sound right, but we never have a rational basis for confidence in prophecy.

The problem is not solved by treating predictions as probabilistic. Exactly the same considerations apply to predictions worded in terms of probability. Unless one specifies a probability, the estimate of success of a prediction can range all the way from zero to one, and we have a situation analogous to prophecy. If all we know is that the probability that a prediction is correct is greater than zero and less than one, the prediction is never wrong. Unless the predictor can specify the numerical probability of success, we have no rational basis for any confidence in the prediction. With prophecy, even when it is stated probabilistically, there is no basis for specifying the probability of success. One cannot escape the problems of prophecy by making prophecies probable rather than certain.

When not attempting to prophesy, the orientation to phenomena represents a misplaced emphasis. The objective of conditional prediction requires the formulation of conditions under which predictions are applicable. Simply listing factors which affect a phenomenon, without discriminating those factors that operate as scope conditions, does not promote the formulation of conditional predictions. For the generalizing orientation, successful prediction is one way to evaluate a knowledge claim. That is, we can view derived observation statements as predictions about the results of observations in a particular study; but these predictions apply only if the scope conditions are realized. The orientation to phenomena, by ignoring scope, often attempts to test a prediction in an inappropriate situation; hence, without the explicit formulation of scope, an investigator testing predictions is likely to come up with contradictory results. The particularizing orientation, by its failure to regard issues of scope, is unlikely to contribute to the production or the evaluation of knowledge claims.

Some readers will object that predictions of phenomena and events are made all the time by social scientists and others. Polls predict the winners of elections; government officials predict voter turnout; economists predict the growth of the economy; meteorologists predict the weather. Is it fair to consider all such efforts as prophecies? Furthermore, doesn't the nature of modern society compel such predictive efforts? If a city is going to solve its transportation problems, isn't it necessary for that city to predict its future population?

It is true that the better of these efforts are more than prophecy and do provide some basis for evaluation which is absent from prophecy. For these better efforts let us use the term forecasting. We reserve the term *forecasting*, however, for those efforts that are conditional. Forecasting, it should be clear, is not a scientific activity; it is more appropriately regarded as engineering. Even election polls have moved from prophecy to forecasting. Instead of reporting that Jones will win the election, polls now report, ''If the election were held today, Jones would win the election.'' This is not really very helpful—the election is not held today. This condition is tantamount to saying that if all the factors which may affect the outcome of the election were to remain constant from poll day to election day, the poll's prediction would hold. To be sure, the closer to election day the poll is taken, the more justified that assumption.

Forecasting based on the condition that all important factors remain constant may be called *persistence forecasting*. In its extreme form, persistence forecasting is like predicting that tomorrow's weather will be like today's. For very short time spans, or for isolated systems, persistence forecasting can be useful, as long as its major assumption is recognized. The user of the forecast, upon recognizing some dramatic event, is alerted to throw out the forecast.

A more sophisticated form of forecasting is one that develops alternative forecasts based on differing sets of conditions. For example, shortly after World War II demographers attempted to forecast population growth in the United States for the ten years to follow. They formulated three alternative sets of conditions based on differing assumptions about birth rates, death rates, and migration rates, and came up with three different forecasts of population size. These alternative sets of conditions provided guidelines for the user to choose among the alternative forecasts. It turned out that all three forecasts underestimated the population growth; but when it became apparent that the assumed conditions for all three forecasts did not hold, it was possible to revise the forecast.

From this discussion it should be clear that forecasting is a necessary and legitimate activity. Furthermore, forecasters who recognize the need for conditionalization will come up with more useful forecasts. However, nothing that has been said about forecasting alters the view that the orientation to phenomena is not optimal for generating scientific knowledge. Forecasts may use scientific knowledge as well as historical knowledge, but the task of the forecaster is primarily an engineering task. As we have said earlier, the particularizing orientation is quite appropriate to engineering tasks, and some features of the orientation to phenomena may be useful for forecasting a phenomenon, provided that the necessity for conditionalization is recognized.

Naturally, it would be extremely beneficial if we could scientifically predict the future, but such prophecy, except in very limited circumstances, is not possible. Nevertheless, forecasting in engineering tasks and conditional prediction in science are extremely important and powerful tools. The ability to forecast what would happen under certain conditions allows the engineer to design his work to meet those conditions. Solving a practical problem involves predictions of the kind, "What would happen if _____?" Hence, for engineering tasks, forecasting the behavior of a phenomenon is often the most important objective for solving the problem.

The generalizing strategy, however, treats prediction as a means to an end, not as an end in itself. Conditional prediction is the means for evaluating knowledge claims; it is the instrument for provisionally accepting or provisionally rejecting an idea. To be sure, we have argued that for an idea to be empirical there must be a way for the empirical world to modify that idea. The failure of a conditional prediction can bring about such modification. Nevertheless, conditional prediction is a tool and empirical evaluation is only one type of evaluation of a knowledge claim.

Since we have argued that single studies can witness the operation of unique and idiosyncratic factors, we also must argue that the single failure of a conditional prediction may be due to unique and idiosyncratic causes. Thus a single failure of a prediction derived from knowledge claim, even when the scope conditions are met, is not sufficient grounds for rejecting the knowledge claim. At a minimum, scientists require a reproducible failure of a prediction. While the reproducible failure of a prediction is necessary, it is not sufficient for us to reject an idea. Before we can understand why it is not sufficient, however, we must examine other criteria for the evaluation of knowledge claims. In the next chapter we will look at concept formation and consider criteria for the evaluation of concepts. Since knowledge claims are statements of relationships between concepts, the evaluation of concepts plays a role in the evaluation of knowledge claims.

SUGGESTED READINGS

Phillips, D.C., *Holistic Thought in Social Science.* Stanford, Calif.: Stanford University Press, 1976.

In a brief treatise, Phillips dissects both classical and contemporary arguments for holism.

Popper, Karl. *The Poverty of Historicism.* New York: Harper and Row, 1961.

Although this a polemic against Marxist historicism, it does provide useful ideas for differentiating between particularizing and generalizing strategies. Some of Popper's critique of Marxist historicism applies to those who confuse particularizing and generalizing orientations without being Marxists.

will exactly mean to the researcher in the future. Yet it is for certain that new issues will come up and new "myths" will be created. Therefore, it is important to the development of science for the researcher to have a discerning mind equipped with a thorough understanding of the scientific undertaking. For a beginning researcher who is determined to become a genuine social and/or behavioral scientist, it might be disappointing to be given the information on both the bright and the dark sides of this enterprise. Being prepared for all the imperfections of the undertakings, however, you can put yourself in a better position to make your intended contributions. Once you know what you are doing, keep going. After all, the old saying has it: "Where there is a will, there is a way."

SEVEN

Concepts, Definitions, and Concept Formation

We have introduced several different types of statements and illustrated their usage in research. Furthermore, we have emphasized the requirement that statements must be intersubjectively evaluated using reason and evidence; for this to be possible, there must be intersubjective agreement on the meanings of statements. In order to obtain such agreement, we must consider the problem of how one constructs statements or sets of statements. In this chapter we focus on the basic ingredients of any statement (reserving for a later chapter the problem of putting together several statements in a systematic way).

A statement such as "Power varies with status" relates terms—in this case the terms "power" and "status." But terms are merely labels for concepts; hence, concepts are the basic ingredients of any statement. Our statement relates some idea of power to some ideas of status. The problem is that there are many ideas of power and many ideas of status. Intersubjective agreement on the meaning of the statement "Power varies with status" requires that we specify which ideas of power and of status we are talking about.

Just as statements are ideas, concepts are ideas. In forming concepts one makes use of statements to specify the ideas involved. This is what we mean by defining terms—that is, spelling out the idea for which the term is a label and providing explicit guidelines for when it is appropriate to use the term. What differentiates statements used in defining concepts from knowledge claims is that definitional statements are not problematic: they are not true or false in the way that knowledge claims can be true or false.

While no one has yet discovered a recipe that guarantees good concepts, there are certain properties of concepts and certain issues in concept formation which should help us recognize good concepts. In the next section of this chapter we will discuss three important properties of concepts. Subsequent sections will examine three objectives central to concept formation: the objective of precise communication, the objective of trying ideas to the empirical world, and the objective of fertility (i.e., of forming concepts that enable us to generate a wide variety of statements). Since concepts are defined using language, we must also consider some features of definition; we will present a typology of definitions and indicate how each type relates to our three objectives of concept formation.

PROPERTIES OF CONCEPTS

Concepts are abstractions from shared experience. This is not adequate as a definition, however; it turns out to be very difficult to define *concept*. Indeed, definition is a more difficult problem than most of us believe. Hence, rather than try to define the term concept, we can try to promote shared understandings by talking about properties of concepts.

In the first place, a concept is an idea—not a thing. To be sure, ideas may refer to things but they are distinct from the things to which they refer—the *referents* of the concept. If we have a concept of a table, our idea is not a perfect representation of any existing table. In forming the concept of a table, we leave out many properties of existing tables. We also exaggerate or emphasize some crucial properties that presumably are present in all or most things to which we would apply the concept. For example "legs" may be a property central to our concept of tables; yet we would recognize that some things we would call tables have four legs, some have five, some have three, and some may have only one. Our concept of a table may slide over these differences as unimportant in our usage of the idea of a table. As long as we recognize the distinction between a concept and its referent, we should have little trouble because we will be aware that the referent can always be described in more ways than we have captured in our formulation of the concept. Such awareness will guard against reifying the concept—that is, confusing the idea with its referent.

The danger of confusing an idea with its referents is even greater in dealing with sociological concepts, because sociological concepts are typically abstract in that their referents are not concrete objects, but the referents themselves are ideas. Consider, for example, the term "power." The term may have as its referents ideas of physical force, ideas of persuasion and influence, ideas of control of valued resources, and others. When we assert that power varies with status, we usually do not have all of these ideas as referents of power. Not many sociologists worry about the exercise of physical force when they talk about power. If we talk about powerful, high-status individuals, we do not usually mean that high-status people are physically strong. But unless we specify the ideas that are the referents of the term power, we leave other people free to infer any set of referents they desire, and we generate no end of argument about the meaning of our statements employing power. The point of defining power is to limit explicitly the ideas that can be associated with the term. When we ignore those explicit limitations, or when we argue that the limitations leave out the properties that really represent the idea of power, we are blurring the distinction between the concept and its referent.

In the case of abstract concepts such as "power," where the referents are themselves abstract ideas, one cannot capture the totality of possible ideas any more than one can capture the totality of properties of all concrete tables in a concept of a table. An idea and its referent are not identical. An idea can contain properties not shared by all of its referents, and the referent can contain properties not represented in the idea. There is not, and cannot be, a perfect correspondence between a concept and the referents to which it refers.

Since concepts are abstractions from experience, and since there is not a perfect correspondence between a concept and its referents, it follows that concepts are neither true nor false. Abstraction is necessarily a selective process. With any phenomenon there are an infinite number of ways to select. The analogy of cutting a pie into eight pieces fits this situation. There are an infinite number of ways to divide a pie into eight slices because there are an infinite number of starting points on the circumference of the pie, and that is true even if we insist on eight slices of equal size. And it follows that there is no right way to cut a pie into eight slices. Similarly, there is no right way to abstract from experience in forming concepts.

Unfortunately, as fundamental as is the idea that concepts are neither true nor false, a great many people do not recognize the principle. We find examples of researchers who are concerned with "validating" their concepts. For example, a researcher may formulate a concept of status which includes the idea that people who occupy statuses differentially evaluate one another. He then shows that people believe that some statuses are better than others, and so he argues that his concept of status is valid. But what could his claim mean? If he simply means that there is some correspondence between the properties contained in his concept and the properties of the referent, then we would have little cause to argue with him; we would also claim that he is not saying very much. It stands to reason that there would be some correspondence between an idea and its referent. Correspondence arises in the process of formulating the concept, and correspondence is guaranteed if the concept is to have any empirical referents.

In arguing that a concept is valid, what is usually meant is that there is a unique way of capturing properties of reality in an idea. At least, the idea of a valid concept implies that there are some invalid concepts, but the notion of true concepts implicitly involves the error of reification, in this case the belief that the idea can be identical with its referents. As long as one grants that an idea does not capture all of the properties of its referents, and that there is "conceptual license" to magnify some properties and diminish others in forming a concept (part of the selectivity inherent in the process of abstraction), then one cannot argue that there are true and false concepts. One mode of abstraction may be useful for a particular purpose and not useful for another purpose, but the concept is not true because it is useful or false because it is useless. The only way we could justify a criterion of "true" for concepts is if we believe that an idea captures the totality of its referents. And that is a belief which we emphatically reject.

Consider two conceptions of "power." Lenski (1966) defines power as "the probability of persons or groups carrying out their will even when opposed by others." Collins and Guetzkow (1964) define power in the following way: "When the acts of an agent can (actually or potentially) modify the behavior of a person or group of persons, the agent has power over that person or group of persons." One cannot argue that one of these definitions is right and the other one is wrong. Furthermore, while there is some overlap in the ideas captured by the two definitions, they are by no means identical. For example, Lenski includes the idea of opposition, but Collins and Guetzkow do not. In addition, Collins and Guetzkow formulate a more relational idea than Lenski does; that is, they want to limit their usage to an agent having power over another person or group of persons, whereas Lenski's notion is not limited to the relationship of two sets of actors.

One can form different kinds of statements using these two different conceptions of power. For example, Lenski asserts that power will determine the distribution of nearly all of the surplus possessed by society, and Collins and Guetzkow assert that high-power persons will be less affected by the efforts of others to influence them. The Collins and Guetzkow statement is irrelevant to the Lenski concept (and probably vice versa). As long as we recognize that both concepts of power can be useful (both the Lenski and the Collins and Guetzkow knowledge claims can be true) and we recognize that both concepts are partial, we can appreciate that different modes of abstracting must be evaluated, not in terms of truth or falsity, but in terms of their contribution to generating true knowledge claims.

Although we argue that concepts cannot be true or false, this does not mean that concepts are totally arbitrary. We focus on the objective of using concepts to generate true statements, and then the objective constrains how we formulate our concept. As our example of power illustrates, the relational notion of Collins and Guetzkow, who talk about power of one agent over another, would not be very helpful in making statements about the consequences of power for the distribution of goods in society, because it would be cumbersome to look at society as the aggregation of all pair relations between agents. While it does not make sense to argue whether power is a relationship of one agent to another or simply a property of the agent itself, it is reasonable to ask whether the ideas in one conception are appropriate to the knowledge claims in which the concept is to be employed. Asking the question this way demonstrates that, for a particular collection of knowledge claims, some abstractions are appropriate and others are inappropriate. That being the case, one cannot argue that the mode of abstraction is arbitrary. This is especially true from the perspective that we have adopted—namely, that one begins with an idea, transforms it into a knowledge claim, and makes the idea more precise by defining the terms in the knowledge claim.

A word must be said about the issue of arbitrariness. Some writers claim that one is free to define a term in any way he chooses so long as he is explicit in his definition and consistent in his usage. To put it another way, since concepts are abstractions and there are an infinite number of ways to abstract, no one of which is more true than any other, one is free to choose any abstraction as long as his choice is explicit and his usage consistent. This position is usually defended with the argument that, since we cannot know usefulness in advance but only after the fact, it does not matter how we begin. In short, the starting point of concept formation is arbitrary. Much in this position is sound, but the position ignores some constraints on concept formation which limit the arbitrariness of forming concepts.

Let us briefly discuss two of these constraints. If an individual is working by himself with a single isolated concept, and with no concerns for communicating to others, we could live with the idea of arbitrariness. However, once we are concerned with shared abstractions on the one hand and systems of connected concepts on the other, we have constraints which limit arbitrariness. If one is concerned with intersubjective understanding and usage of a concept, one is concerned with maximizing shared abstractions. Therefore, a totally idiosyncratic way of looking at the world and of talking about what one sees will not do. In forming a concept one must be concerned that others can perform the same mental processes as the inventor of the concept. "I feel this idea in my innermost soul" is not acceptable. Similarly, a totally idiosyncratic use of words to define a concept will defeat the objective of shared usage. Hence, the inventor has some limitations on both his thought processes and his use of language in defining his concepts.

The constraint that a system of concepts imposes on the formation of one concept is perhaps best illustrated by looking at arithmetic. The definition of the idea of "one" is not arbitrary if the definer wants to build a system of arithmetic. The definition of "one" is constrained by the definition of "two," the definition of "plus," the definition of "equals," and the sentence "One plus one equals two." Not every definition of "one" would be consistent with that system. Since our orientation is to constructing statements, the formation of concepts must be done with an eye on the statements in which we might embed the concept; that is, no concepts exist in isolation. Concepts must be used in sentences, and this potential usage in sentences limits the arbitrariness with which we can define a concept. If, for example, we intend to relate the concept of power to the concept of status in a sentence, we cannot choose any old concepts of power and status.

Since scientific concepts are shared abstractions—shared ways of looking at phenomena—concepts must be communicable in order to promote these shared understandings. Furthermore, science uses evidence to evaluate statements containing concepts; hence scientific concepts must be tied to empirical phenomena. This requirement is called the requirement of *empirical import*. Finally, as the discussion of arbitrariness indicated, concepts must be part of a system; concepts are used to form statements in which the concept may be embedded. This is called the *fertility* of the concept. Thus there are three primary objectives in concept formation; we want to form concepts that (1) communicate precisely, (2) have empirical import, and (3) are fertile in generating knowledge claims.

The Objective of Precise Communication

We have all had the experience in heated discussions of reaching the point where someone says, "Define your terms." Such comment is a signal that the argument has broken down because we are no longer communicating to others what we think we are communicating. Particularly in abstract arguments, the communicator and the audience may have widely different associations to the words used in a discussion. If, for example, someone is arguing that high-status people have more power than low-status people, the term power can mean widely different things to individuals in a group all of whom would be willing to agree to the statement. To some, power might mean the ability to affect the outcome of a series of decisions; to others, it could mean greater ability to influence other people's opinions; to still others, it might mean greater access to the arenas where decisions are made; to some, it might mean all of these things. The significance of the statement "High-status people have more power" is not the same for all of these meanings of power.

This difficulty in communication centers around the problem of *surplus meaning*. Concepts are defined using words, and ordinary English words are never precisely enough defined to insure that two people have the identical set of associations and connotations when they use the same word. Almost any English word has more meanings than are intended in a particular sentence. The use of ordinary words may thus set off a chain of associations that go far beyond the intention of the person who uses the word, and it is these extra associations and connotations that represent its surplus meaning. Scientists develop special languages in an attempt to minimize surplus meaning; such attempts represent the principal motivation for scientific jargon. While sociology is often criticized for its jargon, some of this criticism is mistaken because it fails to recognize the role of a specialized, limited language in facilitating precise communication.

The problem of surplus meaning is, however, particularly acute in sociology, where we use terms like power, status, influence, and role, that have a general usage which may conflict with its intended sociological usage. When a sociologist uses a term like power or status, rarely does he or she intend to convey all of the meanings that a layman might attach to those words. Insofar as the audience hears more (or less) than the sociologist intends, we have a problem of communication. This problem of communication exists not only between social scientists and the general public, but between social scientists as well.

In order to evaluate any knowledge claim, it is essential that all potential evaluators have similar understandings of the meaning of the knowledge claim. In other words, intersubjective evaluations depend upon precise communication of meanings. If we use terms like power and rely on others to automatically understand what we intend, we are very unlikely to come up with shared evaluations of knowledge claims.

The first objective, then, is to consider forming concepts and defining terms in such a way that we maximize shared understandings, in part, by controlling surplus meaning. The problem of precise communication is fundamental to the construction of statements. As we shall see later, there are ways to maximize precise communication.

The Objective of Empirical Import

Providing concepts with empirical import distinguishes science from other types of thinking activities. In sociology, it is especially important to emphasize this objective because the history of thinking about social phenomena is filled with highly abstract arguments that remain totally insulated from all evidence and observation. From Hobbes's remark, "Society is a war of all against all," to Durkheim's comment, "One function of punishment of crime is (consciously or unconsciously) to enhance the solidarity of society," we are dealing with ideas that have very elusive referents. While both of these statements have strong intuitive appeal, on analysis they pose serious problems. What observations would allow us to decide that society is *not* a war of all against all? What evidence would allow us to reject the view that punishment is designed consciously or unconsciously to enhance the solidarity of a society? The problem in part results from the lack of empirical referents for "war of all against all" and "function for society."

Because science deals with abstract ideas, there is always the danger that discussions can go on in a vague way, almost empty of any content. For this reason, scientists need to be especially sensitive to ultimately linking their abstract ideas to the world of sense experience. Fortunately, most scientists are sensitive to this issue; most recognize that a concept without empirical import is excess baggage because there is no way for the phenomenal world to affect statements containing that concept. Indeed, scientists may be too sensitive to this issue and not sensitive enough to other objectives of concept formation.

The reaction among scientists against vague, private concepts and introspective ideas has been so strong that useful theoretical concepts have suffered the same attack as vague, abstruse ideas. Some writers would even exclude theoretical concepts entirely, admitting only those concepts which can be directly defined in terms of observables. Thus, for example, these writers would rule out the Collins and Guetzkow idea of power because we cannot observe "potentially overcoming the resistance of others." The position of these writers, known as *operationism*, insists that an idea must be defined by an explicit set of physical operations in the phenomenal world which anyone can perform—namely, by an *operational definition*. An operational definition of "power," for example, might be "the number of decisions a person wins in a discussion." Clearly, that definition limits the idea of power to something that can be observed and so maximizes the objective of empirical import. But it maximizes the objective of empirical import at the expense of the objective of fertility (and also, as we will argue later, at the expense of the objective of precise communication—though operationists claim that operational definitions facilitate precise communication).

To raise this criticism of operationism might seem inconsistent when we have insisted that scientists must ultimately link their abstract ideas to the phenomenal world. The key word is *ultimately*. The requirement of empirical import does not mean that all concepts must be formulated in terms of observable sense experiences; it allows indirect linkages between ideas and the phenomenal world. By defining a concept using observables, one establishes a direct link between the concept and sense experience, and limits the meaning of the concept to the explicitly stated, concrete examples. The alternative to establishing direct links is to tie the concept to the world of sense experience through a chain of reasoning that involves other concepts and statements relating concepts. At some points in this chain of reasoning, there are concepts tied to observables. This is what we mean by indirect linkages. The chain of reasoning contains theoretical concepts (i.e., concepts defined in terms of abstract ideas) that are related to other concepts directly linked to the phenomenal world; these theoretical concepts obtain their empirical meaning through their relationships with the directly linked concepts.

Consider the Collins and Guetzkow concept of power: 'When the acts of an agent can (actually or potentially) modify the behavior of a person, or group of persons, the agent has power over that person or group of persons.'' This is a theoretical concept. We can indirectly link this concept to observables with a statement of initial conditions: ''Winning decisions in a discussion is an instance of actually modifying the behavior of another person or group of persons.'' This provides empirical import to their concept without putting their ideas in a straitjacket. Our formulation calls attention to the fact that their concept has more meaning than is represented by the observable instance, and it allows them the freedom to think about these other meanings.

When we emphasize the use of a chain of reasoning to provide empirical import to theoretical concepts, we once again call attention to the necessity for scientists to deal with systems of concepts. It is not possible to deal with a single concept in isolation from other concepts. Communicating an idea precisely and giving the idea empirical import immediately involves us in spelling out the relationships among ideas systematically, that is, in a system of concepts and statements.

The Objective of Fertility

If we keep in mind that the goal of forming concepts is to generate knowledge claims using these concepts, then the objective of fertility follows directly from this goal. We want concepts that will allow us to formulate a large number of true knowledge claims—that is, knowledge claims that will be supportable by evidence. A fertile concept, then, is one that can be embedded in a variety of different types of statements—knowledge claims, scope conditions, initial conditions — providing us with new relationships to test and the intellectual tools with which to test them.

A concept that is so tightly constrained that it fits only one statement is not a useful concept. While the objectives of precise communication and empirical import impose constraints on the way concepts are formulated, we must be careful not to constrain our ideas so much that they are no longer fertile. This is one of the difficulties with operational definitions of concepts—they are too constraining. Compare, for example, the operational definition of power given above with either the Lenski or the Collins-Guetzkow concepts of power. In thinking about the idea captured in the operationally defined concept, we have many fewer associations than we have with either theoretical idea of power. But these associations in thought are the beginnings of how we relate one concept to other concepts; if a concept is too constrained, it inhibits thought. It is difficult to relate "winning decisions in a discussion" to ideas of status, authority, or prestige. It is less difficult to associate these ideas to theoretical concepts of power. Concepts, therefore, must be sufficiently open to allow new relationships.

Openness, however, can lead to vagueness. In fact, if we were only concerned with generating new ideas and not with their intersubjective evaluation using reason and evidence, vague concepts would probably be the best. The concept that means all things to all people—like the sayings in the fortune cookies—generates more relationships to other similarly vague ideas than is possible with any precisely defined, constrained concept. Maximizing fertility clearly has a cost in the limiting case, the "most fertile" concept is not communicable at all and has no empirical import. (We say "most fertile" in the previous sentence because, given our objective of generating true knowledge claims, such vague concepts would not be most fertile.)

Even when defining fertility from the perspective of generating knowledge claims, it turns out to be impossible to maximize all three objectives—precise communication, empirical import, and fertility—simultaneously. As we move toward more concrete and specific definitions of concepts, we may increase empirical import, but we certainly decrease fertility and we may or may not increase communicability. As we move toward more asbstract ideas, particularly when we label the concepts with ordinary English terms, we may increase fertility, but we certainly make precise communication more and more difficult and we may make linkages to empirical phenomena more problematic. Let us look again at the problem of surplus meaning, which we pointed to as being especially serious in sociology. Surplus meaning is a two-edged sword. On the one hand, it creates problems of communication; on the other, it provides the fuel for new ideas. It is just those extra associations and connotations of terms like "power" that create insights into new relationships between power and other concepts. Scientific concept formation confronts a fundamental dilemma.

Scientists and lay people both must recognize and appreciate this dilemma. The lay public must realize that scientific concepts cannot be stated wholly in ordinary everyday language, because ordinary language does not facilitate precise communication. Moreover, in the case of abstract ideas, ordinary language does not provide the necessary linkages to observables. Scientists for their part must recognize that technical concepts can never be completely divorced from ordinary language because communication depends in part (and our thought processes depend in large part) on ordinary language. A technical jargon may have empirical

Bibliography

Aron, Arthur, & Norman, Christina C. (1997), *Study Guide and Workbook*, for use with Aron, Arthur, & Aron, Elaine N. (1997), *Statistics for the Behavioral and Social Sciences: A Brief Course*, Prentice-Hall, Upper Saddle River, NJ.

Babbie, Earl (1995), *The Practice of Social Research*, 7th Ed., Wadsworth Publishing Company, Belmont, CA.

Bentler, P.M. (1985), *Theory and Implementation of EQS: A Structural Equations Program*, BMDP Statistical Software, Los Angeles.

Bentler, P.M. (1989), *EQS: Structural Equations Program Manual*, BMDP Statistical Software, Los Angeles.

Blalock, H.M., Jr. (1962), "Four-variable causal models and partial correlations", *American Journal of Sociology*, 68:182-94.

Blalock, H.M., Jr. (1964), *Causal Inferences in Nonexperimental Research*, University of North Carolina Press, Chapel Hill, NC.

Blalock, H.M., Jr. (1970), *An Introduction to Social Research*, Prentice-Hall, Englewood Cliffs, NJ.

import and may appear to be precisely communicable. But if that jargon is totally separated from ordinary language, it will be sterile for generating new ideas.

The only resolution to the dilemma involves a series of trade-offs. We want to control, but not eliminate, surplus meaning. We want to communicate precisely enough so that our ideas are shared, but allow enough openness so that our audience can extend our ideas. We want to tie our ideas to observables, but not to restrict them to the extent that they have no meaning beyond the specific concrete instance. We want to develop ideas that can be embedded in a variety of knowledge claims, but not make them so loose that they mean all things to all people. In short, we must keep all three objectives in mind in formulating concepts, must be aware that we cannot maximize all objects at once, and must be ready to sacrifice a little of one objective to gain ground on another. This is one more reason why we emphasize that research is a process of developing, evaluating, and modifying ideas. From this perspective, one can start out with a vague idea but cannot stop with a vague notion. The idea must continually be developed and fine-tuned so that it communicates more precisely, has greater empirical import, and yet retains the capacity to relate to other ideas.

Our central thesis is that concept formation and definition must address the objectives of precise communication, empirical import, and fertility. The task is to strike a balance among these objectives. If we recall that science deals with systems of concepts, we are aided in that task. In a system of related concepts, every concept need not be equally bound by all three objectives; some can carry the load of precise communication, others can be attuned to the objective of empirical import, and still others can emphasize the objective of generating new ideas.

Perhaps the most important aspect of concept formation is the act of definition. The next section examines general problems of defining concepts and introduces three types of definition. In the final section of this chapter, we will return to the three objectives of concept formation to see how the act of defining relates to these objectives.

DEFINITION

A major purpose of definition is to facilitate communication and promote shared usage of terms. When we define a term, we introduce a new idea or usage and specify the new idea by means of already familiar ideas. Let us introduce some terminology. Let us call the new idea, the *definiendum*, which is simply a Latin name for the "thing to be defined." The Latin word *definiens* will stand for "the familiar ideas used to specify the meaning of the definiendum." Thus, for example, if we define "status" as "position in a social structure," then "status" is the definiendum, and "position in a social structure" is the definiens.

One point must be emphasized. We require that the definiens contain already familiar ideas. To put it more precisely, the definiens is an expression whose meaning is already determined. Unless this requirement is met, definition cannot accomplish its objectives of facilitating precise communication and shared usage. In our example, if "position in a social structure" is not widely understood and used in the same way, we have not contributed to our goals by using that expression to define status. If we are vague or unclear about the meaning, and particularly the limits of usage, of "position in a social structure," we will also be vague and unclear in using the term"status." For example, some of us associate the idea of a hierarchy with positions in a social structure. Our example takes no stand on whether the idea of hierarchy is to be included or excluded from the concept of status. Hence, it is extremely important that the definiens be an expression for which there is already shared understanding.

Our insistence that the definiens be an expression whose meaning is already determined raises another problem which, in a sense, is a version of the chicken-or-egg problem. How does one begin the process of explicit definition when there are no expressions to begin with whose meanings are already determined? We have all experienced something of this problem. One can go to the dictionary to look up a new term. In so doing, we often find that the definition of that term contains other terms with which we are unfamiliar; we look up those terms, and some of them contain terms whose meaning we do not know. If the process continues long enough, one of the definitions leads us back to the term we started with; that is, the definiens of some term contains the term which was our original definiendum, and we have gone in a circle.

The experience of going in a circle illustrates a very important feature of definition: we cannot explicitly define all of the terms and ideas that we use. Some ideas must be *primitive*; these ideas are not explicitly defined, but are used as definiens to explicitly define other terms. Thus, one builds a system of definitions by defining ideas in terms of a limited number of primitives. Any new idea is explicitly defined either by using as the definiens a previously defined idea or a primitive idea.

If we are to achieve our goals, we must be careful in selecting those ideas which will serve as primitives. If we choose as our primitives ideas which are vague, have widely varying usage, or which are empty of content, then explicitly defining other ideas in terms of these primitives will not generate a useful set of concepts. If, for example, we define "role" as the "dynamic component of status," and treat "dynamic," "component," and "status" as primitives, we have not moved very far toward precise communication and shared usage. On the other hand, if we offer a definition such as "Two positions differ in status if they are differentially evaluated," the primitive idea of "differential evaluation" seems reasonable—since most people will associate to the term "differential evaluation" the idea that one thing is better or worse than another. In short, "differential evaluation" seems to meet the requirements for a useful primitive term, while "dynamic component of status" does not.

Let us briefly review the principles we have introduced. The purpose of explicit definition is to facilitate precise communication and shared usage by indicating the limits of meaning and usage for any new term. But terms usually reflect ideas;

hence, the definiens delineates the boundaries of a new idea in terms of familiar ideas. But not all ideas can be explicitly defined; one must begin with a small number of primitive ideas, ideas for which there are no explicit definitions within the system. We use these primitives as definiens for explicitly defining the ideas we want to introduce. Furthermore, any idea that is explicitly defined is defined either in terms of primitives or in terms of ideas previously introduced by explicit definition. Finally, great care must be exercised in selecting primitives. These are the foundation of the system of concepts, and a weak foundation can cause the collapse of the system. To insure a firm foundation, we should choose as primitives ideas for which we believe there is a wide agreement on meaning and usage.

We should recognize that a system of concepts is not introduced once and for all; rather, formulation of ideas and their definitions undergo revision and refinement. It often turns out that some definitions which initially seem fruitful generate more problems than they solve, and then it is necessary to reformulate the concept. Sometimes we are sure that a primitive is widely shared, but we discover that it has a much broader range of meanings than we imagined; hence we recognize that it is inadequate as a primitive. Sometimes, we are forced to use a term as primitive, recognizing its inadequacy, because of our inability to explicitly define it. We do so with the faith that eventually we will be able to provide an explicit definition.

Two points deserve special emphasis. First, concept formation is a process which involves conceptualization and reconceptualization, formulation and reformulation. If we deal with a system of explicit definitions, its very explicitness reveals both the strengths of the system and its problem areas, directing attention to those ideas that need further thinking. The second point to emphasize is that the greatest difficulties arise from attempting to formulate a single concept in isolation from other concepts. By emphasizing the distinction between primitive terms and explicitly defined terms, we hope to draw attention to concept formation as a systematic process. Defining one term requires us to use other terms in the definiens. The selection and analysis of the terms in the definiens cannot be passed over lightly. Careful selection of primitives and the use of other previously defined terms point to the systematic structure of definition.

TYPES OF DEFINITION

We have discussed the general problem of definition. Now it is necessary to introduce some different types of definition in order to make finer distinctions among the purposes definition serves. We will consider three types of definitions: *nominal, denotative,* and *connotative.* As will be seen shortly, most of the preceding discussion referred specifically to connotative definition.

A *nominal definition* is a purely conventional agreement to let the definiendum be synonymous with the definiens. Nominal definitions are frequently very convenient. For example, in algebra we often wrote:

$$\text{Let } y = x^3 + 2x^2 + 16x - 44$$

We then proceeded to develop a proof in terms of y, rather than write and rewrite the complicated expression on the right. When we did, we were providing a nominal definition for y. In talking about social science ideas, it is often convenient to substitute a single term for a complicated phrase; this we introduce by stating, for example, "Let attitude mean a response toward or against an object." Rather than constantly writing "a response toward or against an object," it is much more convenient to write "attitude." This is providing a nominal definition for the term attitude. Nominal definitions are extremely useful, but they do not add meaning or specification to a concept. In our example, we have not limited the concept of "response toward or against an object" in any way by letting it be synonymous with "attitude"; nor can we be sure that the phrase has an already well-established meaning. However, if our readers interpret "attitude" only as a substitute for "response toward or against an object," we will have greatly simplified our writing task. It demands a lot of the reader to keep his thinking within such strict boundaries. Despite our explicit nominal definition, people will associate other ideas with the term attitude—for example, strong positive or negative feelings—and our purpose of having a convenient single-word substitute for a complicated phrase will be defeated.

At this point we can see a virtue for technical jargon. The coining of new terms in the social sciences is frequently an object of criticism, but there are circumstances in which such jargon is not only justifiable but absolutely necessary. If we coin a new term as a nominal definition, we can be sure that excess connotations of the term would not transfer with the limited meaning we intend. Hence, instead of using "attitude" as our nominally defined term, suppose we said, "Let 'axet' be 'a readiness to respond to or against objects.'" That would provide us with convenience without running the risk that extra meanings would be read into our term "axet." Of course, it would be possible to choose more pleasant sounding coined terms than "axet," but that should not obscure the point.

A *denotative definition* provides a partial definition by pointing to examples to which we would attach the term in the definiendum. Thus, we could denotatively define "attitude" by saying that each of the following is an example of an attitude: feelings toward war, prejudice toward minorities, favorability toward abortion, and dislike of criminals. Very often, the best we can do is denotatively define a concept because we are not ready to establish the boundaries of that concept, although we may be quite at home in picking out instances where the concept is applicable. Many very useful concepts in sociology are only denotatively defined.

The final type of definition is what we shall call *connotative definition*.[1] By connotative definition we mean a definition which specifies in the definiens properties of the concept being defined. Again, consider the concept of "attitude."

---

1. From Plato's time on, this has been discussed as *real* definition. Plato regarded real definition as capturing the essential features of the thing to be defined. We resist the notion of "essential features of the thing to be defined" and the implication that there is one best definition of a concept; for these reasons, we prefer the term *connotative*.

Allport in 1935 (reprinted in Fishbein, 1967) offered a definition which attempted to formulate properties of his concept when he defined "attitude" as "a mental and neural state of readiness exerting a directive influence upon the individual's response to all objects and situations with which it is related."[2] While Allport's intention is to specify properties which limit the usage of his concept of attitude, it should be clear that his attempt is only partially successful. The problem again is that the terms in the definiens are not terms for which meaning has already been clearly established. For that reason, Allport's definition in practice becomes quite difficult to work with. It is hard to know, for example, what people think of as "mental or neural states of readiness." Unfortunately, instead of providing us with clear guidelines for the usage of the concept of attitude, "mental and neural states of readiness" serves merely as a physiological metaphor.

If we apply our two tests to Allport's definition, we find that it does not facilitate precise communication; on the other hand, it is exceedingly fertile. A great many more instances than those in our simple short list can be considered as attitudes fitting Allport's conception when we denotatively define the concept. In fact, Allport's concept may be too fertile—it may exclude little of human behavior. Nevertheless, Allport's objective of formulating a connotative definition must be applauded.

The Allport definition has one important consequence for research. Since its infancy, attitude research has been controversial. Some critics, noting that people say one thing and do another, have questioned the wisdom of doing attitude research at all. The controversy has simmered ever since La Piere's classic study in 1934. La Piere traveled across the United States with a Chinese couple, stopping at hotels, motels, and restaurants. They were refused service at only a few places, although this was a time when there was considerable prejudice against Chinese. At the end of his trip La Piere mailed a questionnaire to the managers of all the hotels and restaurants he had visited. The vast majority of respondents indicated they would refuse service to Chinese people, providing a dramatic example of the discrepancy between attitudes expressed on a questionnaire and behavior in a real situation.

Yet, it is too easy to use La Piere and other such studies to condemn attitude research. Moreover, Allport's concept of attitude provides a useful perspective on this controversy. If an "attitude" is a "readiness to respond," we should not always expect a correspondence between attitude and behavior. Sometimes this "readiness" is triggered and overt behavior occurs; but sometimes the "readiness" is turned off. In Allport's view, an attitude is a predisposition that may not always result in the corresponding behavior. His concept enjoins the researcher to look for factors which intervene between the predisposition and the behavior in order to discover those factors which transform the predisposition into overt behavior and those which turn off the predisposition. In the La Piere study, perhaps

2. Although this concept was formulated more than fifty years ago, it is very similar to conceptions of attitude in current use (McGuire, 1985).

attitudes toward La Piere as the white Anglo-Saxon host of the Chinese couple intervened to turn off prejudiced predispositions against Chinese.

Applying a connotative definition to the issue of the discrepancy between attitudes and behavior illustrates how connotative definitions can help resolve such controversies. Unfortunately, this controversy is still unresolved, in part because there is no shared connotative concept of attitude. Researchers mean different things when they argue about "attitudes," and some attempt to deal with the issue of discrepancy with no connotative concept at all. Hence there is no shared intersubjective understanding of what the argument is about. Resolution of such controversies depends on shared usage of key concepts, and it is our contention that the intersubjective usage of concepts can only be achieved if at least some concepts are defined in terms of the properties captured in the abstraction which the concept represents.

Our analysis of Allport's definition documents our claim that the three objectives—precise communication, empirical import, and fertility—conflict with each other. In formulating concepts we are involved in a trade-off, attempting to achieve a reasonable balance among these objectives. The analysis also points to one further problem. One of the major difficulties with Allport's formulation lies precisely in the fact that it does not pay sufficient attention to embedding a concept in a system. Furthermore, his definiens introduces several terms, all of which are treated as primitives. If some of these terms had been explicitly defined previous to their use in the definiens for "attitude," the definition would have been much more useful. In other words, a conceptualization of attitude based on a system which contained a small number of primitives and a few explicitly defined terms, all of which build up to a definition of attitude, would have as a result fewer difficulties in our shared understandings of the meaning and usage of attitude.

We have talked about the importance of embedding a concept in a system of concepts. We have also emphasized building up the definition of a complex idea from primitive terms and previously defined concepts. We should illustrate these two important aspects of concept formation and the benefits that result from pursuing concept formation in the way we suggest.

Let us try to formulate a concept of attitude following the strategy we recommend. While a complete conceptualization of attitude is beyond the scope of this book and would take us into too many technical details, we can go far enough to illustrate important gains that result from this approach.

Let us choose as primitive terms the following: person, object, attribute, situation, positive, negative, evaluation, perception, and action. With these primitives, we can introduce some explicit definitions. First, let us assert that an object has many attributes that a person may or may not perceive. In other words, a person may react to the president of the United States as a totality or may distinguish the president's personality, his international policies, and his domestic policies as three separate attributes. Needless to say, an object as complex as "president of the United States" could be perceived in terms of a long list of attributes.

We can let the term *object* stand for any person, thing, or situation which a person may perceive or evaluate. Now let us define the term *consistent evaluation:*

A person has a consistent evaluation of an object if, and only if, that person's evaluation of the perceived attributes of the object is either all positive or all negative.

Here, what we are suggesting is that objects are complex. The typical object can generate both positive and negative feelings—a person can strongly support the president's foreign policy and be strongly opposed to the president's domestic policy.

We now define the term *consistent action:*

Any action of a person toward an object is a consistent action if, and only if, the object or some third person perceives an evaluation of the object that corresponds to the person's evaluation of the object.

In the example, La Piere would be a third person perceiving that serving the Chinese couple was positive evaluation and that refusing to serve them was negative evaluation. Knowing the questionnaire response of the motel manager, La Piere could judge whether the action was consistent or inconsistent. We are now ready to define *readiness to respond*, and our definition will have the virtue of providing an empirical test of the appropriate use of the concept:

A person has a readiness to respond to an object if, given that the person perceives the object, the person has a consistent evaluation of the object, then the person performs a consistent action toward the object.

This definition does not restrict us to situations or objects where there is a consistent evaluation; that is, there can be a readiness to respond where there is not a consistent evaluation. However, there must be a readiness to respond which leads to action when there is a consistent evaluation. The definition both eliminates the vagueness of Allport's "readiness to respond" and gives empirical import to the concept. In other words, having established that a mental state (person's evaluation) leads to an action in the pure situation (consistent evaluation), we are then free to use the idea in the inconsistent situation where a person's evaluation is mixed. We have put limits on the usage of *readiness to respond* because, if a consistent action does not appear in the pure situation, we have no license to use the concept.

We are now ready to give a connotative definition of attitude:

An attitude of a person toward an object is a positive or negative evaluation of one or more attributes of that object that generates a readiness to respond by means of positive or negative action toward that object.

In this definition, we have made use of both primitives and our previously defined terms. We have increased the empirical import of tbe concept of attitude by building in an empirical test of *readiness to respond*. At the same time, we have preserved a high level of fertility for the concept. Our definition of *attitude*, for example, allows there to be both positive and negative evaluations of different attributes of the same object. In so doing, it calls attention to the balance of positive and negative evaluations as a researchable issue. For example, in thinking about this definition, one would not expect the potential for action to result in action when the positive and negative evaluation of the object are action roughly equal. Approaching attitude research with a concept something like this (of course, it would need to be more fully developed) gives us the realistic expectation that attitudes will not always result in overt behavior. Hence, the fact of the discrepancy between attitude and behavior is no longer a damaging criticism of attitude research; rather, it is what is to be expected. It follows then that the interesting research question becomes: Under what conditions do attitudes lead to consistent action apart from the situation in which there is consistent evaluation? "Consistent evaluation leads to consistent action" is true by the definition of *readiness to respond*. (We will not call something an attitude unless, when there is consistent evaluation, there is also consistent action.)

We argue that this way of formulating the concept of attitude is exceedingly fertile. We can introduce other ideas, such as intensity of attitude, and refer them to the balance of positive and negative evaluations of attributes of an object. A natural proposition that arises from thinking about the problem is that the greater the intensity of an attitude, the greater will be the readiness to respond to an object, and the greater the likelihood of there being an action consistent with the attitude. The ideas that have been illustrated are not new. Bits and pieces of them and many others abound in the literature. The virtue of developing a system of concepts is that it takes these bits and pieces and brings them together in a way that should stimulate attitude research and bring some clarity to the circumstances in which one should expect attitude and behavior to correspond, and to situations in which one should expect a discrepancy between attitudes and behavior.

TYPES OF DEFINITION AND THE OBJECTIVES OF CONCEPT FORMATION

In this concluding section of the chapter we should bring together some ideas that were mentioned in the previous two sections. In a system of concepts, each type of definition has a role to play; our system will contain some primitive terms, some nominally defined terms, and some connotatively defined concepts. Furthermore, when we attempt to relate our ideas to empirical phenomena, we are often involved in denotative definition. The choice of type of definition, then, depends very heavily on the role that the concept plays in the system.

Furthermore, we argue that formulating a concept explicitly imposes constraints on its usage. The different types of definition impose different constraints that are suitable under different circumstances. If we summarize these constraints, we will see some of the advantages and disadvantages of each type of definition and also how each type relates to the objectives of precise communication, empirical import, and fertility.

Clearly, nominal definition is the most constraining type of definition. The definiendum means only what the definiens explicitly says. Neither the user nor the audience has any right to think about anything else when a nominally defined term is used. If the meaning of the definiens is well established, then nominal definition should facilitate precise communication, for the definiendum is just a simpler substitute for a more complicated expression and it neither adds nor subtracts meaning. If the definiens states observables, then a nominally defined concept may have empirical import; on the other hand, nominal definition can be stated solely in theoretical terms. For instance, defining the power of Person *A* over Person *B* as the number of times *A* influences *B* is a nominal definition with empirical import; defining "power" as "ability to overcome resistance" is a nominal definition with little empirical import. Finally, the fact that this type of definition is most constraining means that it is the least fertile. By conventional agreement, nominal definition severely limits associations and connotations; by neither adding nor subtracting meaning, it forecloses the possibility of generating new ideas. In fact, in those cases where new ideas arise from nominally defined concepts, they arise because the rules for using nominal definition have been violated.

Denotative definitions are typically not very effective in promoting precise communication. Defining by example may allow us to share concrete examples but there may be completely idiosyncratic ideas underlying the examples. Suppose we define "prestige" denotatively by saying that doctors have it, lawyers have it, and so forth. One person's "it" may be wealth and another's may be respect. Even if we could agree on a list of examples of prestige, we might have difficulty adding new examples without any specification of properties that the examples share. However, simply by raising the question of what properties the examples in our denotatively defined concept do share, we can initiate a very productive activity. Thus denotative definition can be very fertile.

Denotative definition is also a way to give concepts empirical import that is closely akin to stating initial conditions. In this way, denotative definition enables us to conduct research even when we are unable to formulate a connotative definition capturing the properties of the idea we are studying. In the process, our ideas become clarified and we can move toward a connotative definition. Suppose we could agree on a set of examples of high prestige and a set of examples of low prestige; we could then study each set to uncover properties that were shared within each set and those that were sharply different between the sets. In such an investigation, we could not get started if we required a connotative formulation of prestige in advance, but the investigation itself could promote the development of a connotative concept. It should be clear, then, that denotative definition serves a useful and important purpose in the development and evaluation of ideas.

Our insistence on a system of concepts would not make much sense unless the system contained some connotatively defined concepts. It is the explicit spelling out of properties of an idea that enables us to generate relationships among ideas. The definiens of connotative definitions spells out properties; furthermore, the expressions that state these properties must be expressions whose meaning is already determined. To meet this requirement, the definiens is constructed from both primitive terms and previously defined ideas. In other words, constructing a definiens that formulates properties of an idea requires embedding the idea in a system of primitive terms and other defined terms. An interesting mutual dependence emerges; we cannot have a system of concepts without connotative definitions, and we cannot have connotative definitions without a system of concepts.

Connotative definitions are less effective, by themselves, in facilitating precise communication. In the last analysis, every connotative definition can be traced back through the definition chain to the set of primitive terms, and thus every connotative definition depends on undefined ideas. As undefined ideas, however, primitive terms do not limit associations and connotations, even if there is a widely shared understanding of the primitives. Thus, the chain of definitions always contains some openness to varying interpretations of the concepts, and this openness increases the likelihood of miscommunication and misunderstanding of ideas. The same openness provides the possibility of new interpretations and fruitful new insights. Moreover, connotative definitions are not totally open and unconstrained, since the meanings are somewhat limited by the relation to other concepts and by the sentences in which the connotatively defined concept is embedded.

In principle, it is possible to connotatively define a concept with a definiens whose expressions all refer to observables. Such a concept, in and of itself, would have empirical import. But restricting the meaning of a concept to a specific set of observable properties drastically limits the usefulness of that concept and defeats the purpose of connotative definition. Usually, connotative definitions do not have direct empirical import, but obtain empirical meaning through links to observables such as statements of initial conditions.

Connotative definition chiefly addresses the objective of fertility. Formulating properties of a concept suggests things that should be related to that concept and, therefore, to sentences which relate that concept to other concepts. If ''ability to overcome resistance'' is a property of ''power,'' and we start thinking about things which might enhance the ability to overcome resistance, we can come up with a number of knowledge claims relating power to other ideas.

An important conclusion emerges from this analysis. No type of definition accomplishes all three objectives. This is not surprising, since we have already noted that it was not possible to maximize all three objectives simultaneously. But the significant point is that the analysis of types of definition shows the way in which the types of definitions complement one another. Furthermore, this complementarity gives additional support to our belief in the desirability of a system of concepts. In order to get the best trade-off among the objectives of precise communication, empirical import, and fertility, it is necessary to have a system of concepts employing all three types of definition.

Boudon, R. (1965), "A method of linear causal analysis: Dependence analysis", *American Sociological Review*, 30:365-74.

Campbell, Donald, & Stanley, Julian (1963), *Experimental and Quasi-Experimental Designs for Research*, Rand McNally, Chicago.

Chen, Sheying (1996), *Social Policy of the Economic State and Community Care in Chinese Culture: Aging, Family, Urban Change, and the Socialist Welfare Pluralism*, Avebury, Brookfield, VT/Aldershot, Hampshire.

Chen, Sheying (1997), *Measurement and Analysis in Psychosocial Research: The Failing and Saving of Theory*, Avebury, Brookfield, VT/Aldershot, Hampshire.

Coombs, C. H. (1964), *A Theory of Data*, Wiley, New York.

Cook, Thomas D., & Campbell, Donald T. (1979), *Quasi-Experimentation: Design and Analysis Issues for Field Settings*. Rand McNally, Chicago.

Davis, James A. (1985), *The Logic of Causal Order*, Sage University Paper Series on Quantitative Applications in the Social Sciences, 07-055. Sage, Newbury Park, CA.

Dometrius, Nelson C. (1992), *Social Statistics Using SPSS*, HarperCollins Publishers, New York.

Donald, Cathy A. and Ware, John E., Jr. (1984), "The measurement of social support", *Research in Community and Mental Health*, 4:325-70.

Duncan, O.D. (1966), "Path analysis: Sociological examples", *American Journal of Sociology*, 72:1-16.

Flinders, David J., & Mills, Geoffrey E. (eds.) (1993), *Theory and Concepts in Qualitative Research: Perspectives from the Field*, Teachers College Press, Columbia University, New York.

215

SUGGESTED READINGS

Hempel, Carl G. "Fundamentals of Concept Formation in Empirical Science." *International Encyclopedia of Unified Science* 2, no. 7. Chicago, Ill.: University of Chicago Press, 1952.

Although difficult and somewhat technical, this small volume sets out the fundamental issues of concept formation in a clear and concise manner.

Lachenmeyer, Charles. *The Language of Sociology.* New York: Columbia University Press, 1971.

A critique of the present state of sociology, this work argues that sociology is not a science and attributes its failure to the fact that sociological language is more like conventional language than a scientific system. The author, however, believes that sociology can and should be a science.

EIGHT

Tying
Concepts to
Observations

The previous chapter emphasized the necessity for developing systems of concepts. It introduced three main objectives of concept formation, presented three types of definition, and indicated how these types related to those objectives. Further, the discussion demonstrated the impossibility of maximizing the objectives simultaneously and, therefore, the necessity for a trade-off among the objectives of precise communication, empirical import, and fertility. The analysis concluded that constructing systems of concepts could achieve the best trade-off among these objectives.

This chapter looks more closely at the problem of tying concepts to observables. Although the discussion of the objective of empirical import pointed to the importance of tying concepts to the world of sense experience, it is necessary to go deeper into that aspect of concept formation. In the first place, the problem is crucial because the empirical character of science depends on it. Second, the implication of our emphasis on a system of concepts is that, rather than a piecemeal approach to tying concepts to observables, a strategy is needed that attacks the problem in the context of the other two objectives. Third, we must examine the problem closely because the approach suggested here runs counter to much current practice in the social sciences.

In this chapter we will present a strategy for tying concepts to observables—that is, for achieving as high a level of empirical import as is consistent with also achieving high levels of precise communication and fertility. We will call it the *strategy of indicators*. We will discuss the strategy, the direction it provides to the researcher, and two criteria for the evaluation of its success. The two criteria to be examined are *reliability* and *validity*. While they have been widely discussed in the methodological literature, their place in the overall strategy puts them in a new light.

Before presenting the strategy of indicators, it is instructive to analyze an alternative approach to the problem of tying concepts to observables. The next section will discuss the approach known as *operationism*. Since operationism is prevalent in the social sciences, an understanding of its basic ideas and their implications is important. We will examine what is perhaps the best-known and most striking example of operationism in action, the IQ-testing movement. A case study of the IQ will reveal some of the principal scientific and practical consequences of the operationist approach.

OPERATIONISM: THE CASE OF THE IQ

The key idea of the operationist position is the *operational definition.*[1] An operational definition specifies the meaning of the definiendum in terms of a set of physical measuring operations; the definiens constitutes a set of directions for physical acts of the observer. Thus, for example, the concept "leader of a group" might be operationally defined as the person who receives the most votes in answer to the question, "Who is the leader of the group?" Such a definition implicitly directs the observer to perform the physical act of asking the question and the physical act of counting up the responses. Similarly, the concept of a person's IQ is operationally defined as that person's score on a particular IQ test.

Operationism has been influential in the social sciences and particularly in sociology. It quite rightly recognizes that scientific concepts must be tied to the empirical world; that is, concepts must be linked in some way to the sense experience of the observers. Furthermore, operationism has performed a useful service as a critique of vague concepts and concepts that have no empirical import. Yet, as a constructive approach to concept formation, operationism fails. It is important for both the researcher and the consumer of research to understand the reasons for its failure.

For the purposes of analysis, we will examine the strictest form of the operationist position (Hempel, 1965). Few operationists adopt this extreme position, but their work reflects its influence. In addition, analyzing the purest form of the position will aid in clarifying the issues. Throughout this analysis, we must keep in mind that many operationists recognize the problems we point to and have somewhat modified the strict operationist approach.

In analyzing the case of the IQ, we will argue that operationism fails for several reasons:

1. It does not facilitate precise communication.
2. It is too stringent in attempting to limit meaning to a specific set of operations, and thereby opens the door to vagueness and ideological dispute.
3. It focuses on a single concept, and therefore does not provide guidelines for thought or investigation.
4. It provides no basis for deciding whether the operations used to define the concept are relevant to the statements people want to make using the concept.

---

1. *Operational definition* should not be equated with "operationalization," which is a general term for relating concepts to observations; operational definition is only one type of operationalization.

The case of the IQ well illustrates some of the main features of operationism. In its strictest form, operationism regards as inadmissible any concept for which there is not an operational definition; it excludes connotatively defined, or what we will call *theoretical*, concepts. Thus a person's IQ is defined as the score on a particular test, and no ideas about the person's intelligence are admissible. The test constructor does not think about intelligence; the user who thinks about intelligence is misusing both the test and the concept of IQ. (As we shall see, it does not work that way; but many test builders argue that it should and that they are not responsible for misuse of the test or the concept.)

Operationism asserts that each different set of operations defines a different concept and that equivalence of operations can only be established by demonstrating a strong empirical relationship between the two sets of operations. Thus, there is a different and noncomparable IQ for each test that is constructed. The only way to substitute one concept of IQ for another, or to compare IQs, is to show that, for a given group of people, their scores on one IQ test are highly correlated with their scores on a second IQ test. However, because most tests are constructed so that 100 is an average score, many users ignore this injunction and compare the scores as if the number has a reality independent of the test on which it was based. For example, we see many instances of comparing the scores of black children who took one test with those of white children who took a completely different test.

Operationism deals exclusively with empirical relationships between operationally defined concepts—in our terms, exclusively with observation statements. Thus, IQ scores predict performance in school, operationally defined by other tests; or IQ scores distinguished between brain-damaged and normal patients, where brain damage is operationalized in terms of physical procedures like lack of hand-eye coordination. "Theorizing" is done empirically by categorizing observation statements. For example, differing mental abilities are asserted by showing that there are sets of questions on an IQ test such that: within a set there are high correlations among the questions, and between the sets there are low or zero correlations. But these are observation statements and are singular. From our earlier discussion of singular statements, there is no reason to expect them to hold in other times, places, or historical circumstances. It is no wonder that the number of such abilities ranges from 2 to 120.

With this brief sketch of the operationist approach to IQ, let us turn to analyzing the scientific and practical consequences of such an approach. Underwood (1957) asserts that "operationism facilitates communication among scientists because the meaning of concepts so defined is not easily subject to misinterpretation." In the case of IQ, anything but precise communication has occurred. Although there is no meaning to IQ other than score on a particular test, no one thinks about IQ in this way. If the meaning of IQ were in fact restricted to its operational definition, we would not have seen any of the recent controversy over the use of IQ tests. Laymen and many scientists have regarded these tests as measures of intelligence. Not only is the intelligence interpretation widespread, but people read their own meanings of intelligence into the test results. Needless to say, there is wide variation in the usage of ideas of intelligence in the population.

Consider, for example, the controversy over genetic or racial differences in IQ scores. It is a repeated observation that, on the average, blacks score lower than whites on IQ tests. It is not justifiable, however, to conclude that blacks and whites differ in intelligence, or that these differences in average test scores are hereditary. In order even to think about these differences in intelligence, one must have a concept of intelligence, and that concept must have some properties. In other words, much of the controversy and discussion of racial differences on IQ tests is based on implicit connotative definitions of intelligence. Since these connotative definitions are never made explicit, often the parties to the controversy are not talking about the same things. It is important to explain or interpret differences in test scores, but it is not valid to leap from an observation of average differences on a paper-and-pencil test to a vague, unspecified, idiosyncratic notion of intelligence.

It is clear, for example, that connotations of the word *intelligence* range far and wide. Some people associate ''intelligence'' with the ability to solve problems, the ability to learn, creativity, insight, or some combination of these notions. If one specified an explicit connotative definition of intelligence, it is not obvious and automatic that any particular IQ test would be related to that definition. However, such specification would have two virtues. First, it would indicate directions for establishing that a test is related to the definition—an activity known as *validating* the test. Second, it would clarify what the parties to the controversy are arguing about. It makes no sense to argue that intelligence is or is not hereditary without some specification of the thing which is or is not hereditary. Certainly no one is prepared to argue that the collection of checkmarks on the answer sheet to a particular paper-and-pencil test is genetically determined; yet the same people who would grant the absurdity of that position have no hesitancy in masking the huge inductive leap from ''average scores on such tests'' to ''hereditability of intelligence''; and this is true for both proponents and opponents of the genetic interpretation. Thus, instead of precise communication, operationally defined IQ provides a license for ideological and idiosyncratic interpretations of the observations.

That operationally defined IQ is the center of controversy is not an accident; it is inherent in the operationist approach. The attempt to limit meaning to a specific set of operations is too stringent. It so inhibits thinking and communication that it creates a void that cannot be filled *within* the operationist approach. People are going to think about ideas and are going to communicate ideas. Unless concepts are explicit enough to provide guidelines and constraints, there will not be intersubjective usage and precise communication. The history of IQ suggests that we delude ourselves if we believe that we can limit people's thinking to a set of physical operations like those involved in computing a score on a test.

The operationist approach to IQ focuses exclusively on a single concept, the IQ. Other concepts, also operationally defined, are introduced only after empirical investigation—that is, after showing, for example, that IQ predicts grades in school. But how does a researcher decide to use an IQ test to predict school grades—or anything else, for that matter? The operational definition of IQ provides no guidelines for such investigation. It does not even provide guidelines for thinking about how to construct an IQ test. Obviously, the researcher does think about what to include in the test and what other variables might be related to performance on the test. The test constructor does not throw together random assortments of test questions and does not collect random observations. According to strict operationism, however, we have no right to think about IQ in this way—our thoughts add meaning beyond the operational definition.

Strict operationism, taken literally, so inhibits thinking that research is almost impossible. It rules out the speculations out of which science is made—the attempt to imaginatively generate statements which guide the collection of observations and the formulation of theories.

Despite the commitment to operationism, the IQ-testing movement has not worked according to its injunctions. Researchers have been forced to step outside of the approach. When they do think about concepts and relationships, they do so privately, almost illicitly or with a sense of guilt. When they step outside of the operationist approach, they go from being overly constrained to being totally unconstrained. IQ tests are built, investigated, and used on the basis of implicit assumptions, unrecognized biases, and subjective interpretations. The absence of explicit definitions, assumptions, and guidelines for interpretation creates serious problems for intersubjective evaluation and usage, and prevents the collective application of critical reason.

Consider the fact that IQ tests place a heavy emphasis on vocabulary. Why? What justified the inclusion of vocabulary questions on the test? Certainly not the operation definition of IQ. In the United States, there is the implicit assumption that because English is the common language, every American has similar experience with some set of English words. This ignores the vast subcultural differences within the United States—differences among ethnic groups, among rural versus urban residents, among regional groups, and among social classes. The assumption of *equal* familiarity necessary to justify the inclusion of any vocabulary is untenable. Even the procedure by which tests are standardized does not help. The fact that when the test is given to thousands of people, 90 percent can answer "correctly" does not justify the inclusion of the item—unless the other 10 percent are randomly scattered across all the social groups which might have varied experience.

In general, any vocabulary test depends upon the testees' having learned the vocabulary. Vocabulary is culture-bound; no one would think of giving the average American a vocabulary test with the words in Russian. We have only recently realized just how culture-bound IQ tests are, and this recognition is largely due to the protest that IQ tests are discriminatory against minority groups.

The critics, however, have fallen into the same trap that has ensnared those whom they criticize.[2] Some of these critics have attempted to substitute a ghetto vocabulary that asks about words like "blood" and "square"; others have attempted to construct "culture-free" tests using pictures. They have failed to realize that it is the whole operationist orientation to IQ, with its unanalyzed assumptions, that is the problem. They too have developed IQ tests which depend upon learning and experience, and the relation of learning to their implicit notions of "intelligence" remains unanalyzed.

Consider the following example. The question on the test presents picture of a rat, and a child is asked to match that picture with the one of three other pictures to which it is most alike. The three other pictures are of a bird, a squirrel, and a cockroach. The slum dweller will likely match the rat and the cockroach as two pests which invade his home; the suburbanite, if he is not stymied completely by the question, will probably match the squirrel with the rat, as two rodents. Previous experience and previous learning determine the response to the question. Any inferences about "intelligence" from answers to questions such as these remain unjustified.

The focus on a single concept rather than on a system of concepts obscures the fact that the construction of IQ tests and the use of IQ in research and practice involves many ideas, ideas about properties of the tests, about relationships between these properties and other variables, and about the consequences of using these properties—for example, to decide that a testee is mentally retarded. Unless these ideas and their systematic relationships are made explicit, there are no guidelines for intersubjective evaluation.

Our final point is that operationism provides no basis for deciding whether the operations used to define the concept are relevant to the statements people want to make using the concept. Why, for example, is knowledge of vocabulary relevant to claims that intelligence is, or is not, hereditary? Perhaps relevance is justified by the implicit assumptions that one property of intelligence is the ability to employ previous experience and learning, and that the vocabulary questions measure that ability. If so, the simple act of making these assumptions explicit allows us to intersubjectively evaluate the claim of relevance. In this case, we would have to conclude that these assumptions are inadequate to justify relevance. Even if intelligence is hereditary, and if one property of intelligence is the ability to employ previous experience, and if vocabulary questions measure that ability, we still could not conclude that the test was relevant to the claim that intelligence is hereditary. We would still need to assume that the vocabulary questions were part of the common experience of the testees. We might be willing to assume that every testee was exposed to a common set of experiences (although that is a sweeping

2. Even Ehrlich and Feldman, who recognize many of the issues raised here, write: "Intelligence, then, is whatever the tests measure, and the adequacy of tests resides in the ability to make accurate predictions" (1978). They proceed to criticize the ways in which IQ has been used in what is a strong critique. But even these authors do not lay any blame at the door of operational definition.

assumption); but exposure does not guarantee that the testees had common experience. If someone is presented with new vocabulary and fails to learn it, the person may have been unmotivated or the teacher may have been incompetent. While making assumptions explicit is necessary, it is not sufficient. We would also have to think about the conditions under which the assumptions hold. In short, deciding the issue of relevance is a theoretical activity which involves choosing among alternative assumptions on the basis of an explicit set of shared criteria. As long as one remains within the operationist framework, the issue of relevance of the measurement and the issue of the conditions under which relevance can be justified are swept under the rug. Rather than precisely communicating a shared usage, operationism opens the door for arbitrary and unconstrained assertions of relevance. While the recognition of the cultural bias of the operationist is a step forward, proponents of "culture-free" IQ tests fall victim to the same failure, because they have formulated neither a connotative definition of intelligence nor a theory that explains the development and consequences of an individual possessing more or less of the properties formulated in such a connotatively defined concept.

It is our contention that operational definition creates more problems than it solves; it neither facilitates precise communication nor allows for fruitful application of a concept by embedding it in statements which relate concepts. Although there have been some attempts to formulate theories of intelligence, these theories still rely heavily on the IQ test. It may well be that the operationalization of IQ exerts too strong an influence on the theorizing. It may well be that we would be better served to theorize about intelligence and use those theories in an effort to devise ways to measure and observe intelligence. The result of such a strategy may be quite different from the current IQ test. Certainly, as long as the controversies about race and heritability revolve around current tests, there is unlikely to be an intersubjective resolution, for the operational definition of IQ allows everyone to hold fast to the interpretation that is most ideologically congenial.

THE STRATEGY OF INDICATORS

The strategy of indicators is an alternative strategy to operationism which allows us to promote precise communication, generate fertile concepts, and link our ideas to empirical phenomena. An indicator is a set of empirical procedures (operations) which generates an instance of a concept. At first glance our concept of indicator may seem identical to the concept of operational definition; but there are extremely important differences in the way indicators are used and in the attitude that a researcher takes toward indicators.

The careful reader will have recognized that an operational definition comes close to being a nominal definition. One could regard operational definitions as purely conventional agreements to substitute a term for a set of operations, adding no meaning to that term. (One could, of course, question how conventional the conventional agreement is for any particular operational definition.) Indicators, on the other hand, come very close to being denotative definitions. The operations used in creating an indicator represent an example of the concept which does constrain the meaning of the concept, but only partially constrains it. The crucial difference is that while an operational definition defines a concept, the procedures in forming an indicator merely give an example. Thus, if income is used as an indicator of social class, the user recognizes that income differences may correspond to class differences; but those income differences do not capture all that is intended by the use of the concept "social class." Social class differences may involve differences in life chances, consumption patterns, social power, or more. None of these other differences are excluded from the concept because one has chosen income as an indicator.

Perhaps the most crucial difference is that a concept can have many indicators but only one operational definition. This fact alone enhances the fertility of concepts which are linked to empirical phenomena through indicators. At the same time, indicators enhance precise communication. Once a researcher has chosen one indicator of a concept, he is not free to choose a second indicator without considering the relationship of the second indicator to the first. Thus, a researcher who uses income as an indicator of social class cannot arbitrarily choose religious affiliation as an indicator of social class unless he thinks carefully about the relationship between variation in income and variation in religious affiliation. Usually such thinking seeks the common underlying properties. One could perhaps make a case that religious affiliation and income are both appropriate indicators, but, in making that case, the researcher is involved in relating the concept "social class" to other concepts—which is precisely what we mean by fertility. The fact that the researcher cannot arbitrarily choose two indicators signifies that the meaning of the concept "social class" is partially constrained.

The strategy we have sketched points toward the formulation of connotative definitions. In thinking about the underlying ideas that relate indicators, we are in effect thinking about properties which make up a connotative definition of a concept. In our example of social class, an underlying idea that members of different social classes differentially evaluate one another may be the common idea that justifies using income and religious affiliation as indicators of social class. Here, then, is another crucial difference between operationism and the strategy of indicators. Operationism rules out theoretical meaning to concepts; indicators, on the other hand, promote theorizing about concepts. With operational definition, the meaning of a concept is restricted to what is essentially an observation statement; using indicators not only asserts that there is meaning to a concept beyond an observation statement but also facilitates the formulation of theoretical statements.

The strategy of indicators does not solve all problems. Indicators may be misused—that is, used uncritically. Investigators may, and sometimes do, choose indicators more or less arbitrarily without considering the relationship among indicators. These investigators may hope to solve all problems empirically. They may simply decide to treat a set of observed variables which correlate as indicators of the same concept, without giving it any more thought than that. While the strategy can be misused, it does contain within it the possibility of generating fertile concepts tied to empirical phenomena, and it does enhance communication by at least partially restricting the range of meanings that can be read into the concept. This strategy promotes the use of critical reason; using it in an unthinking way, however, defeats the objectives of concept formation.

EVALUATING INDICATORS: RELIABILITY AND VALIDITY

The strategy of indicators emphasizes the relationship between ideas and observations—in contrast to operationism, where the observation *is* the idea. From the perspective of the strategy of indicators, observations do not have an independent status; they are not treated as having a reality of their own. Recall that observations are taken in a particular place at a particular time; thus, any statements describing a set of observations must be singular. If we want to use such statements—observations statements—to evaluate knowledge claims, we must be very careful. If the observation statement describes only one time and one place and is not *reproducible*, it is not likely to help us evaluate universal knowledge claims. Furthermore, even if the observation statements are reproducible, we must consider whether or not they are *relevant* to the knowledge claim we wish to evaluate.

The reproducibility and relevance of observation statements depend upon the reliability and validity of the indicators we use. Problems of reliability and validity are widely discussed in the social science literature, but are usually approached as if observations can be reliable or valid in themselves without reference to the ideas that generated the observations. Our emphasis on a system of concepts requires us to look at reliability and validity in the context of the ideas that direct the collection of observations. Let us consider reliability first.

We define *reliability* as follows:

Reliability is the stability of a set of observations generated by an indicator under a fixed set of conditions, regardless of who collects the observations or of when or where they are collected.

Consider, for example, using a questionnaire item asking the respondent's occupation as an indicator of the respondent's social class. We would take each reported occupation and decide which social class it represented according to rules we would spell out in our concept of social class. For example, we might decide that someone who reported "bank clerk" was "lower middle class," while someone who said "assistant bank manager" was "upper middle class." Such decisions depend upon the stability of the respondents' answers. If the questionnaire sometimes elicited the response "bank clerk" and sometimes "assistant bank manager" from the same person, our indicator of social class would be unreliable.

Fox, William (1992), *Social Statistics Using Microcase*, Microcase Corp., Seattle, Washington.

Garfinkel, Harold (1967), *Studies in Ethnomethodology*, Prentice-Hall, Englewood Cliffs, NJ.

George, Darren, & Mallery, Paul (1995), *SPSS/PC+ Step by Step: A Simple Guide and Reference*, Wadsworth Publishing Company, Belmont, CA.

Ghiselli, Edwin E. (1964), *Theory of Psychological Measurement*, McGraw-Hill Book Company, New York.

Goldstein, H. (1990), "The knowledge base of social work practice: Theory, wisdom, analogue, or art?" *Families in Society*, 32-43.

Grimm, James W., & Wozniak, Paul R. (1990), *Basic Social Statistics and Quantitative Research Methods*: *A Computer-Assisted Introduction*, Wadsworth Publishing Company, Belmont, CA.

Horst, Paul (1966), *Psychological Measurement and Prediction*, Wadsworth Publishing Company, Inc., Belmont, CA.

Jacoby, William G. (1991), *Data Theory and Dimensional Analysis*, Sage University Paper Series on Quantitative Applications in the Social Sciences, 07-078, Sage, Newbury Park, CA.

James, L.R., Mulaik, S.A. & Brett, J.M. (1982), *Causal Analysis: Assumptions, Models, and Data*, Sage, Beverly Hills, CA.

Jöreskog, K.G., & Sörbom, D. (1989), *LISREL 7: A Guide to the Program and Applications*, 2nd Ed., SPSS Inc., Chicago.

Kahane, Howard (1992), *Logic and Contemporary Rhetoric*, Wadsworth, Belmont, CA.

More generally, to be a reliable indicator, our occupation question should produce the same response whether asked by a Marxist or an anti-Marxist; whether asked in an industrial society or in a developing society; whether asked during a depression or during prosperity. Furthermore, asking the question repeatedly should produce the same answer every time. Those who focus exclusively on observations would agree on these requirements for a reliable indicator. Our perspective, however, requires a closer look at this example.

Notice that we have made a number of implicit assumptions. We have assumed that Marxist interviewers, anti-Marxist interviewers, industrial societies, developing societies, times of depression, and times of prosperity all represent identical fixed conditions for the collection of observations. Some of these assumptions may be justified some of the time, and some may even be justified all of the time. The justification of these assumptions, however, depends on the context of ideas in which the indicator is used. It may be that our example is far-fetched and that there is no context that would justify all of these assumptions. Even if that were the case, it would not help to ignore the assumptions involved because the problem would not go away. Treating the indicator as if it had a reality of its own is equivalent to ignoring the implicit assumptions involved in any observation procedure. For this reason, we argue that it is impossible to assess the reliability of an indicator apart from the context of ideas in which the indicator is employed.

The definition of reliability sets out criteria, but the extent to which any particular indicator used in any particular study meets these criteria depends on the assumptions of the investigator, and these assumptions tie directly into the system of concepts underlying the particular indicator. If we are dealing with an indicator of social class, properties of our concept of social class and ideas about social mobility may help resolve whether we can assume that the conditions of our example constitute a fixed set of conditions. Spelling out properties of our concept of social class would help us decide what were equivalent occupations in industrial and nonindustrial societies, or in depression and prosperity. For example, should our indicator classify a postal clerk in a developing society in the same way as it classifies a postal clerk in an industrial society? For some purposes, variations in the opportunities available in the two situations could be disregarded—as for instance, if we were using our indicator of social class to study the relationship between class position and consumption patterns. On the other hand, if we were using our indicator to study social mobility, changes in class position would depend on available opportunities. We would not then want to assume that differences in available opportunities represented fixed conditions. In addition to making assumptions about fixed conditions, we must make them consistent with the purposes of our investigation.

Assessing reliability depends on deciding what make up the "fixed set of conditions." This decision cannot be made solely by looking at the set of answers to the question; it involves the context of ideas of which the question is a part. Assessing reliability requires intersubjective agreement, and intersubjective agreement is facilitated by making the system of concepts explicit.

For an indicator to be reliable, it must generate a consistent set of observations every time it is used when the fixed conditions hold. From one point of view, every time and place is unique, and so no two situations can be said to meet the same set of fixed conditions. If one takes that position, however, one operates with a particularizing orientation to which reliability is irrelevant. It may be difficult to assume that an industrial society and a developing society, regardless of the unique properties of each, both meet the same set of fixed conditions, but we cannot, a priori, say it is impossible. The generalizing point of view selects a small set of conditions regarded as important, and abstracts these from the concrete uniqueness of each society. If both the industrial and the developing society meet this small set of conditions, that is sufficient for asserting the reliability of an indicator, regardless of the other ways in which two societies may differ.

The importance of reliability cannot be overemphasized. If one is going to use indicators to make inferences about statements, then an unstable set of indicators will generate one set of inferences on one occasion and another set, perhaps contradictory, on another occasion. If an indicator of occupational prestige sometimes ranks clerical positions above semiskilled jobs and sometimes vice versa, then inferences about the relationship between prestige and other characteristics of these occupations will be inconsistent. Hence, an essential precondition for any empirical evaluation of knowledge claims is the stability of the indicators of the concepts contained in the knowledge claim. As we will see shortly, reliability is a necessary condition, but not a sufficient condition. Furthermore, one cannot trust to luck that indicators are reliable; one must assess the degree of reliability of every indicator used. Unfortunately, much sociological research does not pay sufficient heed to the requirement of assessing reliability. In other sciences, an investigator would not think of reporting observations obtained with an indicator without also mentioning an estimate of the error contained in that indicator. When such estimates of error are reported, they can be interpreted as measures of degree of instability in the indicator, given a fixed set of conditions.

The failure to assess reliability of indicators means that any observation statements based on those indicators rest on a very weak foundation. In a public opinion poll, for example, we cannot just assume that the responses to a question like, Whom do you favor for president in the November election? are automatically reliable. For example, pollsters make much of a 2 percent shift from one candidate to another in two successive polls. But such a shift may be smaller than the unreliability of the question. We know that slight changes in question wording— like changing the order of presenting the names of the candidates—or minor variation in the way the interview is conducted can produce as much as a 10 percent shift in response. Such shifts are usually signs of the unreliability of the question as an indicator of political preference, because there is no evidence of an actual shift in preference and no reason to assume that such minor alterations of the observation procedure could cause a change in preference. Hence, a good deal of fluctuation in public opinion surveys during a political campaign may well indicate unreliability of the question as an indicator of political preference. One could be misled in interpreting trends in public opinion if one is totally unaware of the amount of unreliability in the indicators used to measure public opinion.

In our public opinion example, the problem is to disentangle true changes in public opinion from unreliability of the indicators used. This is not a trivial problem; moreover, it cannot be solved empirically. One cannot simply give the question to large numbers of respondents under a wide variety of circumstances, and correlate different sets of answers. If we think again about the definition of reliability, we recognize that what constitutes a "fixed set of conditions" is a conceptual problem. Hence, in order to assess the reliability of an indicator, we are forced to make explicit assumptions concerning what constitutes a fixed set of conditions for that indicator. In our political example, we could assess the reliability of a question like, Whom do you favor for president in the November election? by defining a fixed set of circumstances, such as those time periods in which no appreciable campaign activity took place. Then, by administering our question to the same sample of individuals at two points during this period, we could obtain an estimate of unreliability by looking at the amount of change in responses to the question that cannot be attributed to any external influence, such as a campaign. Of course, we might well want to require certain things other than the absence of campaign activities; but the point here is that assessment of reliability is essential, and it can only be done by assuming what constitutes fixed conditions.

In order to successfully tie our ideas to observables, we must construct reliable indicators of our concepts; while reliability is necessary, however, it is not sufficient. It is possible to have highly reliable indicators that are totally irrelevant to our ideas. In a society where social class depends on family ties, a question about occupation may produce stable responses under fixed conditions but may have nothing whatsoever to do with our concept of social class. Hence the strategy of indicators requires procedures to demonstrate the relevance of what we observe to our ideas. Such procedures are known as *validation*, and the general issue is called the validity problem.

The idea of validity is difficult to define satisfactorily. One traditional approach, for example, argues that "an indicator is valid if it measures what it is supposed to measure." The major drawback of this position is that its proponents never provide any suggestions for deciding how you know that an indicator measures what it is supposed to measure. Despite this obvious problem—it is almost equivalent to saying an indicator is relevant if it is relevant—there have been few attempts to develop the formulation far enough even to provide direction for solving the problem. If we understand the reasons why there have been few attempts to develop the idea of validity, it may help us to formulate a more useful conception.

To be sure, the problem of validity is difficult and complex. When we considered simple knowledge structures, we avoided the problem by simply assuming that our observations were relevant to our ideas. That, after all, is the significance of asserting an initial condition such as, "In place P at time T, occupation is an instance of social class." Such assumptions may be a useful way to begin research and may even be necessary, because we cannot solve all problems *before* we do any empirical research. At some point in the development of knowledge, however, we must confront such assumptions and evaluate them. At that point, we come face to face with the validity problem.

The inadequate development of the notion of validity stems from two sources. In the first place, there is a naive identification of validity with "truth." When we give people a questionnaire, we can naturally wonder whether or not they are giving truthful answers. If we believe they are telling the truth, there is a natural tendency to regard an indicator based on their truthful responses as valid. All our respondents could be telling the truth, however, and their responses could still be irrelevant to the ideas we want our indicator to reflect. For example, there is a recent critique of research on occupational prestige (Villemez, 1977) that uses this argument. Such research typically asks respondents to evaluate the "general standing" of a list of occupations, and uses the responses for each occupation as an indicator of the prestige of that occupation. The critic claims that people give truthful answers, assigning high general standing to high income occupations; but he argues that such answers are irrelevant to prestige, because his concept of prestige contains the notion of deference, rather than reward, as a crucial property. With this critic, and with us, relevance of the responses to the concept is more important than the simple notion of truthful responses.

As a second source, inadequate treatments of the idea of validity occur when validity is considered an empirical property of an indicator of and by itself. IQ tests are described as valid, and much effort goes into validating them. But IQ is a score on a test. What, then, does it mean to say an IQ test is valid? Since no "idea" of IQ is formulated, there is no legitimate way of deciding whether it measures what it is supposed to measure. Validity of IQ tests, then, must have some other meaning; it is not too overstated to say that, in this context, validity means successful prediction. The typical way of validating an IQ test supports this view: validating a particular IQ test consists of showing that it predicts scores on another IQ test that has been validated. Such validation procedures do not resolve any of the controversies surrounding IQ, primarily because such procedures do not deal with the central issue of validity. In these controversies, we do not want to talk about predicting other tests or performance in school; we want to talk about ideas of intelligence. Since successful prediction may or may not be relevant to these ideas, prediction cannot be the defining criterion of validity.

The central issue of validity is the question of the relationship between the meanings abstracted in a concept and the properties of the indicators of that concept. This view of validity enables us to use observations to talk about the ideas formulated in our concepts, and directs us to validate indicators by demonstrating that the properties of an indicator are relevant to the properties that define the concept. We can now present a provisional definition of *validity*:

Validity is the degree of correspondence between the defining properties of a concept and the observed properties of an indicator of that concept.

Although this definition is still not adequate, it does provide some useful ways to think about validity and some direction for validating indicators. It suggests, for example, that the definition of the concept offers a basis for deciding what the indicator is supposed to measure. It exposes the fact that validity involves both conceptual and empirical issues. It emphasizes that, while empirical procedures are necessary in the validation of indicators, they are not sufficient.

In this formulation, validity only concerns connotatively defined concepts. If properties of a concept are not specified, we cannot talk about correspondences between properties of the concept and properties of the indicators, and the issue of validity is beside the point. Once we have formulated a connotative definition, however, we then must analyze the properties of potential indicators. Our analysis leads to assertions or statements about relationships between potential indicators of the same concept, or to statements relating indicators of two different concepts, that may be used for validation purposes. These statements are testable, and the empirical evaluation of the validity of a set of indicators involves testing them. But when the statements are supported, we do not have a finding, because we have assumed the truth of these statements; validation is not discovery.

What has been said so far implies that it is impossible to validate one indicator by itself. In order to form an assertion one must have at least two indicators to relate. Our expectation that two indicators of the same concept should be related derives from our analysis of the properties of the concept, the properties of the two indicators, and the claim that both are indicators of the same concept. Take *power* as a concept, for example. Let us define it as Emerson (1962) has:

The power of actor A over actor B is the amount of resistance on the part of B which can potentially be overcome by A.

Thinking about this definition will help us choose some potential indicators of the concept of power and will also provide direction for validating these indicators. Suppose we assume that power in a group is exercised when a group has to make decisions, so that an actor overcoming resistance of others means persuading them to go along with the actor's own position. Further, let us assume that overcoming resistance in the group is accomplished through discussion by talking sufficiently to persuade others to adopt an actor's own position. For the sake of simplicity, let us also assume that there are no boomerang effects; the more an actor talks, the more he overcomes the resistance of others. These arguments lead to a set of statements which can be sketched as follows:

1. More talking leads to more resistance overcome.
2. More resistance overcome leads to more decisions won.
3. Therefore, more talking leads to more decisions won.

These statements suggest that we observe the amount of time each actor talks as one indicator of the actor's power, and the number of group decisions that correspond to the actor's own position as a second indicator of power. Further, if we assume that statement 3 is true, we should find a correlation between these two indicators. Our expectation, however, is based on our assumptions, statements 1 and 2. Given some alternative assumptions, we might not expect a correlation. Furthermore, the correlation will never be perfect for the reasons that, first, our indicators contain some unreliability and, second, indicators have other properties besides those that are relevant to the properties of our concept of power. For example, actor *A* may win some decisions because all *B*s have zero resistance.

Validation, then, is both a theoretical and an empirical activity. It depends upon analysis of concepts, making assumptions about our indicators, forming statements, and testing these statements empirically. We emphasize the point made earlier about the kind of empirical testing involved in validation. When we form a statement to validate two indicators, and the behavior of the indicators is consistent with our statements (is what we expect), we have not discovered anything about the empirical world. We have not tested a knowledge claim. We are using the assumed truth of our statement as a criterion for evaluating the relevance of our indicators to our concept. If we did not assume that our statement was true, we would have no standard against which to test our indicators. Because we are assuming that our statement is true, we do not establish a knowledge claim.

Our example represents only a beginning of the process of validating indicators. One set of observations supporting one hypothesis would hardly be sufficient. But we have illustrated the important features of the process: formulating an argument, assuming its assertions are true, and choosing indicators that behave empirically in correspondence with the theoretical assertions.

There is one more aspect of the example that should be noted. Validation could involve testing two indicators of the same concept or two indicators of two different concepts. The assumptions sketched above could apply to either approach. In fact, many researchers would prefer to treat participation, or talking, as a separate concept from power. The only change this would necessitate would be to consider "amount of time each actor talked" as an indicator of participation rather than of power. In validation, we would still assume the truth of statement 3 and use the empirical correspondence of the indicators as validating one indicator of power and one indicator of participation. The crucial element is that in order to evaluate the correspondence of properties of indicators with properties of the concepts to which they refer, we must assume the truth of a set of assertions and select indicators that behave as the assertions require. By assuming the truth of some assertions, we validate tools which allow us to investigate other knowledge claims that we do not assume to be true in advance. Clearly, the process is a bootstrap operation.

The formulation we have just presented does not solve all the problems of validity. But by formulating validity as an issue of the relationship or relevance of an indicator to a concept, we do provide a validation strategy, an approach, and a set of standards by which some progress can be made toward intersubjective agreement on validity of indicators.

SUGGESTED READINGS

There is a voluminous literature on IQ and the controversy over the heritability of IQ. Three works that present some of the issues in this controversy are:

"Environment, Heredity and Intelligence." In *Harvard Educational Review*, Reprint Series, no. 2. Cambridge, Mass. Harvard University Press, 1969.

Ehrlich, Paul R., and S. Shirley Feldman. *The Race Bomb: Skin Color, Prejudice, and Intelligence.* New York: Ballantine, 1978.

Gould, Stephen Jay. *The Mismeasure of Man.* New York: Norton, 1981.

A summary discussion of contemporary views of reliability and validity may be found in:

Kerlinger, Fred N. *Foundations of Behavioral Research,* 2d ed. New York: Holt, Rinehart & Winston, 1973. Pp. 442–51, 456–69.

NINE

The Special Problems
of Quantitative Concepts
and Quantitative Indicators

To complete the discussion of forming concepts and tying them to observables, we must consider quantitative concepts and indicators because of their importance and the controversies that surround quantification. The issues divide sociologists into opposing camps, qualitative sociologists versus quantitative sociologists. Some extreme views in each camp entail total rejection of the work of those on the opposing side.

From the point of view of this author, this is meaningless controversy which arises in large part from a misunderstanding in both camps of the nature of quantitative concepts. Aside from the extremists, most people would agree, on the one hand, that some very informative research does not involve any quantification and, on the other hand, that some research cannot be informative without quantification. These people would also acknowledge that there are many examples of misplaced precision and/or mindless quantification. We can see such examples not only in research but in many arenas of society.

Amusing examples of misplaced precision occur in the sports field and particularly in football. A chain exactly ten yards long is used to determine whether a team has made a first down by advancing the ball ten or more yards from the starting point of the previous first down. In most cases, the referee can judge the question without using the chain because the call is not at all close—the team has either succeeded or failed by a large margin. On close plays, however, the officials solemnly carry the chain onto the field, stretch it to where the ball has been placed and then the referee makes the judgment according to whether the nose of the football extends past the stick at the end of the ten yard chain. This "precision" presumably allows the referee to make an "objective" judgment except for one thing—where the ball rests at the end of a play depends totally on the subjective placement by the officials. What makes this a case of misplaced precision is that the "inches" by which the ball lies before or beyond the end of the chain are much smaller than the errors associated with the subjective placement of the football, and the errors are not always random.

The passion of Americans for rating anything and everything from their own happiness to feminine beauty provides numerous examples of mindless quantification. Many of these are quite harmless but some are not. The "quizzes" in popular magazines by which readers can get a numerical score for their happiness, well-being, self-esteem, sexual prowess, and the like are probably quite harmless, but it is doubtful if students would agree that the consequences of assigning numbers in grading exams are totally benign. Not all grading is mindless, of course, but when grading occurs without regard for whether or not the performance being evaluated meets the assumptions underlying the assignment of numbers, then we would call it mindless. The problem does not occur only with assigning scores to essay answers; consider, for example, a true-false test in which each correct answer counts one point and the points add up to the test score. The procedure assumes that all questions are equivalent, and this assumption justifies assigning the same point value to every answer. Yet, many professors include some especially difficult items (and some include trick questions) to increase the power of the test to discriminate among students. That action clearly violates the assumption that justifies totaling the number of correct answers at one point each.

Unfortunately, some social science research can be faulted on the same grounds as this grading procedure—lack of adequate regard for the assumptions underlying the assignment and use of numbers. Considerable misunderstanding of the nature of quantitative concepts exists among both quantitative and non-quantitative researchers. In this book, we cannot hope to clear up these misunderstandings, because the field of measurement covers a wide range of issues and requires a high level of technical sophistication. To alert potential producers and consumers to the problems, however, we must consider two central issues.

Both issues relate to our discussion of validity of indicators. First, there is the general question of the relevance of numerical properties of indicators to the abstract properties of the concept; a year is an appropriate unit of chronological age but it may or may not be an appropriate unit of age as a sociological concept. Second, we must examine, albeit superficially, the more general issue of how one justifies assigning numbers to observations.

CONCEPTS AND IRRELEVANT NUMBERS

Some concepts are not quantitative; clearly, quantitative indicators for such concepts are irrelevant and inappropriate. Despite its obviousness, this fact is sometimes overlooked in research. Various processes may operate to blind the researcher to this obvious fact; they include the need to combine different observations, the desire to use sophisticated statistical techniques, or the use of fine discriminations to compensate for the inability to make large distinctions. In grading essay questions, we have the problem of combining the evaluations of answers to several questions to arrive at an overall grade; the most convenient way to combine is to assign points to each essay and then add them up. But how many graders have concepts of performance that correspond to the fine distinctions of their numerical assignments? Perhaps graders can conceive of four or five distinct levels of performance from "excellent" to "very poor," but the assigned numbers imply more finely honed and precise concepts of performance. The student who complains about missing a "B" by one point has a justifiable case.

There is no question that when a concept has quantitative properties, the construction of quantitative indicators enhances the researcher's power in analyzing data. The range of available techniques is broader for quantitative than nonquantitative observations; furthermore, greater efficiency is possible with quantitative measures in the sense that one can draw stronger conclusions from fewer observations. But assigning numbers solely for the purpose of using these techniques can lead to uninterpretable, or even nonsensical, results. For example, considerable research involves people giving numerical ratings to some property generating, for example, "leadership scores"; while the scores could be related to many other variables, it is not at all clear what these relationships might mean. This is not surprising because the underlying concepts are often not quantitative. We can think of people having more or less of leadership ability or leadership potential or even perhaps leadership strength, but it is difficult to think of people varying in degrees of leadership. In general, when using techniques which ask people to make ratings of some property, researchers must make sure that they and their raters have a clear understanding of the underlying quantitative concept.

One study in which the investigators attempted to compensate for the fact that their sample contained few middle-class people provoked considerable controversy. This argument centered around the relevance of quantitative status scores to the concept of social class. Chiricos and Waldo (1975) attempted to test the proposition from Chambliss and Seidman (1971) that "when sanctions are imposed, the most severe sanctions will be imposed on the lowest social class" (p. 475). Chiricos and Waldo examined the sentences received by 10,488 prison inmates for a total of seventeen offenses. Since the prison populations did not contain many middle or upper class people, the investigators decided to treat socioeconomic status as a continuous quantitative variable. They found no relationship between the status scores and severity of sentence and regarded their findings as evidence against the Chambliss and Seidman formulation.

Critics such as Hopkins (1977) have challenged the result, claiming that the status score is not relevant to the concept of class in Chambliss and Seidman, that the finding is only that variation *within* the lower class is unrelated to severity of sentencing, and that status must be treated as a dichotomous variable for an adequate test of the formulation. Hopkins, although acknowledging imprecision in the original formulation, argues: "The theory. . . requires a comparison between those who have some degree of power and those who have none—between, in other words, the lower class on the one hand and the middle and upper classes on the other" (p. 176). Hopkins claims that the theory treats class as a nonquantitative concept so that status scores are inappropriate for testing the claim.

Kim, J. and Mueller, C.W. (1978), *Factor Analysis: Statistical Methods and Practical Issues*, Sage University Paper Series on Quantitative Applications in the Social Sciences, 07-014. Beverly Hills, CA.

Kington, J. (1990), *Agendas: Alternatives and Public Policy*, Little, Brown & Co., Boston.

Krueger, Richard A. (1994), *Focus Groups: A Practical Guide for Applied Research*, 2nd Ed., Sage, Thousand Oaks, CA.

Li, Pei-liang (1988), *Shehui Yanjiu de Tongji Fenxi* (Statistical Analysis in Social Research), *Juliu* Books Co., Taipei. (in Chinese)

Lin, Nan (1976), *Foundations of Social Research*, McGraw-Hill, New York.

McArdle, John J. (1996), "Current Directions in Structural Factor Analysis", *Current Directions in Psychological Science*, 5(1):11-18.

McIver, John P., & Carmines, Edward G. (1981), *Unidimensional Scaling*, Sage University Paper Series on Quantitative Applications in the Social Sciences, 07-024, Sage, Beverly Hills, CA.

Mauch, James E., & Birch, Jack W. (1993), *Guide to the Successful Thesis and Dissertation: A Handbook for Students and Faculty*, 3rd Ed. (revised and expanded), Marcel Dekker, New York.

Meier, Scott T. (1994), *The Chronic Crisis in Psychological Measurement and Assessment: A Historical Survey*, Academic Press, San Diego.

Morrison, Donald F. (1990), *Multivariate Statistical Methods*, 3rd Ed., McGraw-Hill, New York.

Neuman, W. Lawrence (1997), *Social Research Methods: Qualitative and Quantitative Approaches*, 3rd Ed., Allyn and Bacon, Boston.

Norusis, Marija J. (1988), *SPSS-X Introductory Statistics Guide*, SPSS Inc., Chicago.

In a rejoinder to several critics, Chiricos and Waldo (1977) defend their quantitative status score and cite two other propositions from Chambliss and Seidman which imply a quantitative concept for class, including the following: ''Where laws are so stated that people of all classes are equally likely to violate them, the *lower the social position* of an offender, the greater the likelihood that sanctions will be imposed on him (p.282; Chiricos and Waldo added the emphasis). Since ''lower social position'' implies degrees of status, these researchers insist that their quantitative status scores are appropriate and reject the argument that they are dealing only with variation within the lower social class.

This is not a controversy we can settle, especially since it involves other facets we have not discussed. We can, however, underscore the problem. The status scores in this study are not self-justifying. Both the original formulation and the empirical study can be faulted, the former for not explicitly and consistently specifying the properties of social class and the latter for what this author suspects is making a virtue of necessity, that is, defending a procedure in which assigned numbers mostly reflect within class variation because there were few middle or upper class inmates. The explicitly formulated properties of the concept provide the best justification for the relevance of measurement procedures. Even when connotative definition is not possible, social scientists have some obligation to specify whether a concept has quantitative properties, because quantitative indicators are only appropriate to quantitative concepts. If researchers and consumers would recognize these limitations, many controversies could be avoided and we could eliminate some instances of the meaningless use of numbers.

QUANTIFICATION

Quantification—assigning numbers to observations—serves several purposes. When done appropriately quantification: (1) increases intersubjective agreement on the nature of the observations, (2) increases the amount of information about the observations that can be communicated in a concise way and (3) increases the ability to manipulate and summarize observations by making possible the use of powerful tools of mathematics and statistics. These potential benefits provide strong incentives that sometimes blind researchers to the conditions that must be fulfilled for the appropriate assignment of numbers.

We have no quarrel with researchers who, for heuristic reasons, knowingly violate the underlying assumptions in assigning numbers to their observations. As we have noted earlier, a researcher cannot solve all problems before doing a study so that tentative quantifications are frequently necessary prerequisites for data collection. Subsequently, however, the researcher needs to assess whether there are systematic errors due to measurement; this assessment requires an evaluation of the procedures for assigning numbers.

We do fault those investigators who violate the underlying assumptions involved in quantifying their observations out of ignorance of these assumptions or on the basis of the incorrect belief that "significant results"[1] are sufficient justification for the procedures by which numbers are assigned to observations. The remainder of this section will consider a few of the basic assumptions of quantification, but first let us analyze the often heard justification that measurement procedures which produce results "can't be too bad" or there would be no relationship between the measured variables. This faulty view results from failure to distinguish between random and systematic error.

If assigning two sets of numbers to a collection of objects to "measure" two properties of the objects were completely random, then there would be no correlation between the measures of the two properties—each assigned value would in effect be random error. If there is a systematic bias in assigning numbers to objects, however, that bias could result in a correlation between variables even if there were no relation between the two sets of properties. In short, systematic error could account for an observed result so that observing a significant relationship in itself cannot justify the validity of the quantification procedure.

Some uses of the *Likert scale* exemplify the problem we are discussing. Named for Rensis Likert, who first introduced the technique (Likert, 1932), the scale consists of a series of statements to which a person responds using one of, for example, seven categories such as *strongly disagree, disagree, slightly disagree, no answer,*[2] *slightly agree, agree, strongly agree.* Responses are quantified by assigning numbers ranging from -3 for *strongly disagree* to $+3$ for *strongly agree,* with zero for no answer. Suppose there are two statements like the following:

1. People who discriminate against women should be punished.
2. Affirmative action is a good remedy for discrimination.

A respondent who strongly disagreed with both statements would be given a score of -6 whereas one who strongly agreed to the first and agreed to the second would get a score of $+5$. A correlation between the responses to the two statements could result as an artifact of the way numbers are assigned to the response categories. Suppose some respondents resist using the end categories (the *strongly* responses) while others prefer them. That alone could produce a significant relationship even when the *content* of one statement bears absolutely no relation to the content of the other. Since such verbal habits are well documented, one cannot conclude that observing a correlation between the responses to the two statements either justifies the assigning of numbers or represents a substantive finding. At the very least, the researcher must test and reject the argument that the relationship is an artifact.

The purpose of our example is not to condemn all uses of Likert scales; rather it is to suggest that such procedures are not self-justifying. At some point, the researcher who uses such a procedure is called upon to validate the assignment

---

1. This usage usually refers to statistically significant findings.
2. The "no answer" category is usually not presented explicitly but scored if the person does not choose one of the other alternatives. It is treated as "neutral" and so given the middle score of zero.

of numbers. One way to do this is to show that the relations between objects correspond to the relations between the numbers assigned to these objects. To appreciate what is involved requires us to look at the properties represented in the use of numbers.

Relations Underlying the Assignment of Numbers

There are essentially three types of numbers representing three types of relations. Although terminology varies, we will refer to them as nominal numbers, ordinal numbers, and cardinal numbers. We will briefly discuss each.

The relation underlying the assignment of *nominal numbers* is equivalence; all objects assigned the same number are equivalent to one another with respect to the property which is the basis of the assignment, and any objects assigned distinct numbers differ from one another. The underlying relation is sameness-difference and nothing more. Student identification numbers are nominal numbers uniquely designating individual students. If we code female respondents *1* and male respondents *2*, we are assigning nominal numbers. In fact, nominal numbers do not operate the way we have learned to deal with numbers; we cannot do arithmetic with them and we could use another set of distinct symbols to designate classes whose members were equivalent, like *F* and *M*. Nominal numbers give rise to what is known as the nominal level of measurement, where counting members of each class constitutes the only permissible quantitative operation. If we assume, for example, that every true-false question is equivalent, then we can count the number of correct answers. On the other hand, if we assume that there are hard questions and easy questions, we can count the number of correct answers in each nominal category (hard and easy) and give students more credit for correct answers to hard questions.

The most important thing to remember about nominal categories symbolized by nominal numbers is that the only relation between categories is one of nonequivalence. None of the familiar properties of numbers—for example, additivity—apply to this level of measurement. Some properties are inherently nominal, such as gender, geographic location, and religious affiliation. Other properties must be treated nominally because we have not developed conceptual or observational tools justifying other levels of measurement. One should not conclude that little can be done quantitatively with nominal categories, however; classification and counting combined with recently developed statistical techniques can be quite powerful.

The second type of numbers, *ordinal numbers*, signify order relations. If we have three objects to which we assign three *ordinal numbers*, these objects have different amounts of the property that is the basis for the assignment. If we say that university presidents have more power than deans and deans have more power than professors, then we have formed three ordered categories of power to which we can assign the numbers *1, 2,* and *3*. Here we have not only the relation of sameness within a class and difference among classes but also the relation of more or less between classes. Note that we can assign the number *1* or the number *3* to the class of presidents, treating *1* as the highest or lowest category, because the significant property of ordinal numbers is the relation between the numbers and not their absolute value. Furthermore, nothing in this assignment of numbers says how much more power presidents have than deans and nothing indicates that the difference in power between presidents and deans is the same as the difference between deans and faculty members.

Ordinal numbers have two important properties; they are *asymmetric* and they are *transitive*. If, *a, b* and *c* are three ordinal numbers, then asymmetry means that *a* cannot be equal to *b* but must be either greater or less than *b*. Any president assigned to power class *1* cannot have the same power as a dean assigned to class 2. The definition of transitivity is: if *a R b* and *b R c*, then *a R c* where *R* is some relation such as *greater than*. If presidents have more power than deans and deans have more power than professors, then presidents have more power than professors. If the statement is true for all presidents, deans and professors, we say that the relation "have greater power than" is transitive. If we find one professor who has more power than one president, then we say transitivity has failed. There are many examples of relations where transitivity frequently fails; in most sports, the relation "defeats" is not transitive. It frequently happens that team *A* defeats team *B* and team *B* defeats team *C* and then team *C* turns around and defeats team *A*.

Cardinal numbers have the properties of ordinal and nominal numbers and additional properties as well. These are the numbers we are familiar with from arithmetic. For these numbers, the *difference* between any two numbers is meaningful because there is meaningful *unit difference* and the operation of addition is defined. These properties give rise to what are known as *interval* and *ratio* levels of measurement where the latter requires a definable zero. Since a year is a meaningful unit, chronological ages and years of schooling are examples of sets of cardinal numbers. Annual incomes of individuals and gross national products are cardinal numbers that depend on the meaningfulness of the dollar as a unit. With the Likert scale it is difficult to argue, and even more difficult to demonstrate empirically, that the difference between *strongly agree* and *agree* is equal to the difference between *slightly agree* and *agree*. We must recognize, however, that the assignment of +*3* to strongly agree, +*2* to *agree*, and +*1* to *slightly agree*, makes that assertion.

Relations between values of a property—for example, degrees of power—must satisfy additional conditions beyond equivalence within classes, asymmetry between classes, and transitivity in order to justify the assignment of cardinal numbers. Discussing all seven of these requirements would involve us in technical matters that are beyond the scope of this book, but we will consider one of these conditions to illustrate what must be considered in assigning cardinal numbers. One formal condition requires a defined operation of "combining" such that combining *a* with *b* is equivalent to combining *b* with *a*. Putting dollar bills in a pile satisfies this condition: Placing five one-dollar bills on top of a pile that already has ten in the pile is equivalent to placing ten bills on top of a pile that already has five. This illustrates a physical operation that provides a meaningful basis for addition. But what is the meaningful operation that allows us to combine a person's *strongly agree* responses to two statements like "Most murderers deserve the death penalty" and "Children should be punished for telling even little lies"? And what operation allows us to treat an *agree* response to the first combined with a *strongly agree* to the second as equivalent to a *strongly agree* to the first combined with an *agree* to the second? It is too easy to forget that when we add numbers we are saying something about the objects whose quantities we are combining.

Alternatives to Measurement by Fiat

Assigning numbers without theoretical justification or without empirical demonstration that the properties of the objects being measured correspond to the properties of the numbers assigned is known as *measurement by fiat*. The researcher simply asserts—by fiat—that the relations hold. Frequently we measure by fiat as a way of getting started in a new area of research as part of the "bootstrap" nature of research. As we have noted, at some point in the process of developing our knowledge we need to examine our measurement procedures to decrease measurement error and to eliminate the possibility that our findings are artifacts of the way we assign numbers.

The main problem with measurement by fiat is the possibility of erroneous inferences about the nature of relationships among measured variables. Large random error components to assigned numbers may obscure existing relationships; systematic bias in assigning numbers, as we have already noted, may lead to erroneous inferences about relationships that are simply artifacts. Researchers need to guard against these sources of error, and consumers must be aware of possible pitfalls that might arise from measurement by fiat.

What are the alternatives to measurement by fiat? General measurement models exist and these have clear procedures for testing the applicability of their assumptions. Sociologists need to make more use of these models, and while detailed examination of available models is beyond the scope of this book, we do want to illustrate very briefly how such models work. For this purpose, we will look at a model for ordinal level measurement, the *Guttman scale,* named for its developer, Louis Guttman (1944).

Measurement involves two distinct types of activities: constructing or calibrating the instrument and then using the instrument to assign values of the property being measured to objects. Defining and calibrating a ruler to measure length exemplifies the first activity; using the ruler to measure the dimensions of the rooms in a building illustrates the second. Let us call the first *scaling* and the second, *scoring.* Some models separate scaling and scoring; the Guttman scale, however, combines them. In this model, there are a set of *items* to be scaled and then used to assign an ordinal score to each of a set of *individuals.* Items may be attitude questions, arithmetic problems, national income levels, frequencies of usage of different grammatical forms whereas individuals may be respondents to a questionnaire, students taking a test, nations in the world system, or a set of historical manuscripts. The technique uses the individual *responses* to the items to scale the items and then uses the items to assign scores to the individual.

The Guttman model, like most measurement models, makes assumptions and also provides procedures and criteria for testing some of their consequences. It assumes a cumulative ordering of items which is best illustrated with the analogy of weight lifting. If we have a set of unequal weights and a set of individuals of unequal strengths, then not all individuals should be able to lift all weights. Any individual who lifts a heavy weight except for random errors should be able to lift all weights that are lighter. This is the cumulative property; in an arithmetic test, for example, a student who can do the hardest problem should be able to do all easier problems—again excluding random errors. In weight lifting, we have an external standard that allows us to order the weights, but suppose the only way to determine the order of heaviness of the set of weights was to derive it from the weight-lifting behavior of the individuals. Suppose also that we have no independent way of assessing the strengths of our individuals but must use the "discovered" order of the weights to order the individuals. These two suppositions illustrate the problem the model proposes to solve.

By making some assumptions, the model provides a solution to the problem as well as tests of whether the model fits the data obtained. For example, we must assume that strength is distributed among our individuals so that not every person can lift every weight. In fact, we must assume that as strength increases, the number of people having that level of strength decreases. With that assumption, we can use the proportion of individuals "passing" an item—in this case lifting the weight—to produce an initial ordering of the weights. The heaviest weight is the one that fewest people can lift and the lightest weight is the one that most people can lift,[3] and the others are arrayed in between according to the proportion of individuals that can lift each weight. So far, the ordering is by fiat—we have defined ordinal weight to be equal to the order of the proportion of successful individuals—but the definition does fit with our intuition that as the task becomes more difficult, the number of people able to do the task becomes smaller.

The cumulative assumption allows us to go further and test whether the responses of individuals are consistent with the ordering, that is, whether the relations among the individual behaviors have the requisite properties for the ordering we have assigned to the weights. According to the cumulative property, an individual who lifts a given weight should be successful with all weights that are less heavy. This property allows a test of transitivity which we earlier indicated was a necessary condition for ordinal numbers. Given a heavy, a medium and a light weight: if people who lift the heavy weight can lift the medium weight and people who can lift the medium weight can lift the light weight, then people who

---

3. The model excludes weights that everyone can lift and those that no one can lift.

can lift the heavy weight should also be able to lift the light weight.[4] This idea can be applied to questions on an exam—people who answer a difficult question should answer easier questions—and to attitude statements: people who endorse an extreme statement should also endorse less extreme positions.

Suppose we have five weights—*a,b,c,d,e*—ordered from heaviest to lightest and six individuals with different strengths ordered from strongest to weakest: *1, 2, 3, 4, 5,* and *6.* If we let a + signify passing the item and *0* not passing, we should have the picture:

	a	b	c	d	e
1	+	+	+	+	+
2	0	+	+	+	+
3	0	0	+	+	+
4	0	0	0	+	+
5	0	0	0	0	+
6	0	0	0	0	0

This diagram represents a *scalogram* pattern for a *perfect*—that is, error free—Guttman scale. (In an actual case there would be many replications of each row of the picture.) One test of the model is how closely the actual scalogram pattern conforms to the perfect scale. To put it another way, with five items, there are thirty-two possible types; the six shown above are *scale types* and the remaining twenty-six (e.g., + + + +0, +0+00, etc.) are *error types*. In any actual scale, there are always a number of individuals who are error types and there are criteria for deciding whether the frequency of error types is acceptable.[5]

As we have illustrated for one type of ordinal scale, measurement models contain assumptions about quantitative relations among the things being measured and provide procedures for testing the fit of the model. Models which allow us to assign cardinal numbers are considerably more elaborate; at the interval level of measurement, the models contain assumptions about meaningful units and provide tests to determine whether differences required to be equal are in fact equal. For ratio measurement, models include, in addition to interval level assumptions, procedures and tests for a meaningful zero. In assigning numbers, one needs to consider the consistency of the model's assumptions with the substance of the concept to be measured. Years may be equal and meaningful units for chronological age but not for every sociological concept of age; in some cases, a few ordered categories (e.g., young adults, the middle aged, and seniors) may be more compatible

---

4. As the reader will recognize, ordering the weights by proportion of people successfully lifting each does not force transitivity at the individual level—some people who can lift the heavy and the medium weight can fail to lift the light weight. While this is difficult to imagine with weight lifting, it is quite easy with respect to questions on a test.

5. For the Guttman model, the criteria for acceptable levels of error are somewhat arbitrary. They are, however, explicit guidelines for deciding whether the set of responses fit the assumptions of model and meet the requirements of ordinal numbers.

with the conceptualization. Assigning equal points to exam questions of deliberately varied difficulty is a clear violation of the underlying measurement model. Researchers and consumers both need to be aware of the possibility that measurement procedures introduce distortions that seriously conflict with the substantive nature of the concept being measured. Systematic measurement error can conceal relationships that might otherwise be observed or can produce artifactual relationships due to constant biases in the assignment of numbers.

We must emphasize that assigning numbers involves the explicit or implicit use of a model. If the model is explicit, then we can examine the compatibility of the model's assumptions with the conceptualization of the ideas and we can test empirically whether the relations among objects assigned various numbers reflect the relations among the numbers of a given level of measurement. Social scientists have developed many diversified models for concepts that are unidimensional; that is, the property varies over a single continuum and a single number can represent each value of the property. These are the ordinal, interval, and ratio scales that we briefly discussed. In addition, models are available for multidimensional scaling where key properties vary simultaneously along a number of continua and thus cannot be characterized by a single number. Researchers should be encouraged to look over these models to see if any one is applicable to the problem at hand. If an ordinal concept is formulated, for example, a number of different abstract models are available that involve alternative procedures for meeting the transitivity requirement; a researcher needs to consider these procedures in terms of their feasibility and their compatibility with his or her substantive ideas.[6] We believe that more use of measurement models combined with careful analysis of the properties to be measured would improve the reliability and validity of our indicators.

If existing models are not applicable, as frequently will be the case, then a researcher can either construct a new model or fall back on some form of measurement by fiat. Constructing new measurement models requires major investments of time and resources, and so this alternative is not practical for the average researcher. Measurement by fiat need not be totally arbitrary, however, and there are major differences between totally mindless assignment of numbers and carefully considered procedures.

The first step in any considered procedure requires the recognition that assigning numbers depends upon the formulation of the properties of the concepts for which quantitative indicators are contemplated. Explicating these properties will often provide guidelines for acceptance or rejection of a particular way to assign numbers. If a concept of social class includes homogeneity of social position for members of a given class, then one would reject numerical assignments that yield varying numbers for members of the same social class. If a formulation of organizational complexity involves several dimensions such as level of technology, degree of specialization of roles, and multiplicity of different environments to which it

---

6. The suggested readings at the end of this chapter provide detailed discussions of many of the available models.

must adapt, then any unidimensional assignment of numbers to form a single indicator of complexity would be rejected.

Another salutary step would be to distinguish between scoring and scaling—measuring the lengths of rooms and creating the ruler. Suppose, for example, a researcher wanted to assign *delinquency scores* to adolescents on the basis of the frequency of committing each of a series of acts of deviant behavior such as unexcused absence from school, shoplifting, arguing with parents, reckless driving, or drug abuse. The items of behavior would be the "ruler" which would be used to characterize the adolescents. This "ruler" could be scrutinized in advance: one could ask critical questions about whether the items are comparable, whether one incident of reckless driving is equivalent to one arrest for drug abuse, whether arguing with parents has varying significance for differing ethnic groups in the population. A researcher could pretest the "ruler" on a small sample before doing a large study or, if that is not feasible, the investigator could ask a group of experts to judge the items in terms of questions similar to the three examples given and to evaluate the acceptability of each item.

A final recommendation for improving the quality of quantitative indicators involves minimizing the number of untested assumptions entailed in the assignment of numbers. Although a complete discussion of all of the assumptions of each level of measurement has not been possible, what we have presented should indicate that we can order levels of measurement in terms of the number of assumptions involved as follows (from most to least): ratio, interval, ordinal, and nominal. Twelve assumptions underlie ratio scales whereas nominal scales only assume the equivalence (with respect to the property being classified) among members of the same class and nonequivalence between members of different classes. Furthermore, the levels of measurement form a cumulative scale; ratio measures involve all the assumptions of interval, ordinal, and nominal levels plus one additional assumption.

Consider the following possibilities. A researcher who presents a series of statements with Likert alternative responses to a sample of respondents can make the maximal assumptions by assigning ratio numbers or can make the minimal assumptions of a nominal scale. To assign ratio numbers ($+3$ to -3), one must assume that the difference between, for example, *strongly agree* and *agree* is the same as the difference between *strongly disagree* and *disagree* and that these differences are equal for all statements and for all respondents answering the questionnaire. We discussed earlier the difficulties of assuming equal differences when statements vary in the extremeness of their content. It becomes even more problematic when one must assume that the same relations hold across all individuals. While we may be willing to assume that the *order* of the response alternatives is invariant across statements for a given person, we may be more hesitant to assume that the ordering is invariant for all respondents. One person's *agree* could be a stronger endorsement than another person's *strongly agree*. Most of us would probably be willing to make the least demanding assumptions: (1)all levels of *agree* (*strongly agree, agree,* and *slightly agree*) are equivalent and differ from all levels of *disagree* and (2) this equivalence holds for all statements and all respondents.

With Likert scales, data can be gathered in a way that provides the researcher with a choice in what assumptions to make in assigning numbers to the observations. Although respondents may answer the questions in terms of *strongly agree* to *strongly disagree*, the researcher could combine the three levels of agree responses as *agreement* and three levels of disagree responses as *disagreement*. One could test whether assigning scores of +3 to −3 leads to inconsistencies and if so, fall back on the agree-disagree classification. The same flexibility exists with many other data gathering techniques. Since there are circumstances when researchers feel confident in making maximal assumptions and other situations where the investigator is only comfortable with ordinal or nominal assumptions, such flexibility can be beneficial. Benefits increase when researchers use the data to try to evaluate their assumptions and are prepared to move down to a lower level of measurement should anomalies appear in their data analysis. However, one caution is absolutely necessary in this connection. Once numbers have been assigned, we have a tendency to reify them or to reify the categories to which the numbers are attached. Hence there is a common complaint particularly among novice researchers that combining categories—the three levels of *agree* versus the three types of *disagree*—is "throwing away" information. If the differences between response alternatives are unreliable or not invariant, the complaint represents a reification of the categories. Under these circumstances, the appropriate answer to such complaints is, "You can't throw away what you never had!"

SUGGESTED READINGS

Dawes, Robin M., and Tom L. Smith. "Attitude and Opinion Measurement." In *The Handbook of Social Psychology,* 3d ed., Gardner Lindzey and Elliot Aronson. New York: Random House, 1985. Pp. 519–66.

This handbook chapter presents a critical analysis of both representational and nonrepresentational techniques for measuring attitudes. (Nonrepresentational techniques are similar to what we have termed *measurement by fiat*.)

Torgeson, Warren S. *Theory and Methods of Scaling.* New York: Wiley, 1958.

Torgeson discusses the abstract axioms of measurement and then examines a broad range of social science models. The book is both detailed in its presentation of specific techniques and comprehensive in its coverage. It requires some mathematical sophistication.

Nunnally, J.C. (1978), *Psychometric Theory*, McGraw-Hill, New York.

Orcutt, Ben (1990), *Science and Inquiry in Social Work Practice*, Columbia University Press, New York.

Pedhazur, Elazar J. (1982), *Multiple regression in behavioral research: Explanation and prediction*, 2nd Ed., Holt, Rinehart & Winston, Fort Worth, Texas.

Pedhazur, Elazar J., & Schmelkin, Liora Pedhazur (1991), *Measurement, Design, and Analysis: An Integrated Approach*, Lawrence Erlbaum Associates, Hillsdale, NJ.

Rowland, D., Arkkelin, D., & Crisler, C. (1991), *Computer-Based Data Analysis: Using SPSS^x in the Social and Behavioral Sciences*, Nelson-Hall Publishers, Chicago.

Royse, David, & Thyer, Bruce A. (1996), *Program Evaluation: An Introduction*, 2nd Ed., Nelson-Hall Publishers, Chicago.

Saris, Willem, & Stronkhorst, Henk (1984), *Causal Modeling in Nonexperimental Research: An Introduction to the LISREL Approach*, Sociometric Research Foundation.

Sarle, Warren S. (1996), "Measurement theory: Frequently asked questions", URL: ftp://ftp.sas.com/pub/neural/measurement.html, originally published in the Disseminations of the International statistical Applications Institute, 4th edition, 1995, Wichita: ACG Press, pp.61-66. Revised March 18,1996.

Schon, D. (1983), *The Reflective Practitioner*, Basic Books, New York.

Sewell, W.H. (1941), "The development of a sociometric scale", *Sociometry*, 5:279-97.

Simon, H.A. (1954), "Spurious correlation: A causal interpretation", *Journal of the American Statistical Association*, 49:467-79.

TEN

From Simple
Knowledge
Structures
to Theories

Although many books in the past two decades have dealt with sociological theory and theory construction, there is still no agreed-upon view of what theory is. The word theory is used in many different ways in sociology, and if we were to take the sum of sociological conceptions of theory, virtually the only things that would be excluded are what we have called observation statements. In other words, any idea, speculation, hypothesis, opinion, or belief from someone's point of view would fit under the term theory. Some people justify such an ''anything goes'' attitude on the grounds that sociology is in an early stage of development. Yet this attitude opens the door for *theoreticism*, the promotion of ideas that are not corrigible either by evidence or reason, and fosters a sociology in which one person's opinion is as good as anyone else's. Furthermore, the lack of consensus creates problems for both practicing scientists and consumers. A completely open conception of sociological theory does not provide guidelines for discriminating among ideas, for separating the more useful ones from the less useful, or for understanding limitations as well as extensions of sociological ideas. As long as what constitutes theory is unrestricted, it is hard to tell when to take a sociological idea seriously.

Because of our concern with the evaluation of ideas, we must take a more restricted view of sociological theory. Not every idea is amenable to logical and empirical evaluation. Not every idea is intersubjectively testable. Not every idea is a knowledge claim unlimited by time and space. The consideration of these issues in previous chapters should have already imposed limitations on a conception of sociological theory. In short, ideas must meet certain standards before they are appropriate for scientific evaluation and we must reject an all-encompassing view of sociological theory.

From some perspectives, much of what has already been discussed would be regarded as theory. What we have called simple knowledge structures would be treated as theories by some sociologists. In addition, the definition and explication of a concept are often regarded as inseparable from theorizing. We insist on distinguishing simple knowledge structures from theories because a simple knowledge structure is both more and less than a theory, as will become clear shortly. For the moment, we can say the simple knowledge structures focus more on the process of developing knowledge, whereas theories are the products of this process. A theory may combine universal statements from many simple knowledge structures, but theories never contain specific observation statements.

Similarly, the development of concepts represents an indispensable tool in theorizing. But conceptualization is only a step along the way, not the objective of scientific thought. Too often, sociologists believe that their job is finished when they have defined and spelled out a concept. For example, the concept of status, defined as "position in a social structure," may or may not be useful. Only by using that concept in a theory can we evaluate its utility. In no way does putting forward a definition represent the end of the job.

We cannot overemphasize the process of developing and evaluating knowledge claims. For that reason, the distinctions between simple knowledge structures, conceptualization, and theory are crucial. At the very least, they call attention to different stages in the process.

THEORY

Let us modify slightly our definition of theory from Chapter 4:

Definition:
A theory is a set of interrelated universal statements, some of which are definitions and some of which are relationships assumed to be true, together with a syntax, a set of rules for manipulating the statements to arrive at new statements.

This definition accomplishes several things. It imposes restrictions on the kind of statements that can be called theory and on the number of statements required in order to have a theory. It calls attention to the relationships among statements. Finally, it calls for attention to requirements beyond the content of the statements, requirements dealing with the form as well as the substance of the argument.

Our definition obviously rules out single, isolated statements as theory. Thus, "All societies have incest taboos" or "Political power varies with socioeconomic status" or "A differentiated status structure will emerge in groups whose members are initially of equal status," while all interesting and perhaps appropriate for inclusion in a theory, do not in themselves constitute theories.

Brief reflection will make it clear why we wish to exclude such statements as theories in themselves. In the first place, each of these statements by itself provides no guidelines for dealing with the statement. In the absence of definitions of terms, we can argue that the statements in isolation are meaningless. Anyone is free to read whatever meaning he or she desires into the given terms of each statement. Everyone is free to identify almost any observation with the concepts in these statements.

Once the key concepts for each of these statements are explicitly defined, we begin to have guidelines for dealing with the statements. If we define the concepts "society" and "incest taboo," we begin to have a way of dealing with the statement that all societies have incest taboos. At the very least, we can recognize circularities—that is, situations in which the statement becomes trivially true because the definition of society contains the idea of an incest taboo. If we will not call something a society unless it has an incest taboo, then we are really not saying very much when we assert that all societies have incest taboos. As long as we use single isolated propositions without explicit definitions, we run the risk of implicit circularities in our thinking. The example, "All societies have incest taboos," has a long history of arguments about such circularities. Other single propositions, less controversial, suffer from the same difficulties.

Our definition also excludes singular propositions. Hence, statements like "Americans are becoming more other-directed" and "The prestige of occupations in America varies with the average income level of the occupation" are not properly theoretical statements. Since we are concerned with general knowledge, knowledge whose truth is independent of time and place, the assertions contained in a theory cannot be limited by time and place.[1] Although we may use a theory to explain a concrete situation in a particular place at a particular time, the definition of theory forces us to treat that situation as an instance to which a theory is relevant, rather than as the sole object of the theory. It is perfectly reasonable to want to explain why voters in a particular election may switch their party allegiance in massive numbers. To deal with that situation scientifically, however, means that we cannot have a theory of vote-switching in a particular election; rather, we must connect that particular situation with other situations in different times and places.

The definition of a theory emphasizes the fact that statements must bear some relationship to one another; not any collection of universal statements constitutes a theory. When we require an interrelated set of statements, we impose limitations on the objects of a theory, that is, on what the theory is about. Stressing this point may be belaboring the obvious, and examples that we may choose certainly do appear obvious. Two *interrelated statements* would be the following:

1. Formal power in an organization increases with responsibility.
2. Responsibility increases with status.

These statements are interrelated because they share a common term, responsibility. The following two statements lack this property of interrelatedness:

1. Formal power in an organization corresponds to responsibility.
2. Informal power corresponds to organizational status.

---

1. When writers assert that scientific theory is ahistorical, they are asserting the requirement that theoretical statements must be universal. They are not excluding the application of theory to historical events; nor are they ruling out historical data as either the stimulus for the formulation of theoretical statements or a testing ground for the empirical evaluation of theory.

We will call such statements *disparate statements*. A theory cannot consist of a set of disparate statements. Now we can turn our example of disparate statements into interrelated statements by adding a third statement:

3. Responsibility corresponds to organizational status.

The important things to recognize are, first, the necessity for explicitly including statements such as number 3 to make clear the interrelatedness of the set of ideas and, second, the fact that not every set of disparate statements can be converted into a set of interrelated statements. Our examples are highly oversimplified because they contain so few statements. The problem becomes intensified when there are many statements to be included in a theory.

The distinction between interrelated statements and disparate statements requires our attention, because it suggests that there is no mechanical procedure by which an arbitrary collection of generalizations can become a theory. Recognizing disparate statements implies that the strategy of collecting generalizations about a phenomenon will not lead to theories about that phenomenon. It further suggests that not all phenomena are appropriate objects for theories. Consider, for example, a medical sociologist interested in hospitals. He or she can formulate many statements about hospitals or collect many generalizations about hospital phenomena. What is likely to result, however, is a set of disparate statements which do not belong in the same theory. Consider the following set of statements:

1. Doctors have higher status than nurses.
2. Patients who are told their true prognosis adjust better than patients who are not so told.
3. Private hospitals pay more attention to patient care than public hospitals.

The only thing these statements have in common is that they all deal with hospital phenomena. The example itself is farfetched, since no investigator is likely to be interested in such widely divergent statements. Yet, many descriptive studies of hospitals or of other phenomena present statements that are nearly as disparate as our absurd example. The point cannot be emphasized too much: descriptive efforts which aim at an extensive, wide-ranging description of as many features of the phenomena as possible cannot, by their very nature, generate the kinds of interrelated statements which we require for a theory.

The classic example which illustrates the problem is found in the appendix to the book *Voting* (Berelson et al., 1954), in which the authors present 209 numbered statements (plus some that are not numbered).[2] Someone may argue that this appendix is very rich with material for many theories. The fact that little theoretical development has emerged from the statements in this appendix is testimony to the possibility that it may be too rich. In short, we are suggesting that the

---

2. Recall the discussion of orientation to phenomena of chapter 6.

formulation of interrelated statements requires focusing on limited aspects of a phenomenon, to the exclusion of other aspects which may be interesting but may not fit under a given theory. The attempt to formulate many theories at the same time may lead to no theory at all.

Closely akin to the requirement of interrelatedness are requirements concerning the structure and form of the theoretical argument. In one sense, what we are calling interrelatedness is logical interrelatedness. The student of elementary logic will recognize that our example of interrelated statements employs the logical rule for the distribution of terms in the premises of a syllogism. The student without such background can rely on intuitive notions of interrelatedness and logical structure, and may come to appreciate the need for technical training in logic as a basis for constructing sociological theories.

What we mean by syntax in a theory rests largely on a set of logical rules concerning the form of statements. There are many kinds of possible syntaxes. Often the syntax of a theory need not be made explicit. For many purely verbal theories, the implied syntax is elementary logic, usually from what is known as the calculus of propositions. For other kinds of theories, however, the syntax may be a particular branch of mathematics or a particular computer language. In a very general sense, branches of mathematics and computer languages represent purely formal systems; that is, they are contentless languages whose symbols have no meaning but whose theorems, for example, are rules for the manipulation of the content-free symbols. An algebraic equation has symbols and rules for manipulating those symbols. The equation:

$$ax^2 + bx + c = 0$$

requires the user to identify the letters a, b, and c with numbers and then perform a set of manipulations to arrive at another number for x. The equation, together with the rules of algebra, provide a set of syntactical operations which allow the manipulation of one set of statements ($a = 3$; $b = 2$; $c = 33$) to arrive at a new statement ($x = 3$). The presence of such a syntax quickly reveals erroneous statements. Thus, elementary algebra quickly shows the following two statements to be inconsistent:

$$x^2 - 4 = 0$$
$$2x + 3 = 0$$

In constructing theories, an investigator does not want to say contradictory things. In evaluating theories, the investigator looks for inconsistencies and contradictions. A syntax therefore provides an indispensable tool, because the explicit formulation of statements and the explicit use of a particular syntax allow us to recognize and deal with contradictions.

Consider the following two statements:

1. The higher the status of a member of a group, the freer he is to deviate from the norms of a group. (Hollander, 1958)
2. The higher the status of a member of a group, the more he reflects the norms of the group. (Homans, 1950)

Attempting to put these two statements in the same theory forces us to confront the possibility that they may be contradictory. As we will demonstrate in Chapter 12, the virtue of explicit logical analysis of statements arises from making clear what the problem is and what may be done about it. If we treat the Hollander and Homans propositions as incomplete, we are called upon to reformulate them in such a way as to avoid contradiction. We might reformulate them to avoid contradiction by applying one proposition to an early stage of group interaction and the second proposition to a later stage of group interaction:

1. In newly formed groups, high-status members are more likely to reflect the norms of the group.
2. In well-established groups, high-status members feel free to deviate from group norms.

This is only one possible reformulation. It does away with contradictions by referring the propositions to different points in the history of the group. Other alternative formulations can serve as well. Empirical evidence provides the basis for evaluating which alternative formulation to pursue.

In order to operate with theories, our definition requires that what is the theory be clearly set off from what is a discussion of the theory. In other words, it must be possible to isolate theoretical statements from other text. This runs directly counter to much of current practice in sociology, where theoretical ideas are presented discursively. The typical situation embeds a theoretical assertion in an ordinary prose paragraph, where a claim is made, its meaning is discussed, and its importance is justified, all in one stream of writing. Such presentation makes it practically impossible to know how much is included in the theoretical statement. The reader has difficulty in deciding where the knowledge claim ends and where its justification begins.

Discursive discussion often helps to amplify theoretical ideas and may provide feeling for the source of these ideas and their application. But, as Gibbs (1972) points out,

> the discursive exposition of a theory does serve a purpose: it enables the theorist to make his assertions appear plausible, that is, to make a case for his theory. Hence, sociologists may object to formal theory construction because it excludes argumentation. Although sociologists generally are fond of argumentation, it is difficult to see what rhetoric adds to a theory. If empirical validity should be central in assessing theories, then the outcome of tests—not rhetoric—is decisive. Indeed, should a theorist convince his audience by forceful argumentation, it is nothing more than a personal triumph.

Gibbs uses the idea of formal theory construction as equivalent to our claim that theoretical statements must be clearly set off from other exposition. While he may put the case too strongly, in order to use theory effectively and to evaluate theory intersubjectively, sociologists must indeed move away from purely discursive presentation of theories. This does not require that we give up argumentation; it only requires that we separate theoretical claims from their explanation, justification, and rhetorical promotion. We agree with Gibbs that the evaluation of theories must stand apart from the theorist's ability to persuade others. We believe, however, that there are criteria in addition to empirical validity that play an important part in evaluating theories. Before we turn to these criteria, however, we must examine the components of a theory.

ELEMENTS OF A THEORY

Our definition requires theories to contain interrelated statements. Previous chapters, however, alert the reader to the different types of statements that can make up a collection. Furthermore, as the chapter on concepts explained, statements themselves are made up of different kinds of terms. Here we shall consider three types of statements and two types of terms. The chief components of theories, then, include *assumptions, scope conditions, derived propositions, primitive terms*, and *defined terms*.

Defined terms have already been extensively considered. In the present context, we want only to emphasize two features of definitional statements, that is, the sentences that define what we are calling defined terms. First, definitions differ from knowledge claims, or assertions, in that they say nothing about the empirical world, but merely specify what meanings are contained in the term; that is, they provide rules of usage for the definiendum. Second, the definiens of definitional statements consist of primitive terms and other previously defined terms. As we have indicated earlier, it is not possible to define all the terms used in a theory, but it is important that a theory contain some explicitly defined terms.

Chapter 7 also spelled out the notion of *primitive term*. Since it is impossible to define explicitly every idea, a theory must contain some primitive terms. As we mentioned, it is important that primitive terms be chosen so that we can be reasonably confident of widely shared usage of primitive ideas. It is also desirable to keep the number of primitive terms in a theory to a minimum; if all the terms in a theory were undefined, it would be virtually impossible to intersubjectively evaluate the theory. Primitive terms play two roles: they allow us to build up a system of explicit definitions, and they allow us to form theoretical statements which relate one primitive term to another. The second usage requires further comment.

Embedding a term in a statement which asserts something about that term constrains the meaning of the term, even though the term is never explicitly defined. Suppose we assert that power increases with status, and treat *power* and *status* as primitive terms. (From our earlier discussion, we should not use them as primitives because we cannot feel relatively comfortable about widely shared meanings; nevertheless, they will serve here for an example.) When we posit that power varies with status, we are asserting something about power and something about status. To argue that such an assertion constrains the meaning of the primitive terms is simply to say that not all connotations of power or of status are appropriate if we want to use the terms in our sentence. Thus, we should immediately reject "physical power" as an appropriate connotation and, on reflection, we would probably also reject any meanings of status which did not contain an idea of hierarchy. The point to emphasize here is that using primitives in theoretical statements limits the meaning of the primitives used. Some sociologists not only appreciate this feature of primitive terms but extend the position to say that no definitions at all are required, because usage in assertions can substitute for definitions. This represents an extreme position. The theorist, the evaluator, and the user of a theory need guidelines in order to work with a theory. Explicit definition provides some guidelines. To say that simply making statements constrains the meaning of terms in those statements sufficiently, without the need for definitions, is to hope for too much. When we are able to do so, we should provide explicit guidelines.

We have talked a great deal about theoretical statements without examining their properties. In the literature, what we have called *theoretical statements* have a number of aliases: the terms *assumptions, axioms, propositions, derivations, theorems, assertions, postulates, universal knowledge claims,* and probably some others all refer to theoretical statements. To complicate matters, usage of these terms varies from author to author. We lack a standard, agreed-upon terminology. Some authors use terms like assumptions, axioms, and postulates interchangeably. Some authors make distinctions among assumptions, axioms, and postulates. The same can be said about other terms on our list. This lack of consensus confuses not only the student but the scientist as well. We cannot establish a consistent, conventional usage overnight, but we can recognize that usage varies and attempt to determine in each particular case how an author uses these terms.

Here we will use the term *proposition* as synonymous with *theoretical statement,* meaning any sentence in a theory. We will also treat *assumptions, axioms,* and *postulates* as interchangeable terms. *Assumptions* will mean theoretical statements which are universal knowledge claims that relate two or more concepts by asserting something which can be true or false. Thus, "Power varies with status" can be used as an assumption. To make life more difficult, one sentence can be used in a theory in different ways so that, in addition to the form of the sentence, its usage in the theory determines whether it is an assumption or a derivation. Assumptions cannot be derived from anything else in the theory. They are sentences in which a theorist makes claims, which can then be used to generate other statements through logical derivation. A theory must contain more than one assumption, because it is impossible to derive consequences from a single statement.

The set of assumptions represents the core of a theory. Some writers require that assumptions be true universal statements. The everyday form of this position gives rise to the slogan, "You can't make assumptions unless you have the facts." Such a requirement places impossible limitations on theorizing. In fact, as chapter 13 will demonstrate, it is impossible to know whether or not a universal statement is empirically valid. Authors who recognize the difficulty propose the criterion that assumptions represent universal statements not known to be false. But even this may be too stringent. A theorist may construct a useful set of statements even though he or she may know that one or more of the statements, as they are formulated, are empirically false. Usually, such statements represent simplifying assumptions, which are useful even though they are overstatements of empirical relationships. In the theory to be presented in the next chapter, we will have an example of an assumption which the theorists recognize as too extreme a claim—an assumption which, in the way it is formulated, is literally false but is nevertheless extremely useful.

We want to emphasize that assumptions are not literal truths, laws of nature, or immutable knowledge. They are assertions which, in form, can be true or false; which can be manipulated to generate other statements; and which do not depend on having the facts beforehand. The reader might get the feel of this issue by recalling high school geometry, which assumes that parallel lines never meet. Since we can never get to infinity, we cannot tell whether the assumption is true or false. We know that in concrete applications of geometry, theorems about parallel lines are literally false; yet that does not prevent us from employing geometry to build bridges. We emphasize this point because so many sociologists are so reluctant to formulate assumptions unless they feel fully confident of their truth. Rather than argue about competing beliefs concerning the truth of an assumption, examining the use of the assumption and determining its usefulness is a more appropriate way to look at assumptions.

Assumptions form the core of a theory because they formulate the key relationships that concern the theorist. Assumptions contain the substantive content of the theory, which, roughly speaking, rests with the relation of one idea to another. It is not enough to have an idea about power unless that idea is relational—relating power to other ideas, such as status. Sometimes, assumptions assert the existence of a phenomenon, such as *In all organizations, informal power exists.* But existence statements, while they may be necessary to round out the logical structure of a theory, are less important than relational statements. In working with a theory, we are much more interested in how the theorist views changes in informal power than in the existence of informal power. Hence, we would pay more attention to statements like *Informal power increases with frequency of interaction in an organization* than we would to assertions of the existence of informal power. A theory which in its assumptions posits a number of relationships among its key concepts will allow us to draw out, as consequences, new relationships of which the theorist or the audience may have been initially unaware. It is the ability to generate new consequences which measures the ultimate value of the theory.

These new consequences are typically called *derivations* or *theorems*. Some authors use the term hypothesis for a derived consequence, but we will limit

Sjoberg, Gideon, & Nett, Roger (1968), *A Methodology for Social Research*, Harper Row, New York.

Summers, Gene (ed.) (1970), *Attitude Measurement*, Rand McNally, Chicago.

Timoshenko, S. & Young, D.H. (1962), *Elements of Strength of Materials*, 4th Ed., D.Van Nostrand Company, Princeton, NJ.

True, J. Audrey (1989), *Finding Out: Conducting and Evaluating Social Research*, 2nd Ed., Wadsworth Publishing Company, Belmont, CA.

Van de Geer, John P. (1971), *Introduction to Multivariate Analysis for the Social Sciences,* W.H. Freeman and Company, San Francisco.

Vaux, Alan (1992), "Assessment of Social Support", in Veiel, Hans O.F. and Baumann, Urs (eds.), *The Meaning and Measurement of Social Support*, Hemisphere Publishing Corporation, New York.

Veit, Richard (1990), *Research: The Student's Guide*, Macmillan Publishing Company, New York.

Wright, S. (1934), "The method of path coefficients", *Annals of Mathematical Statistics,* V:161-215.

Yin, Robert K. (1994), *Case Study Research: Design and Methods*, 2nd Ed., Sage, Thousand Oaks, CA.

ourselves to either *derivation* or *theorem,* reserving the term *hypothesis* for another usage. These are statements of relationships between concepts; they result from a correct application of the syntax of the theory to a set of assumptions. A derivation may involve many assumptions of the theory, or only two assumptions. When we talk about correct application of the rules, we mean an application of some logic to manipulate the assumptions to arrive at a new statement. Sometimes, deriving a consequence follows an obvious and straightforward route. At other times, a derivation is very subtle, and its discovery requires ingenuity and imagination. The analogy with mathematics can perhaps clarify the process of derivation. Not everyone can think of new theorems in, for example, geometry. It may take a genius to conjecture a new idea in such an old system as plane geometry. However, once such a conjecture is made, presumably anyone who knows the rules of geometry can determine if the conjecture follows from the assumptions of Euclidean geometry. One evaluates the conjecture as logically true or an incorrect application of the rules—that is, the syntax of the theory.

As an example of derivation, consider the following from Blau (1970). Blau's theory concerns the phenomenon of structural differentiation in organizations. He has observed that large organizations develop specialized subunits, such as a department dealing only with sales, a department dealing only with research, a department dealing only with production, and so forth. He aims to relate the development of these specialized units to organizational size. He first assumes the following:

1. **Increasing size generates structural differentiation in organizations along various dimensions at decelerating rates.**

Although Blau considers this a theoretical generalization, we prefer to regard it as an assumption. Blau then lists three highest-level propositions which he considers as parts of this first assumption:

1a. Large size promotes structural differentiation.
1b. Large size promotes differentiation along several different lines.
1c. The rate of differentiation declines with expanding size.

He then presents a second assumption:

2. **Structural differentiation in organizations enlarges the administrative component.**

Blau then cites as one derivation the following:

Derivation:
The large size of an organization indirectly raises the ratio of administrative personnel through the structural differentiation it generates.

Blau claims that this derivation logically follows, with the following argument: "If increasing organizational size generates differentiation (1a), and if differentiation increases the administrative component (2), it follows that the indirect effect of size must be to increase the administrative component" (p. 204). This derivation is one of many that Blau presents, employing purely verbal arguments to demonstrate the logical truth of the derivations. Implicit in his discussion is an appeal to a syntax—which, in this case, is elementary logic.

One may argue with Blau that his derivation does not rigorously follow, since his purely verbal exposition introduces terms which are not part of his assumptions. But it is a simple matter to "tighten up" his statements so that they would meet the rules of the calculus of propositions. The virtue of his theoretical statement rests with making assumptions and derivations explicit, so that it is possible to apply a syntax and determine what needs tightening up.

Our purpose here is to illustrate the employing of assumptions to generate derivations. Ideally, derivations would follow the strict rules of some logic. As Gibbs (1972) points out, "sociologists rely on the conventions of a natural language rather than formal rules of derivation. Those conventions are putative and notoriously imprecise. Accordingly, when a theorist uses them rather than formal rules, the logical structure of the theory is conjectural" (pp. 104, 105).

We agree that a clear, logical structure represents an important goal of theorizing, but it is a goal which can be reached in stages. One can begin with discursive presentation of ideas. Isolating what the theorist regards as assumptions and derivations can be a second stage. Explicitly using a syntax in a formal, precise, and rigorous way can represent a more advanced stage. Bearing the stages of this process in mind will improve even discursive formulations of theoretical ideas.

The final element of a theory has already been discussed, in that simple knowledge structures also contain *scope statements*. We regard scope statements as extremely important in providing guidelines to both the theorist and the user. Scope statements define the phenomena to which the theory applies. In his theory, Blau does not explicitly present any scope statements. Yet, in his introduction, he presents an example of what we would regard as a scope statement. He argues that his theory applies to work organizations, that is, organizations deliberately established for explicit purposes and composed of employees. Clearly, he intends to direct his audience to a limited class of phenomena for either testing or using his theory of differentiation. But note that his directive is a universal statement. The organizations are not restricted to particular times and places; he states an abstract and universal scope restriction. Unfortunately, his presentation of the restriction is totally removed from his statement of the theory itself (separated by nearly three journal pages). Hence, the reader cannot be blamed for questioning whether or not this restriction is part of the theory. Where theories contain explicit scope statements, however, such questions do not arise. Furthermore, when a theorist formulates explicit scope directives, he must confront a number of questions about the applicability of his theory. We could ask, Does Blau intend the theory to apply to all work organizations? If not, he must put forward additional scope restrictions. He may want to limit the theory to organizations of some minimum size; for example, it is doubtful that a three-member organization would fit his theory.

In spite of our view of the importance of scope statements in a theory, one finds very few examples in the sociological literature that contain explicit scope statements. As a result, sociologists become involved in controversies dealing with whether or not a given study represents an appropriate test of a particular theory. Without statements of scope, however, resolution of such arguments becomes very difficult.

We are now in a position to compare theories with simple knowledge structures. First of all, in the way we have developed the components of a theory we have excluded statements of initial conditions, observation statements, operational definitions, and rules of correspondence—linkages between a concept and observations used to measure that concept. As our conception of theory stands, theories are nonempirical; they lack explicit statements linking the elements of the theory to elements of the phenomenal world. Since we regard empirical import as a necessary property of a theory, we must insist that statements of initial conditions be attached to theories. Whether these statements are part of the theory or auxiliary to the theory is a matter of some dispute.

We prefer to treat initial conditions and observation statements as auxiliary to a theory. We do not want to restrict a theory to one set of initial conditions or one set of observation statements. Both the definitions of the theory and the scope statements provide guidelines in general terms to facilitate the formulation of many different sets of initial conditions and many different observation statements. If, for example, we stated an initial condition such as *Status in work organizations can be measured by the number of people supervised* as an intrinsic part of the theory, we would unduly limit the theory's usefulness. Treating such statements as auxiliary allows us to substitute a wide variety of alternative indicators of status in testing our theory, provided that those indicators have properties consistent with the definition of status in the theory and with the scope restrictions of the theory.

We can now see how theories are both more and less than simple knowledge structures. Simple knowledge structures contain statements of initial conditions and observation statements, which we have excluded from theory. In this sense, a theory is less than a simple knowledge structure. On the other hand, theories contain more than one universal knowledge claim. In fact, they contain at least three such statements—for we have insisted on at least two assumptions, and these two assumptions must have at least one derivation. Furthermore, theories must include explicit definitions, whereas simple knowledge structures need not. In simple knowledge structures, statements of initial conditions may operate to denotatively define concepts, but no further definition is required. For a simple knowledge structure it is enough to say, for example, that observing steam rising is an instance of boiling (recalling our discussion in chapter 4); that is, one need not define boiling. Similarly, we could assert an initial condition like ''Popularity in a high school is an instance of status'' without further conceptualization of status. Hence, theories are more than simple knowledge structures in that they incorporate at least two knowledge claims as well as connotative definitions of key concepts.

Essentially, a simple knowledge structure focuses on the empirical investigation of a single idea. The idea is relational, of course, in that a universal knowledge claim relates two or more concepts which are themselves ideas. Nevertheless, simple knowledge structures have a very limited focus. While the knowledge claim from a simple knowledge structure may become part of a theory, a simple knowledge structure is a tool for use prior to theoretical development. Theories, on the other hand, represent an attempt to integrate a range of ideas in a systematic way. The importance of distinguishing between simple knowledge structures and theories lies in the implications of the distinction for cumulative research strategies, a topic that will be considered in chapters 15 and 16.

WHAT A THEORY DOES

We have stressed the understanding of theory because theories serve several essential functions in the development and evaluation of scientific knowledge. We can list some of these key functions:

1. A theory provides a shorthand for communication.
2. A theory organizes ideas and, in so doing, may uncover hidden assumptions.
3. A theory generates new ideas.
4. A theory may display the complexities of a problem.
5. A theory guides investigation.
6. A theory generates explanations and predictions.
7. A theory may relate what on the surface are different problems.

As shorthand for communication, when scientists are operating with the same theory, they have no need to repeat the rationale and justification for particular pieces of research. As long as it is clear that two different investigators are using the same concepts and assumptions, a great deal of ''common culture'' is understood as buttressing their work. They have no need to go back to first principles or define their terms in every discussion between them.

This shorthand practice often gives the new student a problem. He may approach a study conducted within a particular theoretical tradition without knowing the common culture, and therefore may have great difficulty in understanding what is going on. Where there is a well-established theoretical tradition, it is impossible for a single research report to stand alone and be understood. Scientists operating in a theoretical tradition sometimes create an aura of mystery and an appearance of cultism to the uninitiated. But their theoretical shorthand performs a central role in facilitating economical communication among scientists. If every investigator had to rehash the whole tradition surrounding his work, our journals would require many times the number of pages they presently contain, and readers would waste a lot of time in going over redundant material. The student will avoid the aura of mystery if he recognizes the necessity for this shorthand and the necessity for working through the theoretical background before attempting to understand a new research article.

The organizing role of theory cannot be overemphasized. In attempting to set down explicit assumptions, a theorist often has the experience of recognizing how ideas fit together, where there are gaps in his or her ideas, and where there is excess conceptual baggage. Again looking at Blau's theory, his first assumption relates both the content dimensions of differentiation and the rates of differentiation, organizing these quite distinct ideas into a coherent statement. Furthermore, given his statement number 1, his statements 1a, 1b, and 1c, as they stand, are not really necessary. He himself comments that they are contained in his formulation of assumption 1. In writing down this first assumption, Blau also recognizes something he has been implicitly assuming. He puts forth an additional proposition:

1d. The subunits into which an organization is differentiated become internally differentiated in a parallel manner (p. 204).

Although he numbers this proposition 1d, it does not follow from assumption 1 in the same way that 1a, 1b, and 1c follow. Recognizing this, Blau asserts an additional assumption in his text: ''These generalizations apply to the subunits within organizations as well as to total organizations.'' Even though that assumption appears in the text (i.e., it is not set off explicitly, as are his other assumptions), it is clear that the assumption is very much a part of his theory and has the same status as assumptions 1 and 2. Since Blau wanted to assert 1d, he was forced to rethink his implicit reasons behind that assumption; that is, he was forced to uncover a hidden assumption in his reasoning.

One of the major purposes of theorizing is to generate new ideas. The search for new theorems or new derivations implied by a set of assumptions constitutes a fundamental activity of theorizing. In dealing with complex ideas, one has great difficulty in keeping track of the implications of the ideas while juggling them around in his head. Discursive presentation of theory is analogous to this juggling. Setting down explicit assumptions and systematically tracing out implications constitute a much surer strategy for generating new implications. An illustration of this process will appear in the next chapter, where we look at a theory as a whole.

That a theory may quickly display the complexities of a problem is well illustrated by Davis's *theory of relative deprivation* (Davis, 1959). The basic notion behind this theory is that an individual, call him Ego, evaluates his social situation by comparing himself with others, call them Alters. As a result of this comparison, Ego may feel better off or worse off, relatively gratified or relatively deprived. The basic concern of the theory is to formulate the way in which these comparisons occur. The theory assumes that a population may be divided into categories which reflect differences in desirability. If we limit ourselves to social characteristics that divide the population into only two categories, Ego may then compare himself with Alters in his own category or in the other category. This gives rise to what Davis calls a comparison matrix, as in table 10.1.

Table 10.1. Comparison Matrix (from the Viewpoint of a Given Ego) for a Population Partitioned on Deprivation Only)

		Alter	
		Deprived	*Nondeprived*
	Deprived	*a*	*b*
Ego			
	Non-deprived	*c*	*d*

Suppose that sex is one of the attributes that divide the population into two desirability classes. From a contemporary point of view, some Egos would consider females as the deprived category and males as the nondeprived category. Suppose now we have the attributes of wealth, and consider the poor as deprived and the rich as nondeprived. Instead of table 10.1, we have a comparison matrix that looks like table 10.2.

Table 10.2. Comparison Matrix for a Population Partitioned on Sex and Wealth

		Alter			
		Male		Female	
		Rich	Poor	Rich	Poor
	Male				
	Rich	*a*	*b*	*c*	*d*
	Poor	*e*	*f*	*g*	*h*
Ego					
	Female				
	Rich	*j*	*k*	*l*	*m*
	Poor	*n*	*o*	*p*	*q*

By simply adding one more variable, we have increased the number of possible comparisons in a population from four in table 10.1 to sixteen in table 10.2. If we had divided our wealth variable into three categories instead of two, including a place for the middle class, we would have added twenty more possible comparisons. Davis's theory quickly shows us how complex the problem of formulating social comparisons can be. It is certainly very difficult to juggle each of these comparisons in one's head, as would be necessary if Davis's argument were presented in purely discursive form. Davis goes on to develop additional assumptions which provide him with tools for dealing with these complexities.

A fundamental reason for emphasizing theory is that a theory guides investigations. By now this may be so obvious that it does not need further discussion; yet it might be worthwhile to point out a number of different ways that a theory guides. Most people recognize that the derivations of a theory generate hypotheses that empirical studies are designed to test. But, in addition to guiding research by providing the questions that research should answer, a theory guides in other ways. The concepts defined in the theory point to what features must be observed and measured, and also provide criteria against which to check the quality of the observations. A theory guides us in constructing indicators and in evaluating the reliability and validity of these indicators. Recall, for example, that reliability requires a fixed set of conditions. A well-developed theory formulates what constitutes a fixed set of conditions and thus provides direction for assessing observations. Finally, a theory provides guidance by defining the appropriate research situation for answering the questions posed by the theory's derivations. One of the main reasons for incorporating scope statements into a theory is that scope assertions represent the conditions under which the theorist believes that his derivations would be true. A theory, for example, whose scope was restricted to informal groups, groups without any formal organization, would not be appropriately studied by observing a courtroom. While the example may seem trivial, the problem created by the absence of such explicit scope statements is horrendous. A great deal of controversy would be eliminated if sociologists provided explicit scope statements to guide others in selecting appropriate situations for testing their ideas.

When we say that theory generates explanations and predictions, we must be careful to spell out precisely what we mean. The layman's use of the word explanation has many different senses; to explain something is to indicate why it happened or how it happened. How did the Democrats win the election? is answered by the statement that unions turned out their membership in full force. Why did the unions turn out their membership? is answered by, Because economic issues were central in the campaign. Here we intend explanation to have a more restricted usage, although sometimes it will overlap with these meanings.

We base our view of explanation on what is known as the deductive model of explanation in science (Nagel, 1961). To explain something is to show that it is a deductive consequence of universal knowledge claims. While not all scientific explanations rigorously fit this model, the model does cover a wide range of what are called *explanations* in science. Here again, some Latin terms are helpful. We will call the thing to be explained the *explanandum*, and the premises from which one deduces the explanandum, the *explanans*. Two kinds of explanation occur in science, distinguished by the nature of the explanandum. When the explanandum is itself a universal knowledge claim, we have what is known as the explanation of laws. When the explanandum is an observation statement, we have the *explanation of singular events*.

Suppose we want to explain the observation that air force officers have more influence in group problem solving than air force enlisted men, as in a study by

Torrance (1955). We could explain this observation statement with three universal knowledge claims (KC) and several statements of initial conditions (IC):

KC 1. People who differ in status have different beliefs about their relative abilities.

KC 2. The higher the status of an actor, the more ability he is believed to have.

KC 3. Influence on a task corresponds to beliefs about ability.

IC 1. Rank in the air force is an instance of status, with the higher rank representing the higher status.

IC 2. Beliefs about the task in the Torrance experiment are instances of beliefs about ability.

IC 3. Influence on Torrance's task is influence affected by beliefs about ability.

IC 4. Air force rank is the only status operating in the Torrance situation.

From these statements we can deduce Torrance's result—that is, the statement of his finding—and we can then say that we have explained his finding. In this case we have an example of the explanation of a singular event, the observation statement from Torrance's study.

What is called the explanation of laws is formally identical with generating derivations from a theory. The difference between deriving theorems and explaining laws lies in the direction of the process. In deriving theorems, the principal concern of theorists is with tracing out the implications of their assumptions, and their focus is on the assumptions of their theory. In explaining laws, the universal knowledge claim to be explained represents the theorist's primary concern, and he or she constructs assumptions or uses a theory in an effort to explain the given universal knowledge claim. It is usually not possible to tell from the statements themselves whether we have an explanation or we have the generation of theoretical consequences, since the difference primarily rests with the intention of the investigator. Consider the following universal statement:

A. The greater the division of labor in a society, the greater the solidarity of the society.

We can construct a set of universal knowledge claims from which this statement can be deduced. For example:

1. The greater the division of labor in a society, the greater the interdependence of members of that society.

2. The greater the interdependence of members of a society, the greater the need for bonds of solidarity among members.

3. The greater the need for bonds of solidarity, the more bonds will form— that is, the greater the solidarity of the society.

Simply looking at these four universal knowledge claims does not tell us whether we are dealing with an explanation. If, however, we know that the theorist is primarily concerned with the statement *A,* we would consider it an explanation. On the other hand, if the theorist is primarily interested in statements 1, 2, and 3, then the activity is one of developing a theory by deriving new implications from the theory.

Given either an observation statement or a universal statement, it is always possible to construct many different explanations for the statement. This fact illustrates an extremely important principle:

Explanations in science are not unique.

This principle has great significance for both the scientist's desire to explain and for his strategy in developing theory.

In the first place, constructing an explanation represents no great accomplishment. If the only constraint on the theorist is the explanandum, then any sociologist can formulate an explanans, indeed many sets of explanans, while comfortably sitting in his armchair. The question then becomes, How does one choose one explanation over another? If there is no basis for choice, then the constructed explanations at best constitute an intellectual game, and, at worst, irresponsibility on the theorist's part. A responsible explanation requires constraints on the theorist. If the theorist focuses on his theory, then he is severely constrained; we cannot deduce just anything from a given set of assumptions, and it is readily decidable whether or not either an observation statement or another universal logically follows from the theory.

The idea of responsible explanation accounts for our emphasis on the similarity between theorizing and explaining. An explanation and a theory are not the same thing; but explanations based on theory are more likely to be responsible than explanations constructed solely for the purpose of dealing with a given statement. It is possible, however, to have responsible explanations even without theory. At the very least, though, we require that any proposed explanation have additional consequences beyond the original explanandum. The fact that an explanation has additional consequences allows other statements to influence the acceptance or rejection of a given explanation. In short, an explanation that explains only the original explanandum is nearly worthless. On the other hand, one may begin with a particular explanandum, construct an explanans that has many consequences, and be well on the road to formulating a theory.

The second important aspect of our principle suggests that a major part of the research task consists of providing evidence for choosing among alternative explanations. In fact, from this perspective one can view research as providing the basis for systematic, rational choice: choice among alternative explanations, choice among alternative theories, or choice between a theory and a nontheoretical alternative explanation. We will have more to say about this in chapters 13 and 14.

Just as there are similarities between the explanation of laws and the derivation of theorems, there are also similarities between explaining singular events and making predictions. A prediction, after all, is an observation statement formulated in advance of collecting observations. Just as one explains singular events by joining initial conditions to universal statements, one makes predictions in the same way. The distinction between explanation and prediction usually focuses on whether the observation statement is derived before or after observations are collected. Many writers have placed much more emphasis on predicting, that is, on deriving observation statements in advance, than on explaining. Implicitly, the reason for this emphasis concerns the ease with which explanations can be constructed after the fact. Once you know what the observation statement is, if your only task is to explain that observation statement, you cannot fail. Hence, constructing an explanation is in no way a test. On the other hand, if you do not know the observation statement in advance, and you predict what it will be, it can disconfirm your prediction; if you can be wrong, but are not, then you have accomplished something.

Much can be said for this position, but it does result in an underemphasis on explanation. Furthermore, the exclusive emphasis on prediction results from considering only those explanations that deal solely with one explanandum (and therefore have no constraints), that guarantee that some explanation will result, and that constitute no test of the theorist's knowledge or ability. The idea of responsible explanation, where an explanation must have consequences in addition to the original explanandum, allows us to put explanation and prediction in proper balance as equally necessary functions of a theory. In short, theory allows us to explain observation statements that have been collected in the past, as well as to predict new observation statements from research yet to be conducted.

CRITERIA FOR THE EVALUATION OF A THEORY

We have considered the components that make up a theory and what a theory does. Implicit in our discussion have been several standards for evaluating theories. By way of summarizing much of what has been said, we can make these criteria explicit.

First, evaluation of theory involves both logical and empirical standards. From the way we have described theories, it follows that evidence cannot be brought to bear on a theory without some consideration of the logical structure of the theory. If we are going to use observation statements to test theories, these must be relevant to the theory. But deciding whether or not an observation statement is relevant to a theory involves logical standards. To be relevant, an observation statement must be derivable from the theory joined with a set of statements of initial conditions, When there is an explicit theory and a set of explicit initial conditions, we no longer have to argue about whether a given set of observations is appropriate to test a theory. We have a relatively simple, intersubjective, logical test by which we can decide.

CHAPTER THREE

How You Decide to Conduct Your Project

How to go about your project appears to be a question that had better be answered in considering your research design. Before you go into the technical details of choosing appropriate research strategies, methods, and instruments for your project, however, you need to resolve some general issues and concerns that would affect the basic configuration of your thesis/dissertation.

Matching your objectives with your resources

The importance of the thesis/dissertation is apparent. It means an academic degree for all those who have to write it. For some, this is probably the last or the only serious academic undertaking that he will be engaged on his own. Even for those who will remain in academia and the research circle, the thesis/dissertation will most likely be valued as a major milestone of their career achievement. It is understandable, therefore, that people tend to take their thesis/dissertation project very seriously. Some students are so ambitious about their project that they are prepared to devote whatever time, energy, and other resources needed to complete it.

The students, however, are restricted by a number of factors in conducting thesis/dissertation research. First, they are limited by time. A degree program usually has a maximum length of study allowed to achieve the degree objective. By the time you embark on your thesis or dissertation, you have already spent a significant portion of that amount of time. Although time still does not seem to be a problem, you should be aware that there is another rule of the game. While

Index

aging study, 30
  psychosocial aspects, 31
analysis of covariance, 146
analysis of dependence, 150
analysis of variance, 131, 145
  between-group sum of squares
    (BSS), 146
  interactive sums of squares (ISS),
    146
  total sum of squares (TSS), 145
  within-group sum of squares (WSS),
    71, 123 145
analysis of variance, decomposition
  131, 145
  interaction effect, 146
  joint effect, 146
  main effect, 146
  one-way, 132
analytic unit, 31
  see also unit of analysis, 56
analytical modeling, 136
atomistic fallacy, 59
backup cases, 71

beta weight, 141
binomial sampling distribution, 129
blind unidimensional treatments, 161
budgeting, 87
  equipment, 87
  prevailing rates, 87
  travel, 87
canonical correlation, 145, 148
  canonical loadings, 149
  canonical variates, 148
case, 59
case study, 84, 89
categorical variables, 143
CAUS
  first component (factor) theory, 167
  maximum component (factor) the-
    ory, 166
  practical approach, 169
  summated factors (components)
    theory, 167
causal model, 150
  endogenous variables, 150
  exogenous variables, 150

In order to facilitate logical and empirical evaluation of a theory, then, we propose the following criteria:

1. The theory should be explicit and relatively precise.
2. The theory should contain explicit definitions based on primitive terms for which usage is widely shared.
3. The theory should provide a clear exhibition of the structure of the argument. In other words, the syntax of the theory should allow an examination of the logical skeleton of the theory apart from its content.
4. The theory should provide clear guidelines as to where it is applicable; that is, the theory should contain explicit scope statements.
5. The theory should be testable empirically; when joined to a set of initial conditions, the theory should allow the derivation of observation statements, both to predict and to explain.
6. The theory should formulate an abstract problem.

Most of these criteria have already been dicussed. A few comments, however, are necessary about the last criterion. Theories must contain universal knowledge claims as assumptions and derivations. It is also necessary for the theory to be applicable to situations beyond the situation which gave rise to the theory in the first place. This last point extends our comments about explanation. Just as a responsible explanation must have consequences beyond the initial explanandum, a useful theory must extend beyond the original situation that motivated the theory. Even if our concern is to explain theoretically a concrete singular event, such as the outcome of the presidential election in the United States in 1988, theorizing requires that we get beyond the concrete problem. We cannot hope to deal with all of the factors that affect a particular event such as a presidential election. Theorizing does not mean trying to capture all of these factors; rather, it means abstracting from the concrete event those aspects which are significant not only to the presidential election of 1988 but to other concrete events as well. If we define our problem as "developing a theory of the presidential election of 1988," we are not likely to formulate a useful theory. If on the other hand, we formulate an abstract problem—for example, the relationship between economic interest and political participation—we are much more likely to make progress in theorizing, and would generate a theory that would contribute to our understanding, not only of the 1988 election, but of other events at other times and places. In the next chapter we will illustrate a theory which formulates an abstract problem.

This chapter has presented a number of new and difficult ideas. Students who intend to be consumers of sociological research may question the need for all of this technical detail. To be sure, most of the technical evaluation of sociological theory takes place in the relevant public of other sociologists. Nevertheless, the layman needs some awareness of the technical issues to appreciate the fact that the evaluation of scientific knowledge claims is a far more subtle and complex process than it initially appears to be. In the next chapter we will illustrate the material of this chapter with a particular theory, to provide both an example of the process of evaluation and an understanding of the requirements of useful sociological theory.

SUGGESTED READINGS

Cohen, Bernard P. "On the Construction of Sociological Explanations." *Synthese* 24 (Sept./Oct. 1972): 401–9.

This article is a fuller development of some of the ideas presented in this chapter.

Nagel, Ernest. *The Structure of Science.* New York: Harcourt Brace Jovanovich, 1961. Pp. 15–28.

This chapter deals with patterns of scientific explanation. It illustrates the ambiguity of the question "Why?" by presenting ten different types of answers to "Why?" questions. Nagel also discusses four different types of explanation in science, one of which is the deductive model.

There are numerous books dealing with theory construction in sociology that have appeared in the past twenty years. They present various approaches to the construction of sociological theory ranging from books on "how to do it" to discussions of philosophical issues in theory construction. The orientations represented in these works sometimes overlap considerably with the present author's orientation and sometimes overlap very little. Three books which represent a diversity of both conceptions of theory and approaches to theory construction are:

Blalock, Hubert M. *Theory Construction from Verbal to Mathematical Formulations.* Englewood Cliffs, N.J.: Prentice–Hall, 1969.

Although Blalock shares the present author's concern for rigor, his conception of theory does not contain many of the elements discussed in this chapter. The book, however, has been extremely important in the development of causal modeling in sociology.

Gibbs, Jack. *Sociological Theory Construction.* Hinsdale, Ill.: Dryden Press, 1972.

Gibbs presents many illustrations of formulating theoretical propositions in sociology. He has a similar orientation to the present author but does not consider the issue of conditionalization.

Stinchcombe, Arthur L. *Constructing Social Theories.* New York: Harcourt Brace Jovanovich, 1968.

This book illustrates various theoretical strategies which are commonly useful in explaining social phenomena. Although Stinchcombe's conception of theory is quite different from that presented in the present work, his enthusiasm and excitement for the task of theoretical explanation are shared by this author.

ELEVEN

A Theory
and Its
Analysis

Up to this point we have discussed the need for theory to guide research, the elements of a theory, and criteria by which to evaluate theories. In the course of this discussion, we have presented theoretical ideas and what might be called theory fragments. What is required now is to look at a reasonably well-developed theory and to analyze it to point out its components and to show how they fit together. Furthermore, we can apply the criteria for evaluating theories, presented in the last chapter, to illustrate how these criteria are used. Such an analysis serves two purposes. First, it illustrates the process of evaluation. Second, by seeing how well the theory we choose measures up to these criteria, our analysis should provide some understanding of the kinds of theories needed in sociology.

We have chosen as our example the *theory of status characteristics and expectation states*. This theory is reasonably well developed, has generated a number of empirical studies, and has both scientific and practical implications. The main reason for choosing this theory, however, is that it is relatively simple to present and to analyze. For the same reason, we will discuss an early version of the theory; recent versions are considerably more elaborate and would introduce complexities that would be distractions from our purpose.

While the simplicity of the theory is desirable, the reader should be cautioned that the simplicity is somewhat deceptive. Researchers who have worked with the theory have uncovered complexities that are not apparent from merely studying the theory in the abstract. Such complexity is largely irrelevant for present purposes, however, and it should be sufficient simply to present such a caution.

THE PROBLEM

There is a long sociological tradition that concerns the effect on social interaction of the way in which actors define social situations. The conceptions which actors form of themselves and of other actors have profound effects on the way in which people behave toward one another. In 1908, Simmel noted that "the first condition

of having to deal with somebody at all is to know with whom one has to deal''
(Simmel, 1908, quoted from Wolff, 1950; italics in the original). While a person
might know with whom he is dealing from direct previous experience, Simmel
observed that one might also know it from the individual's status category. Twenty
years later, Robert Ezra Park formulated a conception of interaction in which one
individual classified another by age, sex, race, and social type, and behaved toward
him on the basis of stereotypes associated with these classifications (Park, 1928).
Thus, we are indebted to Park for an important insight: status conceptions organize
the behavior of individuals in social situations. As Cohen et al. (1972) put it:

> Most of us still use Park's formulation: individuals classify themselves and others
> in terms of already established status categories. Such categories function both cog-
> nitively and normatively and provide individuals with information about how they
> should behave. In the absence of such categories, social situations are ambiguous,
> behavior is unpredictable, and individuals in such situations are anxious and tense.
>   The significance of Park's formulation cannot be over-estimated. Imagine what
> it would be like to visit a close friend in the hospital, but not to be able to distinguish
> doctors from other visitors. A strange man walks into the room and orders you to
> leave, or he asks you to step outside and informs you that your friend has some dread
> disease, or he walks in and orders your friend to take off his clothes. Of course,
> the truth is that social arrangements hardly ever permit such situations to arise, or
> if they arise, to persist for long. Most of the time, the required definitions occur
> almost instantaneously, and even largely without conscious thought, so that most
> people are unaware that they are obeying a rather profound sociological law. (P. 450)

We can all think of examples to which Park's insight applies. There is the Nobel
Prize winner whose opinions are taken seriously even on matters far removed from
his expertise. There is the mixed-sex volleyball game in which the men ''hog''
the ball and the women back off and let them. There are the employees who always
agree with their boss even on things that have nothing to do with the job. In everyday interaction, we all meet situations in which beliefs about status organize the
way interaction takes place.

Furthermore, more than forty years of research has been guided, either explicitly
or implicitly, by Park's general idea. Thus, studies have observed the phenomenon in a variety of settings. In a psychiatric hospital, positions in the hierarchy
determine the participation rates in ward rounds: the ward administrator partici-
pates more than the chief resident; the chief resident more than other residents;
the most passive resident more than the most aggressive nurse. In juries, sex and
occupation determine participation, election to foreman, and evaluation of jurors'
competence. In biracial work groups, whites initiate more interactions than blacks,
and blacks talk more to whites than to other blacks. In air force crews, pilots were
more influential than gunners in convincing crew members to adopt their position,
even when it was wrong, and even when the group task had nothing to do with
the work of the air force.

Even after forty years, however, the research is fragmentary and compart-
mentalized so that two investigators dealing with different concrete problems are

often unaware of the ways in which their efforts are related. Although we have illustrations of Park's idea in studies ranging from field observation to highly controlled laboratory experiments, prior to this theory sociologists had not gone beyond Park's original insight. Essentially, we knew that sometimes status conceptions organize interaction and sometimes status conceptions do not organize interaction. What was missing was a theory which codified previous research, explained previous findings, generated new research ideas, and provided guidelines for those who wanted to use this insight to solve practical problems.

Consider, for example, a biracial classroom. Does Park's general idea provide any suggestions for a schoolteacher who has to deal with interaction problems that might occur in the classroom? The answer is that it might. But in order for it to be helpful, several questions must be answered. These include:

1. Under what conditions do status conceptions organize interaction?
2. What is it about status conceptions that affects interaction?
3. How do status conceptions organize interaction?

An answer to the first question provides some guidelines concerning when status conceptions do affect interaction and when they do not. An answer to the second question gives us clues concerning when we may expect status conceptions based on race (or, more familiarly, racial stereotypes) to affect interaction. And an answer to the third question could indicate what kind of interventions our teacher might make to cope with interaction difficulties.

These issues, then, motivated the formulation of the *theory of status characteristics and expectation states*. The objectives of the theory are to systematize previous thinking and research, to explain previous findings, to generate new research ideas, and to provide guidelines for dealing with problems of status and interaction, both scientifically and practically.

It is necessary to systematize previous thinking and research. Although we have a large number of studies, it is difficult to bring them together and it is even more difficult to say what all this research adds up to. If the next researcher is not simply going to produce another example—perhaps with a different observable variable— of what has already been done, the researcher must be able to evaluate what is and what is not known. If the practitioner wants to use previous research to deal, for example, with sex discrimination, the practitioner should be able to bring to bear a number of these studies to gain insight into the practical problem. For there to be a payoff to the practitioner, studies that do not use sex as a variable—but use other statuses—should also provide useful information. In other words, because a study examined the effects on interaction of air force rank does not mean that the study is irrelevant to the effects of sex differences in interaction. Without a theory, however, it is not possible to know which studies can provide useful information or, indeed, when a particular study is or is not useful. Systematizing previous thinking and research means organizing it in such a way that the next researcher or the next practitioner can evaluate what is and what is not useful.

It is virtually impossible to make such evaluations when all one has is a large body of unrelated studies.

If we look at this body of literature, about all we can say is that all the studies involve status differences. None of these studies, however, defines *status difference* explicitly. In fact, nowhere in this body of literature are the relevant variables precisely conceptualized. The studies all deal with other variables (participation, prestige, influence, etc.), but the collection of studies gives little clue as to how these variables are related to one another, except that they are all affected by status differences.

Similarly, the concrete differences in these studies virtually preclude evaluating the state of our knowledge with respect to Park's insight. For example, the kinds of tasks used in these studies range from estimating the number of dots on a card to deliberating the damages to be awarded in a jury trial. Furthermore, the tasks range from those familiar to the participants to tasks totally foreign to them. For example, discussing a patient's case history is a completely familiar task to staff members in a psychiatric hospital, while asking air force crew members to construct a story about an ambiguous picture represents a task totally foreign to most air force personnel. How important are these concrete differences? When do we take them seriously and when can we safely ignore them? If all we have is the collection of separate studies, we cannot answer these questions. An abstract formulation of the problem together with a theory allows us at least to begin to answer them.

Formulating the theory is also an attempt to construct tools for explaining findings. If we can show that the different results in these widely varying situations are all consequences of a small number of principles (knowledge claims), we have provided a means to summarize the studies and to organize our knowledge about the way status affects interaction in terms of those knowledge claims. If we can show, for example, that beliefs about a person's competence at a task (any task, from playing volleyball to solving mathematical puzzles) arise from beliefs associated with that person's relative status, we then have a powerful tool for dealing with a wide range of different task situations. If we can show that beliefs about status generate beliefs about competence only when no other information on which to judge competence is present, then we have guidelines for when to apply and when not to apply our principle.

Since we have also argued that the formulation of an explicit set of knowledge claims allows us to derive new ideas, it follows that the formulation of knowledge claims to explain previous research will allow us to generate new ideas and also new practical applications. Considering only what we have said up to this point, it is already possible to ask new research questions. For example, how much other information about people's ability at a task is necessary before beliefs about status are canceled out and do not operate to organize the interaction?

With this as background, we can examine the conceptualization of the problem of the effects of status conceptions on interaction and that of the *theory of status characteristics and expectation states* as a solution to the problem. Before turning to the theory, however, one warning is necessary. There are some costs involved in achieving the objectives of the theory. One of the principal costs is that the formulation, of necessity, is abstract, and the explicit ideas of the theory must be stated in a way that is hard to relate to everyday experience. We argue that to develop a scientific theory requires a degree of formalism. This formalism is very awkward to people not accustomed to dealing with abstract theories. In fact, it puts off a great many sociologists. But to achieve generality, precision, and rigor, means that we have to sacrifice the smooth style and the apparently easy flow of the argument to which people are accustomed. It would be delightful if we could achieve our objectives and at the same time present the theory in graceful prose, but these two goals are inherently in opposition. One of the reasons that prose can flow smoothly in more rhetorical writing is that such writing is rarely concerned with precise ideas. The implication in this discussion is that one cannot read a scientific theory like a novel. It is necessary to go slowly and to think carefully about each of the ideas presented.

CONCEPTUALIZATION OF THE PROBLEM

The first step in developing the *theory of status characteristics and expectation states* was to conceptualize the problem, going beyond Park's idea that status conceptions organize interaction. The theorists first attempted to formulate a general proposition (knowledge claim) to summarize the diverse findings of previous research. They noted in previous studies on status categories:

Status categories always appear to imply different evaluations of individuals; and always provide the basis for inferring differences in an individual's capacities.

Previous research seemed to involve one or the other of two kinds of status assumptions. One type assumed that status derived from ability to perform the task immediately at hand. Thus, if a task in a study involved solving a mathematical puzzle, and if one of the actors had high mathematical ability and the second actor had low mathematical ability, it was assumed that high mathematical ability (since it increased the likelihood of success in solving the mathematical puzzle) conferred on its holder high status in that interaction. The second type of assumption was that individuals with generally useful abilities were accorded high status in a variety of situations. Thus, whatever kind of problem-solving task was involved in the interaction, a more intelligent person was accorded higher status than a less intelligent person.

From this reasoning, the theorists were led to formulate the idea of a status characteristic and also the idea of a diffuse status characteristic—where a characteristic is any property of a person that has two or more distinct values or states. Mathematical ability is a characteristic; so is hair color, intelligence, gender, race, and so on. For a characteristic to be a status characteristic, all that is required is that the states of the characteristic be differently evaluated. Thus, if it is better to be a blonde than a brunette, hair color is a status characteristic. A diffuse status characteristic is simply a status characteristic from which one infers other general attributes of individuals. If blondes not only have more fun than brunettes, but are thought to be smarter, better leaders, or better at mathematics, then hair color would be a diffuse status characteristic.

The ideas of status characteristic and diffuse status characteristic are important steps in spelling out Park's idea of status conceptions. The conceptualization claims that people have visible properties which trigger off either specific beliefs or generalized inferences in the interaction situation. If the actors in a situation are aware that one of them is good at math, and the task at hand requires mathematical skill for its successful completion, actors will infer that the person with mathematical ability is better at the immediate task and will accord that person high status.

The next step in the development of the theory tackled the question of what happens to the interaction as a result of according high status to the person who has mathematical ability. Answering that question necessitated a conceptualization of the observable interaction in these task situations. The theorists noted that the interaction in these studies had several distinct and recurrent elements:

1. The individual with mathematical ability was often asked for his or her opinion.
2. When asked, the individual either did or did not give an opinion or a suggestion, or some information.
3. Once the individual expressed an opinion, others evaluated it either positively or negatively.
4. Finally, in attempting to decide how to solve the mathematical problem that was the task, sometimes one individual was influenced by another; that is, sometimes one individual changed his or her mind after differing in opinion with another.

Drawing on previous work, particularly studies of small group interaction, the theorists recognized that the four elements they had distinguished were highly interrelated. They decided to formulate the four elements in terms of abstract concepts, which were:

1. *Giving action opportunities to others*, as when one individual asks another for an opinion
2. *Performance outputs*, as when an individual contributes an opinion or a suggestion
3. *Reward activity*, as when others evaluate a performance output either positively or negatively
4. *Influence*, as when one individual succeeds in changing the opinion of another.

It is important to note that these concepts were only denotatively defined. It was possible to give examples of each of these elements of interaction, which turned out to be sufficient definition for the purposes of the theory. Based on the observation that these elements were highly related to one another, the theorists assumed that all four elements reflected different behavioral consequences of one underlying structure and that they were observable manifestations of that structure. Taken together, these four elements of interaction were called *the observable power and prestige order of the group.*

With this conceptualization of interaction, the theorists were able to expand on the "other end" of Park's idea, namely, the idea that status conceptions organized the observable power and prestige order.

Before they could begin to state the theory, one other aspect of previous research had to be conceptualized. We have noted that the body of studies presented different tasks, different interaction conditions, different settings, and so forth. In all previous studies, however, one feature was clearly present. In all cases, actors in interaction shared a common task. Usually, a group had to make a decision. Its members believed that there was a right or a good decision, and their objective was to make the right or good decision, in which case the group was successful. If the group made no decision or made a bad decision, then they had failed at the task and failure was negatively evaluated.

Another feature of these task interactions which the theorists noted was that the tasks typically involved interdependence among the actors. In other words, in order to come up with a solution to the task problem, actors believed that it was both necessary and legitimate to base an opinion on another person's if the actor believed that the other person's opinion was right.

These features of the task were formulated as three *task conditions:*

1. If a task has a right or a good answer that is defined as a success, and a wrong or a bad answer that is defined as a failure, it is a *valued task.*
2. If the task is one in which it is both necessary and legitimate to use whoever's opinion one believes is right, the task is a *collective task.*
3. A group which wants to succeed at a valued task and employs the collective opinions of the group members is a *task-focused* or *task-oriented* group.

We are now able to add a third element to Park's idea: *in task oriented groups,* status conceptions organize interaction. We are now ready to bring together the three aspects of the conceptualization to formulate a much stronger and more precise notion. Berger et al. (1972) believe that this knowledge claim accurately summarizes the main common feature of a large number of studies:

When a task-oriented group is differentiated with respect to some external status characteristic, this status difference determines the observable power and prestige order within the group whether or not the external status characteristic is related to the group task (p. 243).

predetermined variables, 150
  recursive, 150
  residual variable, 150
causal modeling, 136
causal network, 150
causation, 42, 182
  nonexperimental research, 182
central tendency
  mean, 108
  median, 107
  mode, 107
central-limit theorem, 126
centroid, 144
Chen Approaches to Unidimensionalized Scaling (CAUS), 163
Chi-square test, 131
Chronbach's α, 54
classification, 48
  logical requirements, 48
cluster analysis, 153
  Euclidean distance, 154
  hierarchical clustering, 154
codebook, 91, 92
coding, 47, 90
  dummy variables, 93
  missing values, 93
  postcoding, 90
  precoding, 90
  valid values, 93
coefficient of alienation, 143
coefficient of determination, 143
community study, 81
component/factor-based subscales, 160
composite measure, 153, 157
composite variable, 52
computer applications, 104

computer, 191
computer-aided statistics, 129
  data analysis, 190
  using SPSS, 104
computer editor, 103
computer files, 101
computerization, 91
  of qualitative data, 94
concept, 37
conceptual framework, 36
conceptualization, 36, 40, 45
concordant pair, 36, 40, 45
conditional analysis, 116
conditional frequency table, 138
conditional variable, 112
confirmatory factor analysis (CFA), 138, 159
confounding, 158, 159
content analysis, 34, 55
contingency cleaning, 74, 81
contingency questions, 96
contingency table, 97
  column variable, 112
  marginal distribution, 113
  row variable, 112
  symmetrical relationship, 113
control, 144
  experimental, 70
  statistical, 54
control group, 70
control variables, 70
correlation matrix, 137
covariance, 142
data, 143
  assessment, 57
data analysis
  bivariate, 19

In this context, *external* simply means "brought to the interaction from outside." What this formulation argues is that actors who come to interact on a task and who are different with respect to some status characteristic (or some diffuse status characteristic) will create a status hierarchy based on their beliefs about one another, and they will interact in accordance with this status hierarchy. Thus, if a doctor, a nurse, and a patient are interacting about treatment for the patient, the three actors will believe that the doctor has the most ability to solve the problem and that the patient has the least ability. Given these status beliefs, the nurse and the patient will give the doctor the most action opportunities: the doctor will talk the most, the doctor's suggestions will be most positively evaluated, and the doctor will be most influential. But the assertion says something more. It says that even when the status order is based on beliefs that are unrelated to the task at hand, the same interaction patterns will occur—as, for example, if the doctor, nurse, or patient were discussing a television program that they were all watching.

While the knowledge claim summarizes the most important aspect of a body of studies, it still does not answer the questions that we posed about Park's insight: How and under what conditions do status conceptions organize interaction, and what is it about status conceptions that operates to organize interaction? A theory that explains the knowledge claim also provides answers to these questions. The theory contains scope conditions, posits a mechanism or a process by which status conceptions affect interaction, and singles out beliefs that are attached to status conceptions as the feature which "triggers" the mechanism. Hence, the theory answers the questions of how, what, and when. We turn now to the theory.

THE THEORY

The theory involves the following primitive terms which are not explicitly defined but for which meaning is assumed to be widely shared:

Actor	Differential evaluation
Actor as object	Task
Other	Task outcome
Characteristic	

Before we discuss these primitive terms further, it is convenient to introduce some nominal definitions:

$$p = \text{actor}$$
$$p' = \text{actor as object}$$
$$o = \text{other}$$
$$C = \text{characteristic}$$
$$T = \text{task}$$

In addition, we will use subscripts on these letters to stand for states of characteristics or outcomes of the task, or for evaluations attached to task outcomes or states of characteristics. There is nothing mysterious about these symbols; their purpose is simply to serve as a shorthand so that we do not have to write out cumbersome phrases. Hence:

C_a refers to the *a* state of some characteristic. C_b refers to the *b* state of that same characteristic.

For the sake of simplicity, we shall consider characteristics with only two states. (The theory can deal with characteristics that have many states, but that would needlessly complicate our exposition.) Suppose, for example, the characteristic *C* is wealth. Then the *a* state could be ''poor'' and the *b* state could be ''rich.'' If it is better to be rich than poor, we might write C_{a-} and C_{b+} indicating that the *a* state is negatively evaluated and the *b* state is positively evaluated. Similarly, we will deal with tasks that have only two outcomes, T_a and T_b. Suppose *T* is a mathematical puzzle. Then T_a could represent successful solution of the puzzle and T_b could represent failure to solve the puzzle. If it is better to solve the puzzle than not to solve it, we would write T_{a+} and T_{b-}. Sometimes, when it is absolutely clear which outcome we are referring to, we will drop the *a* and *b* and just attach the symbols + and −. With these nominal definitions, we can formulate the meaning of a valued task.

A valued task is any task for which there are two (or more) outcomes and for which there are differential evaluations of the outcomes.

We can now introduce an explicit definition of a valued task.

Valued task is defined as: *T* such that T_{a+} and T_{b-} exist.

Thus, if an actor is estimating the number of dots on a card and believes that it is better to get the right answer than to get the wrong answer, the estimation task is our *T*; getting the right answer is T_{a+}; and getting the wrong answer is T_{b-}.

We have also introduced p' and o. These ideas require some comment. The theory is formulated from the point of view of an actor who is oriented to social objects, himself (p') and/or others (o). Choosing the perspective of p represents a very significant feature of the theory. It is continuous with a long tradition in sociology which emphasizes the importance of the way an individual perceives and defines a social situation. The theory says, for example, that p attributes to p' and o the states of one or more *C*s. Thus, for example, if p believes that he is poor, and he is in a situation with another who is rich, p attributes to p' the state C_{a-}, and to o the state C_{b+}. If *C* were ''hair color,'' C_a could be blonde and C_b all other hair colors; if p believes that blondes have more fun, we would write C_{a+} and C_{b-}.

The theory is concerned with particular types of characteristics, that is, those which p evaluates. In other words, the major concern is with those Cs for which $+$ or $-$ are attached to a and b. It may be that C is not evaluated by p when p enters the situation. For example, p may not evaluate a C (like ability to judge distance) so that neither C_a (high ability) nor C_b (low ability) has $+$ or $-$ attached to it. The theory asserts that such Cs come to be evaluated, and it formulates processes by which p attaches $+$ or $-$ to C_a or C_b.

One way in which a C comes to be evaluated is through p developing beliefs about the relationship of C_a and C_b to T_{a+} and T_{b-}. If p assumes that possessing the state C_a, of some C increases the individual's likelihood of success at a valued task, then the state C_a will come to be positively evaluated. Suppose T is a crossword puzzle, and T_{a+} represents solution of the puzzle while T_{b-} represents failure to solve the puzzle. (Note that the example assumes it is a valued task, because p attaches evaluations to the outcomes.) Further, suppose that C is verbal ability, with C_a representing high verbal ability and C_b low verbal ability. If p believes that C_a leads to T_a and that C_b leads to T_b, we then say that C is instrumental to T from the point of view that p holds; the theory says that if p did not initially evaluate C but views C as instrumental to T, then p will assign evaluations to the states of C_a according to which state leads to the valued outcome. If p believes that high verbal ability leads to success at solving a crossword puzzle and p values success at this task, then even if p did not evaluate verbal ability beforehand, he will come to evaluate high verbal ability positively.

The actor, p, holds expectations for p' and o, beliefs about how they will behave in specific situations. These beliefs correspond to states of C. In our mathematical puzzle example, if p attributes C_{a+} to p' and C_{b-} to o, p expects p' to perform very well in a task situation calling for mathematical ability and expects o to perform less well in that task. The theory distinguishes specific expectations and general expectations. These two ideas, however, are essentially denotatively defined. Specific expectations are associated with definite and particular situations or characteristics, such as mathematical ability, physical strength, reading ability, manual dexterity, ability to play tennis, and so forth. General expectations are not attached to particular situations. To believe that someone is "smart," that someone is a "gentleman," or that someone is "moral" constitutes a general expectation. The reason that these are only denotatively defined is that it is difficult to draw a hard and fast line between what is specific and what is general. Another way to put this is to ask, How general is *general?* In school situations, for instance, beliefs about reading ability are so pervasive (that is, they enter into so many aspects of the school day) that one might want to consider beliefs about reading ability to be general expectations. It is possible, however, to envision a school curriculum where not all teaching and learning depended upon reading—in which case, beliefs about reading would be specific expectations.

When we are speaking of the high state of an ability, we are also speaking of a specific expectation for success where C is instrumental to T. Thus, for specific ability characteristics, an ability state and an expectation for performance will be used synonymously. The belief that somebody has the high state of physical strength is equivalent to believing that that person will perform well in a task in which physical strength is involved. But it is necessary to be careful, since not all characteristics are ability characteristics and the term C will be used both for characteristics like physical strength and characteristics like hair color. We will only speak of C_a as an *expectation* when we are considering ability or performance characteristics like physical strength.

One of the central concerns of the theory is to connect general beliefs or expectations to specific expectations for a given task. Are there circumstances under which the belief of p that o is a gentleman will lead to the belief that o has higher mathematical ability than p'? Although at first sight this example may seem farfetched, on reflection it may not be so bizarre. One can imagine our actor p reasoning that since o is a gentleman and he himself is not, o is likely to be more educated, hence likely to have studied more mathematics, hence be more likely to have the ability to do mathematical problems. If p believes that being a gentleman implies having mathematical ability, then we speak of one characteristic being relevant to another. If one characteristic is relevant to another, knowing the states of the first characteristic, p expects p' and o to possess specific states of the second characteristic. The idea of relevance is a relational idea connecting two characteristics, an idea which will be more rigorously defined shortly.

Often a given state of one characteristic may be relevant to particular states of a number of other characteristics. Suppose we have ''sex'' as our C, with C_a as male. Then the state C_a may be relevant to the high state of ''mathematical ability,'' the high state of ''physical strength,'' the low state of ''emotional sensitivity,'' and the low state of ''artistic ability,'' and so forth. The last four states constitute a set or a collection which we will call the gamma set, and we will use the Greek letter γ to stand for this so that we can talk about γ_a and γ_b. In other words, we will nominally define γ to be the collection of specific beliefs—in our example, the collection to which the state ''male'' is relevant. Formally, this looks like:

$$\gamma = C_{1a}, C_{2a}, C_{3b}, C_{4b} \ldots$$

In our example, the first characteristic C is sex; the a state is male; C_{1a} is the high state of mathematical ability; C_{2a} is the high state of physical strength; C_{3b} is the low state of emotional sensitivity; and C_{4b} is the low state of artistic ability.[1] The idea that this captures is that with a state such as male, there are a collection of specific beliefs or expectations which actors may associate with males, and each γ set, γ_a and γ_b, represents a collection of beliefs.

---

1. It is important to emphasize that these are beliefs that p may hold and which the theory uses. These beliefs, even though widely shared, are not necessarily true; in fact, such generalized beliefs are often false. Evan false beliefs, however, affect behavior.

Let us introduce two more nominal definitions:

GES = general expectation state
D = diffuse status characteristic

We need one more symbol. Sometimes we do not want to refer to a specific state of *C* or of *GES* but want to make statements that apply to one state, either *a* or *b*. In those cases, we use the letter *x* as an index or a dummy, where the letter *x* stands for either *a* or *b* without specifying which. Initially, this notion may seem difficult to deal with and cumbersome, but after one gets used to it, it turns out to be quite straightforward and convenient.

We are now ready to define the key idea of the theory, the idea of diffuse status characteristic.

Definition 1 (Diffuse Status Characteristic)
A characteristic, *C* is a diffuse status characteristic *D* if, and only if:
1. The states of *D* are differentially evaluated,
2. To each state, *x*, of *D* there corresponds a distinct set, γ, of states of specific evaluated characteristics associated with D_x.
3. To each state, *x*, of *D* there corresponds a distinct general expectation state, GES_x, having the same evaluation as the state D_x.

Several points must be made about this definition. First of all, the definition only applies from the point of view of our actor, *p*. *P* must differentially evaluate the states of *D* and have specific beliefs associated with each state of *D*, and there must be a general expectation state attached to each state of *D*. For example, if sex is a diffuse status characteristic for a given actor, if means that the actor regards male as better than female (or female as better than male), believes that males are better at a list of specific tasks and females are better at other tasks, and believes that, in general males are smarter (or females are smarter). As our example suggests, it is possible for sex to be a *D* for two actors, even though one actor regards male as the high state and the other actor regards female as the high state.

The second point to note is that a given characteristic—for example, sex—may be a *D* for some actors and not others. For this reason, the theorists have chosen to formulate the idea abstractly rather than to develop a theory about beliefs associated with sex, race, or ethnicity. The theory does not say that any of these characteristics will be a *D* for all actors in all situations. Indeed, the theory takes no stand on whether a given *C* is or is not a *D* in any concrete situation. That remains to be determined. What the theory talks about is the consequences that follow *if* a *C* is a *D*, if sex is a diffuse status characteristic in a particular situation for a particular set of actors.

One interpretation of attempts to democratize the military is that such attempts are designed to eliminate military rank as a diffuse status characteristic. If such attempts are successful, then most military personnel should not regard being an officer as better than being an enlisted man, should not believe that officers are better leaders, that officers are more moral, or are "gentlemen." In short, success would mean that the connection between each state (officer and enlisted man) and the sets of beliefs and evaluations would be broken.

This definition begins to answer some of the questions about mechanisms by which status conceptions organize interaction. It abstracts evaluations and belief systems as the crucial aspects of Park's idea of status conceptions. Status conceptions operate in interaction because they represent evaluations of actors and expectations about how actors will behave in specific situations or across a variety of situations. In other words, the beliefs associated with states of D provide an actor with information about himself or herself and the other actors in the situation.

If we think about this information for a moment, we realize that an actor has other sources of information besides the beliefs attached to states of D. If you know somebody well, it is likely that your personal knowledge will outweigh the information provided by your beliefs about a state of D. If the president of a university is a friend, then you will react to that person as a friend rather than as president of the university. This suggests that there are situations in which Ds operate and situations in which Ds do not operate, situations to which the theory applies and situations to which it does not apply. We have noted that defining its applicability requires explicitly stating scope conditions for a theory. The theorists assert the following scope conditions:

1. There is a valued task.
2. There is a C instrumental to T.
3. Actors are task-focused.
4. Actors are collectively oriented.
5. D is the only social basis of discrimination between actors.

The reader will recall that we have earlier defined *valued task, collective task,* and *task-focused*. Condition 5 is concerned with ruling out other information that can be used to form expectations, as in the example where the president was also a friend. (This strong condition was relaxed in later versions of the theory.) Let us refer to this set of scope conditions with the symbol S^*, which stands for a task situation that has the five properties listed above.

Even though a characteristic may be a D for p, it may not be significant to him in a particular social situation, even if that situation meets the scope conditions of the theory. The actor may not recognize that there is a difference between himself and other actors with respect to states of D; or, if p notices it, p might put the idea out of his mind. We refer to noticing the difference between p' and o as *discriminating between actors according to the states of D they possess*. If, in addition to noticing the difference, the difference becomes significant to p, we say that D has been *activated*. We want to emphasize the difference between *discrimination* and *activation*. The first involves recognizing that actors in the situation possess different states of D, while the second involves p paying attention to the beliefs associated with these different states. We should point out that activation is an unobservable state that the theorists posit whereas discrimination refers to observable actions. It is possible that p regards air force rank as a D, recognizes that he is an officer and that he is interacting with an enlisted man, and he says, "Well, I am an officer and he is an enlisted man, and I should be superior to him in leading men in a military exercise." P has then attributed the beliefs attached to states of D to himself and o, and it is that attribution of belief that we term activation. On the other hand, if p says, "Well, he is an enlisted man, but he is very special and not like other enlisted men," then we would argue that D is not activated. These ideas are captured in definition 2 and assumption 1 of the theory (noted below). Unless the distinction between discrimination and activation is understood, however, assumption 1 of the theory will appear trivial, though in fact it is an extremely strong assumption. It does a great deal of work in the theory and is quite likely to be an overstatement.

Definition 2
D is activated in S^* if, and only if, p attributes in S^* the states of GES_x and/or the set of states γ_x to p' and o which are consistent with the states of D that they possess in S^*.

(By *consistent* we mean that the states have the same evaluation. Thus, the positively evaluated GES is consistent with the positively evaluated state of D.) In saying that behavior in a specific situation is based on status conceptions, we mean that p in that situation attributes to p' and o the same evaluations and expectations attached to the status categories of which they are members. If p is an enlisted man and believes that officers know more than enlisted men about leadership, then we say that, for p, rank is activated if he believes that officers know more simply because they are officers. We are now ready to present the heart of the theory— that is, the knowledge claims that represent the assumptions of the theory. We will need two more definitions, and will present those when they occur in the theory.

Assumption 1 (Activation)
Given S^*, if D in S^* is a social basis of discrimination between p' and o, then D is activated in S^*.

Assumption 1 baldly asserts that if the scope conditions are met, whenever p believes that D is a status characteristic and notes that p' and o possess different states, he will automatically attribute one or more of the beliefs associated with states of D to p' and to o. That this is a very strong assumption will now become more clear. Assumption 1 says that in S^*, whatever else is going on, it is enough for p to recognize different states of D in the situation to mobilize a whole set of attitudes and beliefs. For example, if a male and female are interacting in S^*, that alone is sufficient to generate all kinds of sex stereotypes on the part of both actors.

The reason the assumption is characterized as a very strong one is that it is easily false. It may well be that p and o recognize differences and simply shrug their shoulders; or, other features of the situation not covered by the theory could easily interfere with the attribution of stereotypes to the actors.

As Berger et al. (1972) comment:

> There have been almost no studies of activation. We know almost nothing about the conditions under which a status characteristic is activated. There are certainly situations in which some status characteristics that might be used to organize interaction are not: Why are they not? Because we do not know, the present formulation gives only one set of sufficient but not exhaustive or necessary conditions.[2] There may be more general activation conditions than those we assume (p. 244).

Assumption 1 does provide interesting questions for extending the theory. To give just one example, what happens in interaction when p and o are status equals, that is, when both possess the same state of D? Are beliefs attributed to p' and o? Note that assumption 1 takes no stand on the question, since assumption 1 says that if there is discrimination on D there is activation, but takes no stand if there is no discrimination on D. This is why we note that discrimination is a sufficient condition, not a necessary condition.

If D is activated in S^*, two situations may exist. First the performance characteristic that is instrumental to the task in S^* may have strong prior connection with D. For example, a male and female may be playing volleyball, and both may assume the male is a better volleyball player because that is the conventional view. Or a male and female may be involved in a cooking task and may assume the conventional view that the female is a better cook. In other words, the beliefs about the task may already be contained in the sets attached to the states "male" and "female." In this case, activating D is sufficient to determine the expectations of p about how p' and o will perform the task.

Second, the performance characteristic instrumental to the task may have *no* prior association with D. For example, a student and professor may be involved in solving a mechanical puzzle, such as taking apart a complicated set of metal keys. For this task, it is unlikely that there are conventional beliefs contained in the γ sets attached to the states professor and student.

The second case is more interesting and less clear. If the C which is instrumental in S^* has no prior connection to D, how will status differences imply anything for behavior in this situation? But the theory intends to describe what happens when a wholly new ability is required to succeed at the task—a C whose states are not part of γ_x or in no way previously related to the states in γ_x. To cover situations of wholly new abilities, the idea of relevance becomes crucial.

---

2. The distinction between necessary and sufficient conditions is easy to remember with the aid of a rough translation. A sufficient condition can be translated as: if A, then B; while a necessary condition is translated: only if A, then B. In other words, *for B to occur, it is enough that A occurs* represents a sufficient condition, while *A must occur for B to occur* represents a necessary condition. Finally, the translation of something that is both a necessary and sufficient condition is: if and only if A, then B.

Definition 3 (Relevance)
An element e_i is relevant to an element e_j if p' (or o) possesses e_i, then p expects p' (or o) to possess e_j.

The relevance relation is defined in the most abstract terms to allow many different substitutions for e_i and e_j. Thus, an element can be a state of a diffuse status characteristic, a state of a performance characteristic, a state of a task outcome, or a state of a *GES*. Any two of these may be in the relevance relation. If, for example, p believes that an o who possesses a high state of mathematical ability will solve a mathematical puzzle, we say that a state of the performance characteristic is relevant to a state of the task outcome. This definition attempts to capture the idea that one set of beliefs leads to another set of beliefs, from point of view of p. Thus, if p believes that blacks are likely to be poor, we would say that the state "black" is relevant to the state "poor."

Our last example brings out two features of the idea of relevance which should be carefully noted. In the first place, relevance is usually a relation between categories or classes—in this example, the class of individuals possessing the state "black" and the class of individuals expected to possess the state "poor." The relevance relation refers to generalized beliefs which may or may not become attached to a particular p' or o. The process of attaching expectations for a class to a specific individual is what the theory calls *assignment*.

The second feature of the relevance relation is that, in general, it is asymmetric. While in some cases the relevance relation might go both ways, those circumstances are special. P may believe that people who are black are likely to be poor, but he will not usually believe that people who are poor are likely to be black. For this reason it is necessary to emphasize the directionality, or asymmetry, of the relevance relation.

The idea of relevance has two "opposites." One idea is "the absence of relevance" while the second is "not relevant." If relevance is absent between D and C, then p does not know whether or not he can make an inference about C on the basis of D. For example, if p does not know whether or not he can infer that a professor has more musical talent that a student, then relevance is absent. On the other hand, p may explicitly believe that two characteristics are *not* relevant to each other. P might believe that musical ability and academic status are independent, and believe that neither he nor anyone else is able to infer from the fact that p' is a professor and o a student, whether p' has more musical ability than o or that o has more ability than p'. There is then a conventional belief that the two characteristics are irrelevant. When such beliefs exist, we use the term *dissociated*.

The idea of relevance plays a crucial role in determining how *GES* affects beliefs about abilities that may be required to accomplish the task in S^*. It deals with those situations in which prior beliefs about the irrelevance of D to C do not exist; that is, where D is not dissociated from C. Assumption 2 does not deal with the case where conventional beliefs assert that being an officer says nothing about ability to play basketball.

Assumption 2 (Burden of Proof)
If D is activated in S^* and has not been previously dissociated from D, then at least one consistent component of D will become relevant to C in S^*.

From the definition of D, its components will consist of evaluated states of D, the γ sets, and an evaluated *GES*. If D is activated in S^*, then p possesses or has attributed to him one set of these components, say D_x, γ_x, and GES_x; o possesses or has attributed to him a second set of these evaluations. If S^* is an air force crew and p recognizes that he is an enlisted man and o is an officer (D); and believes that officers are better leaders and have greater military knowledge (elements of γ_x); and believes that officers are gentlemen (GES_x); and if the crew task is to solve a puzzle; then p will use one of these components to form a belief that either p' (himself) or o has more ability to solve the puzzle.

The above example illustrates what the theorists call the *expansive property* of D; beliefs about D expand to generate beliefs about abilities relevant to the task at hand. Furthermore, this expansive property operates if nothing bars p from seeing D as relevant to C. In this situation, the actors act as if the burden of proof is on showing that D is not relevant. In other words, it becomes relevant unless p holds prior beliefs that D is not a basis for making inferences about C. For this reason, assumption 2 is called the *burden of proof* assumption. The burden is to stop relevance from occurring in such a situation; and relevance does not occur if other beliefs or other information interfere with the expansion of D—as would be the case if p were dealing with a friend who was an officer rather than simply with an officer.

As the authors of this theory note: "This general assumption permits a number of different mechanisms by which D becomes relevant to C: *GES*, the evaluations attached to the states of D, or even the states in γ might become the basis for forming expectations in S^*" (Berger et al., 1972, p. 246).

From the definition of relevance above, when components of D are relevant in S^*, it means that p expects p' and o to possess given states of C. If air force rank becomes relevant to the ability to estimate dots, then p expects the officer to be good at that task, and himself to be not so good. P has assigned states of C (ability to estimate dots) to p' and o and has formed performance expectations for p' and o. As we indicated above, these are beliefs about the quality of performance outputs. Once a component of D is relevant to C, the theory claims that states of C will be attributed to p' and o that are consistent with that component of D. P makes some assumptions about p' and o on the basis of D, and from these he draws further inference about how p' and o are likely to perform in this particular situation. The theory further makes the reasonable assumption that p attributes performance expectations that are consistent with components of D. For example, if p highly evaluates officers and values a task, then p will believe that the officer will have high ability of the task and that he himself will have low ability. These ideas are formulated explicitly in assumption 3.

integral approach, 105, 111, 136
  multivariate, 158
  univariate, 105, 111, 136
data collection, 105, 136
  direct, 73
  indirect, 74
  information recording, 73
data entry, 99
data management
  cleaning, 93
  computer programs and programming, 95
  creation of new variables, 102
  in stream data, 100
  labeling, 102
  logical control, 98
  manipulation, 97
data reduction, 99
data theory, 47, 153, 154
database, 56
  case ID, 91
  definition of variable, 92
  variable name, 92
definition
  nominal, 92
  operational, 47
  theoretical, 47
  working, 46
degree of freedom (DF), 47
dense sample, 130
dependent variable, 61
descriptive analysis, 106
  bivariate, 106, 134
  cross-tabulation, 97
  frequency, 107
  percentage distribution, 106
  Univariate, 111

dichotomous variable, 93, 106
differential weighting, 54
dimension, 66
dimensional analysis, 56
  factor, 56
  latent variable, 56
  principal component, 56
dimensionality, 56, 160
  multidimensionality, 56, 153
  unidimensionality, 56, 127
discordant pair, 57
discovery, 116
discriminant analysis, 43
  discriminant coefficients, 144, 149, 153
  discriminant score, 144
  discriminant variables, 144
  discriminating power, 144
dispersion
  interquartile range, 144
  mean deviation (MD), 109
  quartile, 110
  range, 109
  standard deviation (SD), 109
  variance, 109
  variation ratio, 109, 119
dissertation, 242
distortion analysis, 109
distribution parameters
  central tendency, 31, 138
  dispersion, 107-109
documentary analysis, 107
dummy variables, 74, 81
ecological fallacy, 59, 123, 142
effect size, 167, 185
elaboration, 64, 135
engineering mechanics, 137

Assumption 3 (Consistent Assignment)
If any components of an activated *D* are relevant to *C*, *p* will assign states of *C* to *p'* and *o* in a consistent manner.

What assumption 3 asserts, for example, is: if *p* is a female and sex is a diffuse status characteristic for *p*; and if *p* is working with a male on a task involving the solution of a puzzle; and further, if *p* believes that men are smarter than women, and believes that being smart is relevant to puzzle-solving ability; then *p* will believe that the male she is working with is better at solving the puzzle than she is.

The reader who encounters this theory for the first time may have some difficulty in distinguishing assumption 2 from assumption 3. A simple rule of thumb will help avoid confusion. Assumption 2 deals with categories of people or classes of people. Thus, if air force rank is relevant to ability to write well, it means that there is a belief that officers as a class are better writers than enlisted men as a class. Assumption 3 deals with taking the generalized belief and attaching it to the particular actors in the situation. If *p* believes that officers are better writers, and if he is interacting with Officer Jones, then he will attribute higher writing ability to Officer Jones. When that happens, the states of writing ability (high to Jones and low to self) have been assigned in a consistent manner.

Thus far, the assumptions of the theory have talked about belief systems, how they become related to one another, and how they get attached to particular actors in a situation. The final assumption of the theory links beliefs to behavior in the situation. As we indicated earlier, the aspect of behavior which the theorists have conceptualized, and which is the concern of the theory, is *the observable power and prestige order*. The reader will remember that this is composed of four kinds of behaviors:

1. Giving action opportunities
2. Performance outputs
3. Reward actions
4. Influence

Position *A* is higher than position *B* in the observable power and prestige order if an actor occupying position *A* is: more likely to receive action opportunities; more likely to initiate performance outputs; more likely to have his performance outputs positively evaluated; and less likely to be influenced when disagreeing with another. The final assertion of the theory claims that the actor's positions in the power and prestige order will be determined by the actor's expectations for self and other, that is, the states of *C* which *p* has attributed to *p'* and *o*.

In our puzzle example, our female *p* will ask for advice and suggestions from *o* more than *o* will ask for advice from *p*; *o* will make more suggestions than *p*; *p* is more likely to regard the suggestions of *o* as good ideas (than vice versa); and *p* is more likely than *o* to change her mind about the right answer to the task.

With assumption 4 we present this formally, and with definition 4 we introduce another nominal definition.

Definition 4
The expectation advantage of *p* over *o* is the degree to which the expectations of *p* are higher for *p'* (himself) than for *o*.

Assumption 4 (Basic Expectation Assumption)
If *p* assigns states of *C* to *p'* and *o* consistent with the states of an activated *D*, then the position of *p* relative to *o* in the observable power and prestige order will be a direct function of the expectation advantage of *p* over *o*.

To summarize this process, the theory asserts that, given a task situation which meets scope conditions, if the actors recognize that they possess different states of *D*, they will think about beliefs[3] attached to these states of *D*; that is, *D* will be activated. If *D* is activated, then the beliefs about categories of people will become relevant to beliefs about task performance in the situation; *D* will become relevant to *C*. If relevance occurs, then the actors in the situation will attach beliefs about task performance to the specific actors in *S**—states of *C* will be consistently assigned. If consistent assignment occurs, then the observable power and prestige order will correspond to the expectation advantage resulting from the assignment of states of *C*.

The knowledge claims of this theory are all formulated as "if-then" statements, and they represent suffcient, but not necessary, conditions. In other words, the theory does not claim this is the only way that actors decide who is a more competent puzzle solver in the group. But it does say that if no other information is available about puzzle-solving ability among group members, the diffuse status characteristic will expand to generate such beliefs.

From these assumptions and definitions, several consequences follow. Here we present some of the principal derivations without formally deriving them. (In the next chapter, we will show how to formally derive consequences from this theory.) Given *S** and a *D*, not previously dissociated from *C*:

1. If *D* is the only social basis of discrimination, then the position of *p* relative to *o* in the observable power and prestige order will be a direct function of the expectation advantage of *p* over *o*.
2. If *D* is the only social basis of discrimination and *D* has been activated, then the position of *p* relative to *o* in the observable power and prestige order will be a direct function of the expectation advantage of *p* over *o*.
3. If general expectation states are attributed to *p'* and *o* and are relevant to *C*, then the position of *p* relative to *o* in the observable power and prestige order will be a direct function of the expectation advantage of *p* over *o*.

---

3. Although it is convenient to describe what occurs with phrases like "think about beliefs," the theory does not assume that the process is conscious.

4. If *p* assigns states of *C* to *p'* and *o* that are consistent with the states of *GES*, then the position of *p* in the observable power and prestige order will be a direct function of the expectation advantage of *p* over *o*.

The consequences above represent various ways in which a status situation may be defined; they also vary in how well-defined the status situation is. Number 1 above is a situation that is only minimally defined. Initially, all the actors know is that they possess different states of the *D*, but *D* and *C* are unrelated. Initially, there is no belief linking the status difference to the immediate task. Consequence number 3 deals with one component of *D*, namely the *GES*. But when there are general expectations among the actors that are activated and relevant, the situation is much more defined than if they recognize only that they have different states of *D*. Incidentally, it is possible to substitute other components of *D* and formulate alternatives to consequence 3, for example. Thus, if γ states were attributed to the actors, we could formulate a consequence parallel to consequence 3. The theory claims that regardless of how status differences are defined and regardless of how well-defined the status situation is, the beliefs attached to these status differences will operate in the same way to determine central features of the interaction.

This theory has been used to study work groups in an organizational setting, students from different colleges interacting with one another, biracial groups of adolescents, and male and female interaction. The theory has also been used to explain findings from other research which was not conducted with the theory in mind. Thus, the theory found that professionals were more likely to be chosen jury foremen than were skilled workers and that men were more likely to be named foremen than women. The theory also helped in the design of the Center for Interracial Cooperation (E. G. Cohen, 1976), a summer school program designed to foster more harmonious interracial relations; this represented a self-conscious application of the theory of status characteristics and expectation states. Since the focus of this chapter is on the analysis of a theory, we will not describe in detail the empirical tests of the theory or the applications to which the theory has been put.[4] We will, however, draw examples from this body of research in applying criteria to evaluate the theory. We will now bring to bear the criteria presented in Chapter 10 to analyze and evaluate the theory of status characteristics and expectation states.

ANALYSIS AND EVALUATION OF THE THEORY

The theory contains all the theory elements that were described in Chapter 10: there are *primitive terms, defined terms, assumptions,* and *derivations.* In addition, the definition of situation *S** contains a set of *scope statements.* It should be recognized that these elements have changed and will change; the theory is in

---

4. Berger et al. (1980) review a body of research bearing on this theory.

no sense a completed product. Nor is it without serious problems. Theories are never static; they continue to develop. As old problems with the theory are solved, new ones are recognized, and efforts are made to solve these new problems. Although the theory of status characteristics and expectation states is a reasonably well-developed theory and one that has demonstrated both scientific and practical utility, it still contains unresolved problems to challenge those who work with it.

The description which we have drawn from in this chapter is the fourth published version of the theory (Berger et al., 1972). Its evolutionary character is well illustrated by the fact that earlier versions treated ideas such as *activation* and *relevance* as primitive terms. When it became apparent that these ideas were not well understood by other investigators who read the theory, *activation* and *relevance* were both introduced as explicitly defined terms. Future versions of the theory may define additional terms whose usage as primitives creates problems. In the present version, primitive terms include *actor, social objects, p'* and *o, characteristic, states of a characteristic, specific, general,* and *evaluation.* It is quite clear that a term like *actor* poses few problems as a primitive idea. Most people who approach the theory share a common usage for the term, and this is not just because the term refers to a person; sociologists have used the term to refer to organizations or societies as acting units. Treating societies as actors is perfectly consistent with the usage in this theory, although it does remain to be seen whether or not the theory can be effectively applied to societies as acting units.

The term *general expectation state,* however, is only denotatively defined. There are wide differences in connotations and associations to the term *general.* Again, how general is *general?* While the authors of the theory can agree on some examples of *GES,* they do not always agree on usage of the idea of general. Is "intelligence"a general characteristic or a specific characteristic? Even to the theoreticians, this is an unresolved issue.

The theory also introduces several explicitly defined terms. These are *diffuse status characteristic, activation, relevance,* and *expectation advantage.* The term *diffuse status characteristic* offers a good example of a connotative definition. It provides three properties that must be present in order to call a characteristic a *diffuse status characteristic.* Furthermore, primitive terms of the theory make up the terms of the definiens. That fact suggests that where there are problems with primitives there will also be problems with this defined term. The definition, however, provides the user with clear guidelines for deciding whether or not a given characteristic for a given *p* is to be considered a diffuse status characteristic. Suppose we wanted to know whether or not sex was a diffuse status characteristic for *p.* The definition tells us to determine whether or not *p* differentially evaluates the states "male" and "female," and whether or not *p* has two different general expectation states attached to the states "male" and "female." Sometimes asking *p* about his beliefs about males and females would allow us to decide whether or not sex is a diffuse status characteristic. At other times, investigators may be willing to assume that a given characteristic has the required properties for a set of *ps*. It is not too great a stretch of the imagination to assume that ''race'' had these properties for a collection of white *ps* in the United States in the 1980s. While such an assumption might occasionally be incorrect for an individual *p*, the current arguments about racism in America suggest that nine times out of ten the assumption would be appropriate, and that degree of error is tolerable for most sociological research today.

These examples point to considerable usefulness for this definition in guiding both researchers and consumers. The definition provides the basis for adjudicating disputes over whether or not a particular characteristic is a diffuse status characteristic. This is true despite the difficulties we have noted with the primitives in the definiens. What those difficulties mean in practice is that there will be some borderline cases that are hazy and undecidable. Clearing up the hazy areas would require a great deal of work increasing the precision of the definitions, primarily by clarifying such notions as *general expectation state*. In principle, we know how to proceed. We would make *general expectation state*, for example, a defined term based on primitive terms whose usage was more widely shared. In practice, however, accomplishing that definition may be beyond our present abilities, and the definition of *D* is still quite useful in its present form. Here we have illustrated a general feature of scientific theories. Explicit formulation enables a theory to be used to solve many problems; yet all theories contain unsolved problems for future theoretical work. Paradoxically, a perfect theory would be less satisfactory than a theory which contains scientific problems that remain to be solved.

One other explicit definition should be commented upon. Definition 4 represents a nominal definition introduced exclusively for convenience in the exposition of the theory. The term *expectation advantage* saves us from constantly repeating the cumbersome phrase, ''the degree to which the expectations of *p* are higher for *p'* than for *o*.'' As a nominal definition, it adds no new idea to the theory. The reader might examine the remaining explicit definitions in order to decide what type of definition each represents.

The assumptions of the theory have two major features. One feature concerns the form of the statements; the second concerns the content. Formally, each assumption of the theory is a statement of sufficient conditions. For activation to occur, it is enough that states of *D* be discriminated in S^*. The assumption does not claim that discrimination of states of *D* represents the only way that activation takes place. Other processes could activate the belief systems. The theory takes no stand on what these other processes may be, except to allow for them to occur. If assumption 1 said, ''only if states of *D* were discriminated in S^*,'' then it would be arguing that discrimination represents a necessary condition. Furthermore, the assumptions as statements of sufficiency allow the possibility of other mechanisms accounting for the same result; namely, that the observable power-and-prestige order is a result of some authority, such as a teacher, establishing the expectation advantage of *p* over *o*.

Differences in the form of assumptions have profound consequences for the type of theory that results. An assumption stated as both necessary and sufficient is much more restrictive than an assumption stated as a sufficient condition only. If the theorist wanted to rule out alternative processes and mechanisms and place the entire burden of the theory on, for example, discriminating states of D, then the appropriate choice would be an ''if and only if'' statement. For example, an ''if and only if'' statement would be tantamount to saying that the only way beliefs about males and females would operate would be if p clearly recognized that o was of the opposite sex. But this may be more restrictive than a theorist wants it to be.

What must be emphasized is that the form of the assumption provides an important clue to what the theory asserts. Some assertions are stronger than others and thus would be cast in a more restrictive form. Readers of sociology need to be cautioned to examine closely the *form* of statements, although the sociological audience is much more familiar with the task of examining the *content* of assertions. Still, one feature of the content of assumptions 1 through 4 deserves our attention.

Assumptions 1 through 4 formulate a process, or mechanism, by which beliefs associated with status become attached to beliefs about task performance and thereby affect task-related behavior. Roughly speaking, this theory postulates beliefs that p carries around, beliefs which are triggered by status *symbols*. These status-related beliefs have two properties: (1) they are expansive and (2) they are regulative (they organize behavior). Status-related beliefs expand to cover any areas for which p does not already have an established belief system. Thus, beliefs about what officers and enlisted men can or cannot do, for example, generalize to influence beliefs about totally new situations, such as when a team composed of an officer and an enlisted man is asked to jointly write a story. In a sense, the status-related beliefs become self-fulfilling prophecies that govern how new task situations develop. Consider the case where a prejudiced teacher puts black and white students into slots and then behaves toward them in a way which confirms each student in his or her slot. Believing that black students are less intelligent, the teacher never calls on them; thus it cannot be shown that the black students have intelligent answers. It is all very well to label this a *self-fulfilling prophecy*, but the theory does more than simply provide a name for a phenomenon. The theory asserts when self-fulfilling prophecies should occur and when they should not occur, and the theory describes *how* the situation generates a self-fulfilling prophecy. Hence, the theory provides more understanding of what is going on than any label can provide.

In short, the assumptions of the theory provide an answer to how status conceptions organize interaction, and the scope statements of the theory provide an answer to when status conceptions organize interaction. Answers to these questions are necessary both for the researcher who wants to investigate the phenomenon and for the practitioner who wants to deal with manifestations of the phenomenon.

Let us turn to some of the general criteria for evaluating the theory presented in the last chapter. The theory of status characteristics and expectation states presents explicit and relatively precise assertions. Although there is room for some disagreement, by and large the theorists and their audience can agree on what the theory claims and what it does not claim. Of course, such agreement depends upon careful and close reading of the theory. A casual reader, for example, might believe that when two actors who are good friends are discriminated by a diffuse status characteristic, the theory would claim that the two actors rank differently in the observable power and prestige order. Careful reading of the theory, however, shows that if the two actors are friends the theory does not apply: the theory requires that the diffuse status characteristic be the only social basis of discrimination; but two actors who are friends are likely to know a great deal about each other, including other social bases of discriminating between them. Hence, the explicitness and relative precision of a theory only operate when the audience closely studies the theory. After all, the theory is abstract, and one cannot expect the same degree of shared usage and understanding that occurs with more concrete statements. While it is reasonable to expect an instantaneous grasp and sharing of the statement ''This is a table,'' it is too much to expect such instantaneous shared understanding of abstract ideas; and that is the reason why the criterion is ''relatively precise'' rather than absolute. Using a theory requires creative intellect just as developing a theory does. Too often, the criterion of precision misleads consumers into expecting mechanical recipes for working with a theory. Our criterion of relative precision demands that the user have guidelines, but cautions the user against expecting recipes. The theory of status characteristics and expectation states does not provide recipes, but does provide guidelines.

The second criterion for evaluating a theory is that the theory should contain explicit definitions based on primitive terms for which usage is widely shared. We have already examined the defined terms and the primitive terms of the theory of status characteristics and expectation states. Although the earlier discussion indicated problems in the choice of primitives in the theory, by and large the theory stands up well on this criterion. Relative to other sociological theories, it more clearly separates defined terms from primitives, and it develops a structure of definitions self-consciously, employing a small number of primitive ideas. Here again, it is important to stress the evolutionary nature of a theory. Not all of the ideas in the version we have examined were well explicated when the theory was first presented. Undoubtedly, future versions of the theory will further explicate the basic concepts on which the theory is built.

With respect to the third criterion, the theory of status characteristics and expectation states provides a clear exhibition of the structure of the argument. The syntax of the theory (which will be examined closely in the next chapter) is the calculus of propositions from elementary logic. Hence, there should be high intersubjective agreement among those familiar with the calculus of propositions concerning both the logical structure of the theory and the logical truth of the derivations from the theory.

The theory of status characteristics and expectation states is nearly unique when it comes to presenting explicit guidelines for its applicability. The idea of scope conditions and the need for explicit scope statements are not generally recognized in the social sciences. If the importance of scope conditions should become well understood, such understanding will be of immense value to scientists and practitioners, because scope statements contain directions for the evaluation and use of the theory.

This chapter has not presented, except in passing, the information necessary to evaluate the theory on the fifth criterion—namely, that the theory should be testable empirically. Since this chapter focused on the presentation of the theory itself, it has not dealt with possible sets of initial conditions. The reader will recall that in order to test a theory one must derive observation statements from the theory, and that observation statements can be derived only when the theory is joined to a set of initial conditions. Although we have made reference to passing examples of possible initial conditions (for example, "In the United States in the 1980s, 'race' is an instance of a diffuse status characteristic"), we have not systematically examined sets of initial conditions that have been used in the research testing the theory. Since there are many empirical studies in the literature testing this theory (Berger et al., 1980), there is evidence that the theory is testable; but describing the various sets of initial conditions that have been used is too extensive a task for our present purpose.

The final criterion—that a theory should formulate an abstract problem—is not well understood, and it is controversial. There is a widespread belief that one can theorize about any problem, concrete or abstract. Thus, investigators try to formulate theories of crime in the streets, the development of higher education in Latin America, the voting behavior of American Catholics, the student revolution in the late 1960s, and others. The conception of scientific theory presented in the last chapter rules out such topics as objects of a scientific theory.

Each of the examples must be described by singular statements; but singular statements are not, from the present point of view, a part of scientific theory. Rather, such singular statements belong to sets of initial conditions used to tie a theory to empirical phenomena. A "theory" about the development of higher education in Latin America, for example, would contain statements like "Higher education in Latin America in the twentieth century is restricted to a small proportion of the population." But a theory containing such a statement would violate the requirement that theoretical statements be universal; hence, such a "theory" is incompatible with the formulation that we have considered.

It is possible to take another view of the meaning of a "theory" of the development of higher education in Latin America. Under this less restricted view, the theory would contain universal statements, but a set of initial conditions would tie the theory exclusively to the problem of higher education in Latin America. Moreover, this view violates the spirit of scientific theorizing. If the assumptions of the theory are universal knowledge claims, then they must have consequences beyond any particular set of historical circumstances. To insist on an exclusive set of initial conditions is to close one's eyes to other possible consequences of the theory. The aims of comprehensive theory, and the object of using a theory to lead us from the known to the unknown, would be seriously compromised by insisting on initial conditions based on particular time and space properties, properties which have historical rather than general significance.

Nothing that has been said, however, rules out using an abstract theory to deal with a particular concrete problem. Of necessity, a theory, when joined to initial conditions, should explain observation statements. There is no reason why a general theory could not explain observation statements about aspects of the development of higher education in Latin America. Indeed, it is our hope that general theory will be useful in dealing with problems such as crime in the streets or the voting behavior of American Catholics. Using the theory to deal with problems, however, is very different from making the problem the exclusive focus of the theory. In short, a scientific theory, by its nature, cannot have a concrete historical problem as an exclusive focus.

The reason for insisting on this criterion—that is, insisting on an abstract problem as the object of a theory, and also insisting on the separation of the theory proper from sets of initial conditions—is to foster the formulation of theories that are widely applicable and not restricted to particular historical circumstances. This criterion also directs the scientist to look for abstract common features among widely diverse concrete phenomena.

The theory of status characteristics and expectation states formulates an abstract problem. It is not a theory of the effect of racial stereotypes on interaction in the United States in the twentieth century. It is not a theory of the interaction of officers and enlisted men in the U.S. Air Force. It is not a theory of the behavior of men and women on American juries. Since it deals with these phenomena as instances, the instances could change, leaving the theory intact. What is an instance of diffuse status characteristic now need not be an instance tomorrow. If racial prejudice is eliminated in the United States, "race" may no longer be an instance of a diffuse status characteristic; but the theory would still apply to explaining the effects of a diffuse status characteristic on interaction. In using the theory, one would only have to formulate different initial conditions identifying what characteristics are instances of diffuse status characteristics. But initial conditions which change through time and space are a general feature of scientific research, and are completely compatible with theories that are independent of time and space. When a user approaches the theory, he is called upon to exercise creativity in formulating initial conditions for tying the theory to his particular concrete situation. Theories which formulate abstract problems are the only theories that permit such creative application to a broad range of diverse circumstances.

SUMMARY

In this chapter we have used the theory of status characteristics and expectation states to illustrate our conception of scientific sociological theory. We have shown how this theory contains each of the elements of a theory introduced in chapter

10. We have also applied the criteria for evaluating a theory presented in chapter 10 to that theory as an example of employing a set of standards in assessing theories. Our assessment was for illustrative purposes and not a definitive evaluation of the theory of status characteristics and expectation states. For example, we simply noted that the theory has been tested many times, but did not analyze the nature of the evidence supporting or disconfirming the theory. Any definitive assessment would deal extensively with the empirical research bearing on the theory, but our purpose was simply to indicate that the theory is empirically testable.

The empirical evaluation of theories is of critical importance in any scientific discipline. Empirical evaluation, however, involves complex issues and chapters 13 to 16 will examine these issues. But we have said that empirical evaluation of ideas requires logical evaluation. Hence, we must first look at the logical analysis of a theory.

SUGGESTED READINGS

Berger, Joseph, M. Hamit Fiske, Robert Z. Norman, and Morris Zelditch, Jr. *Status Characteristics and Social Interaction: An Expectation-States Approach.* New York: Elsevier, 1977.

This is the most recent complete statement of the theory of status characteristics and expectation states. Compared to the 1972 version discussed in this chaper, it is more general and includes many refinements.

Berger, Joseph, Murray Webster, Jr., Cecilia Ridgeway, and Susan J. Rosenholtz. "Status Cues, Expectations, and Behavior." In *Advances in Group Processes, 3,* ed. E. J. Lawler. Greenwich, Conn.: JAI Press, 1986.

Since the 1977 version, there have been many extensions of the theory. This paper presents one of these extensions in the area of cues to status characteristics.

Willer, David, and Bo Anderson, eds. *Networks, Exchange and Coercion.* New York: Elsevier, 1981.

The papers in this volume present an explicitly stated theory to which the analysis in this chapter could be applied.

material mechanics, 163
   Stress, 163
epistemological understanding, 164
epistemology, 135
ethnography, 186
ethnomethodology, 85
evaluation, 85
evaluation research, 82
experimental control, 80
experimental design, 111
   experimental group, 43
   experimental mortality, 70
exploration, 71
   consulting, 32
   personal visit, 32
exploratory factor analysis (EFA), 32
external validity, 158
F test, 132
   F quantity, 132
   multiple comparison test, 131
factor analysis, 132
   extraction, 153, 154, 157, 158, 173
   factor score, 155
   rotation, 156
Factor Analytic Simultaneous Equation Model (FASEM), 156
falsification, 159
field research, 135
focal unit of analysis, 81
focus group, 59
frequency distributions, 89
funding, 96
garbage can model, 86
generalizability, 180
generalization, 71
global measures, 84, 162, 194
goodness of fit, 84, 162, 194

group conversation, 160
heterogeneity, 143
heuristic approach, 89
historical/comparative study, 59
homogeneity, 188
human subjects, 80
hypothesis, 59
   formulation, 86
   testing, 39
hypothesis testing, 39, 42
   p-value, 42
   computer output, 128
   computer-aided, 129
   critical region, 129
   critical value, 129
   significance level, 127, 129
   test quantity, 127
independent variable, 127
indicator, 129
inferential statistics, 31
   hypothesis testing, 47
   parameter estimation, 134
interaction effect, 123
intercorrelation, 123
internal consistency, 146
internal validity, 153
interpretation, 54
intersubjectivity, 72, 123
intervention, 137
interview, 41
interviewers, 76
   interviewer effect, 69
isolated test, 75, 88
item analysis, 88
item scaling procedure, 52
   item characteristics, 34
item sampling, 53

TWELVE

The Logical Analysis of a Theory

One of the virtues of explicitly stating a theory is that the theory can then be subjected to a rigorous logical analysis—to insure, for example, that the conclusions of the argument follow from its premises. In the social sciences, where theory is often only loosely formulated, disputes arise concerning whether or not the "derivations" actually follow from the theory or whether a theory actually explains an observation statement. A logic is a set of rules for the manipulation of statements to derive new statements. Using such a set of rules allows us to evaluate the logical validity of an argument, and it allows us to draw out new consequences which are implicit and often unrecognized unless the verbal statements are subjected to a logical analysis.

While we cannot teach a course in logic in this chapter, we can present a few rules from the *calculus of propositions* to illustrate what a logical analysis involves. In this chapter, we intend to develop a small number of tools, and then subject the assumptions of the theory of status characteristics and expectation states to a rigorous logical analysis.

In a way, it is especially appropriate to apply rules from the calculus of propositions to this theory. When the theory first appeared, its authors simply presented the assumptions and scope conditions of the theory. They assumed that their relevant public would be able to derive the consequences of the theory, and they did not want to "talk down" to their audience. But one critic complained that a great deal of attention was devoted to presenting and justifying the theory, but there were no consequences of the assumptions. In this chapter, we will show the formal derivation of the set of consequences presented in the last chapter. In so doing, we will both justify the assertions of the last chapter and give explicit meaning to the idea of derivation. It perhaps comes as a surprise to some readers that it is necessary to spell out the meaning of *derivation*. Unfortunately, however, the term *derive* is used very loosely in the social sciences. When some writers say they derived one idea from another, they simply mean, "First I thought of this, then I thought of that."

THE CALCULUS OF PROPOSITIONS[1]

The *calculus of propositions* is a set of rules for dealing with statements. Our concern in emphasizing that science deals with statements stems in part from the need to employ logic in the evaluation of ideas. While some logics deal with statements that are not English sentences (as, for example, mathematics deals with equations), the calculus of propositions deals with statements in the sense presented earlier, where a statement is a declarative sentence which asserts something that can be true or false.

Statements can be *simple* or *compound.* Simple statements predicate something of a subject. For example, "Juries are task-oriented groups" and "Task-oriented groups are differentiated" are both simple statements. If we assert, "Juries are task-oriented groups and juries are status differentiated," we have asserted a *compound* statement. A compound statement is a combination of two or more simple statements formed by connecting them with *connectives.* Connectives are terms like *and, or,* and *not.*

Because of the tremendous variety of simple statements, it turns out to be difficult to develop rules governing simple statements. On the other hand, it is relatively easy to develop rules for compound statements. The calculus of propositions presents rules for dealing with compound statements. If we assume the truth or falsity of the simple sentences making up a compound sentence, the calculus of propositions provides rules for determining the truth or falsity of the compound statement.

We can develop powerful analytic tools by restricting ourselves to five basic connectives. The connectives and their symbols are presented in table 12.1. The

Table 12.1

Name	Symbol	Translated As
Conjunction	$\wedge$	and
Disjunction	$\vee$	or*
Negation	$\sim$	not
Conditional	$\rightarrow$	if...then...
Biconditional	$\leftrightarrow$	...if and only if...

*In ordinary language, there is an ambiguity between the inclusive "or" and the exclusive "or." If we say, "Next year I will take sociology or next year I will take psychology," we could mean that we will take one and not the other, or we could mean that we will take both. The first is the exclusive sense of "or" while the second is the inclusive sense. When we use the "or" connective, we will use it in the inclusive sense.

truth value of a compound statement depends upon the truth values of its components. The rules for determining how the truth of a compound statement made from a particular connective depends upon the truth of its components can be

1. The development in this section is based on Kemeny et al. (1966), pp. 1–51.

presented in what is known as a *truth table*. For example, the truth table for the connective " ¯ " is as follows:

Let *p* stand for our statement, *T* stand for "true," and *F* stand for "false" (see table 12.2). If *p* is a statement, table 12.2 asserts that if *p* is true, its negation

Table 12.2

p	*p̄*
T	F
F	T

is false, and if *p* is false, its negation is true. The calculus of propositions posits as a fundamental property of statements that any statement is either true or false and cannot be both true and false. Hence, we only need *T* or *F* in our truth table. The truth table for $p \wedge q$, where *p* and *q* are statements, is shown in table 12.3. There are four possible pairs of truth values for the simple statements *p,q*. It will be noted that the conjunction $p \wedge q$ is true only when each simple statement is true. In other words, $p \wedge q$ asserts no more and no less than "*p* and *q* are both true."

Table 12.3

p	*q*	$p \wedge q$
T	T	T
T	F	F
F	T	F
F	F	F

In order to illustrate the logical analysis of our theory, we need the truth table for one other connective, the conditional. (We will not present truth tables for disjunction and biconditional.) In ordinary language, we often do not want to make an outright assertion, but want to qualify it by some condition. For example, let us consider the sentence, "If it rains, then I will take my umbrella." We have qualified the assertion "I will take my umbrella" by using "if it rains" as a condition. As table 12.1 showed, we have constructed a compound statement using the "if. . .then" connective. As we noted in chapter 4, the "if" clause is termed the *antecedent conditions* and the "then" clause is the *consequent conditions*. Assumptions 1 to 4 in the theory of status characteristics are all formulated as conditionals.

It is clear that if both simple statements are true, the conditional is true. If I fail to take my umbrella when it is raining, then the compound statement is false. But what happens if it is not raining? That is, what happens if the antecedent simple statement is false? The truth table for the conditional solves that problem, although the rules may appear somewhat arbitrary. The compound statement of the conditional is always true when the antecedent conditions are false. On intuitive grounds, we could say that if the antecedent is false, we really have no way of testing the conditional; therefore, we give it the "benefit of the doubt." We cannot leave the question undecided when the antecedent is false, because that would violate our requirement that any sentence is either true or false. We then have table 12.4 as a truth table for the conditional.

Table 12.4

p	q	$p \rightarrow q$
T	T	T
T	F	F
F	T	T
F	F	T

USING THE CALCULUS OF PROPOSITIONS

With these simple tools, we can proceed to analyze the logical structure of the theory of status characteristics and expectation states. There are more elegant ways to pursue a logical analysis of the theory, but they would require a more advanced knowledge of logic. First, let us translate the assumptions of the theory into logical symbols. In table 12.5 we assign the symbols p, q, r, s, and t to the simple statements of assumptions 1 through 4. The fact that we use p for both a statement and an actor should not cause any confusion since the usage will be clear from the context.

Table 12.5

Symbol	Stands for
p	D in S^* is a social basis of discrimination between p' and o.
q	D is activated in task situation S^*.
r	At least one consistent component of D will become relevant to C in S^*.
s	p will assign states of C to p' and o in a consistent manner.
t	The position of p relative to o in the observable power and prestige order will be a direct function of the expectation advantage of p over o.

With these translations, we now claim that assumptions 1 to 4 assert that the following conditionals are true:

1. $p \rightarrow q$
2. $q \rightarrow r$
3. $r \rightarrow s$
4. $s \rightarrow t$

This simple step of translating the verbal assumptions of the theory into logical symbols immediately reveals several problems in the way we have verbally presented the theory. Some of these problems are easily disposed of, but others require more thought. For example, our translation of assumptions almost ignores the statement of scope conditions. This is easily remedied. Strictly speaking, we would rewrite our assumptions so that the antecedent of the conditional was a conjunction, as follows:

$$(S^* \wedge p) \rightarrow q$$

This again makes use of the properties of the conjunction asserting that both the scope conditions (represented by S^*) and p must be true.

A more difficult problem emerges if we look at our translation of q. In the verbal statement of the assumptions (see p. 212), the q in "$p \rightarrow q$" and the q in "$q \rightarrow r$" are not the same q. In assumption 1, q is a translation of "D is activated in task situation S^*." But in assumption 2, q is a translation of "D is activated in S^* and has not been previously dissociated from C." Thus, we immediately see a logical problem with the verbal formulation of assumption 2. Implicitly, the phrase "D has not been dissociated from C" is assumed to be a consequence of meeting the scope conditions of the theory, but it is precisely the purpose of logical analysis to reveal such problems; in this case, we find that the argument would not follow logically without making the assumption explicit. Having pointed this out for the sake of this exercise, we can drop out the qualification about previous dissociation. We have already reaped one benefit from our analysis; to retranslate this set of sentences to include this additional idea would needlessly complicate our illustration of the use of truth tables. Let us assume, then, that the translations in table 12.5 are adequate for our purposes.

In the last chapter, we listed without derivation four consequences of the theory. In terms of our symbols, we can write these consequences as:

$$p \rightarrow t$$
$$(p \wedge q) \rightarrow t$$
$$r \rightarrow t$$
$$s \rightarrow t$$

Since the general expectation state is equivalent to one consistent component of d, we can write $r \rightarrow t$ and $s \rightarrow t$, and the latter becomes equivalent to assumption 4 of the theory. Since table 12.3 showed that a conjunction is true if, and only if, both simple statements are true, we can simplify our problems with the following reformulation:

$$p \rightarrow t$$
$$q \rightarrow t$$
$$r \rightarrow t$$

We now want to prove that these derivations logically follow from the assumptions of the theory. To do this, we will make use of truth tables constructed from the truth table of the conditional.

First let us rewrite the truth table for the conditional in reverse order, as in table 12.6. The purpose of rewriting this is to emphasize the fact that this theory assumes that the conditional is true, and this assumption means that we can examine the truth value of the simple statements that make up the conditional.

Table 12.6

$p{\to}q$	p	q
T	T	T
F	T	F
T	F	T
T	F	F

Because we have assumed that $p{\to}q$, the second line of the truth table can be discarded because it is inconsistent with our assumption; that is, it contradicts what the theory is assuming.

Next, we proceed to construct a more complicated truth table which makes use of assumptions 1 and 2. Essentially, this more complicated truth table combines two truth tables, one for $p{\to}q$, and the second for $q{\to}r$, as in table 12.7. Table

Table 12.7

	$p{\to}q$	$q{\to}r$	p	q	r
1.	T	T	T	T	T
2.	T	F	T	T	F
3.	F	T	T	F	T
4.	F	T	T	F	F
5.	T	T	F	T	T
6.	T	F	F	T	F
7.	T	T	F	F	T
8.	T	T	F	F	F

12.7 shows the truth values of the component simple statements p, q, and r for the various truth conditions of the conditionals $p{\to}q$, $q{\to}r$. Again, remember that the theory assumes that $p{\to}q$ and $q{\to}r$ are true. Thus, we want to eliminate those rows of the truth table which contradict what the theory asserts. We can see that the horizontal rows 2, 3, 4, and 6 are inconsistent with the theory. In other words, these rows are logically not possible according to the theory. It is important to emphasize that we are dealing solely with the logical structure of the theory and not with its empirical truth. Empirically, the theory could be wrong, which would be another way of saying that what the theory regards as logically not possible occurs empirically.

We are now left with rows 1, 5, 7, and 8 of table 12.7. This gives us a set of logical possibilities for the truth values of p and r. We use these to construct another truth table for the conditional $p{\rightarrow}r$. We construct this truth table in the usual way (rather than in the reverse), because in this case we are making no assumptions about the truth of $p{\rightarrow}r$. See table 12.8. We have constructed this table using only the logically possible truth values of p and r which we have discovered from table 12.7. These are indicated by the numbers in parentheses, which refer

Table 12.8

	p	r	$p{\rightarrow}r$
(1)	T	T	T
(5)	F	T	T
(7)	F	T	T
(8)	F	F	T

to the rows of table 12.7. First of all, we note that rows 5 and 7 are identical. In table 12.7, they differed with respect to the truth value of q, but we have dropped q from our analysis. The second and most important thing to note is that $p{\rightarrow}r$ is true for all the logically possible truth values of p and r in the theory. What we have proven, then, is that the first two assumptions of our theory have $p{\rightarrow}r$ as a consequence. If we translate this back into the verbal formulation, we find that we have derived a new consequence for the theory which was not among those listed in chapter 10:

If D in $S*$ is a social basis of discrimination between p' and o, then at least one consistent component of D will become relevant to C in $S*$.

We have also done something else that is very important. We have demonstrated the logical relation of implication. Implication is a relationship between pairs of statements. By implication, we mean that one statement logically implies the other one. As we have listed all logical possibilities, then we shall characterize implication as follows:

The first statement implies the second if the second is true whenever the first is true, that is, in all the logically possible cases in which the first is true.

We have shown that for the theory of status characteristics and expectation states, the statements $p{\rightarrow}q$, $q{\rightarrow}r$ logically imply $p{\rightarrow}r$, because $p{\rightarrow}r$ is true in all the logically possible cases, as the truth table demonstrates.

By successfully applying the method we have just illustrated with three truth tables, we can proceed to prove that $p{\rightarrow}s$ and $p{\rightarrow}t$. We would have to apply the method successively to achieve these results. We will not construct the two sets of necessary truth tables, but the reader might want to try it as an exercise: it is simply a straightforward application of the three truth tables presented above, inserting the appropriate symbols for the statements we wish to examine; thus, we would have one set of tables with $p, r; p, r, s;$ and $p, s;$ and we would have a second set of tables with $p, s; p, s, t;$ and p, t.

Using these truth tables, we have demonstrated that we can logically derive consequences of the theory. To put it in another way, the consequences are logically consistent with the assumptions of the theory, and we have made a valid argument.

Using the same technique, we can also derive the consequences $q{\rightarrow}t$ and $r{\rightarrow}t$ (stated verbally in chapter 11). By again applying the same type of analysis, we can prove the validity of these derivations.

Subjecting the theory to logical analysis has revealed two things: (1) there is some "sloppiness" in the verbal statement of the theory, which fortunately can be easily tidied up; and (2) the consequences of the theory logically follow from the theory's assumptions. We have an intersubjective way of deciding whether or not a consequence is implied by the set of assumptions. The same tools can be applied to settle the question of whether a statement is relevant to a knowledge claim conjoined with a set of initial conditions. Hence, we have an intersubjective method for resolving disputes about the relevance of observation statements (evidence) to knowledge claims.

Although we discovered some difficulties in the verbal statement of the theory, we did not uncover any irremediable logical errors in the formulation. If there had been a fallacious argument in the theory, subjecting it to rigorous formal analysis would have revealed fallacies that might have remained hidden in a loose verbal formulation. Suppose, for example, that the theory made the following claims:

1. If D is activated in S^*, one consistent component of D will become relevant to C in S^*.
2. If p assigns states of C to p' and o that are consistent with the states of D, then one consistent component of D will become relevant to C.

With these two assumptions, suppose we then claimed that the following consequence could be derived:

If D is activated in S^*, then p will assign states of C to p' and o that are consistent with the states of D.

Some readers may recognize that our conclusion does not follow from these two premises. On the other hand, it is the nature of verbal arguments that such logical flaws are not always clear. Thus, some readers will be caught up in the flow of the verbal presentation and not see any difficulty, especially if one reads the argument rather quickly. Since intersubjective agreement requires all members of the relevant public to agree either that the conclusion does or does not follow from the premises, and since the reader is temporarily a member of the relevant public, we should analyze this simple argument rigorously. Applying our truth-table analysis, we immediately see the fallacy of the above claim. We begin by translating the statements into symbols from our earlier translation:

1. $q \rightarrow r$
2. $s \rightarrow r$

and we are trying to prove that $q \rightarrow s$. Two truth tables will show us the fallacy. First, see table 12.9.

Table 12.9

	$p \rightarrow r$	$s \rightarrow r$	q	r	s
1.	T	T	T	T	T
2.	T	T	T	T	F
3.	F	F	T	F	T
4.	F	T	T	F	F
5.	T	T	F	T	T
6.	T	T	F	T	F
7.	T	F	F	F	T
8.	T	T	F	F	F

In table 12.9, we eliminate lines from our truth table which are inconsistent with the truth of our assumptions; in this case, lines 3, 4, and 7. We now construct a truth table for q and s for the conditional involving q and s using the allowable truth values for q and s. From table 12.9 we obtain table 12.10.

Table 12.10

	q	s	$q \rightarrow s$
(1)	T	T	T
(2)	T	F	F
(5)	F	T	T
(6)	F	F	T
(8)	F	F	T

If we compare table 12.10 to table 12.8, we immediately see an important difference. In table 12.10, the statement $q \rightarrow s$ is false for one of the logically possible combinations of values of q and s (line 2 of the table). In table 12.8, on the other hand, the statement $p \rightarrow r$ was true for all logically possible truth values of p, r. Thus, by the definition of implication, we have shown that q does not imply s. Hence, the claim that our consequence followed from our two assumptions is false.

Of course, our proof was inelegant. With additional tools, we could have shown the fallacy of the argument in a much simpler fashion. But our point here is not to develop a set of logical tools, but rather to illustrate how a rigorous logical analysis works and what it can accomplish. We have shown that logical tools enable

item sampling adequacy, 53

item variable, 52

lagging correlation, 168

latent structure, 52

least errors criterion, 121

least squares criterion, 185

lessons from science, 153, 158

levels of inquiry
  descriptive, 121
  explanatory, 166
  exploratory, 42, 43

linear correlation, 42, 43

linear equation, 39, 43

linear research logic, 118

linearity, 120

linearization, 33

literature review, 118, 150
  citation, 121, 142
  notes, 28, 30, 32, 35

log-linear regression, 32

logic of a statistical test, 32

logistic regression, 140, 152

longitudinal data, 123

longitudinal data analysis, 145
  panel design, 157
  trend study, 184
  turnover analysis, 185
  cohort study, 185
  Experimentation, 185
  panel study, 68
  trend study, 69

manifest structure, 68

matching, 69

measurement, 158
  composite measure, 72
  indicator, 45, 47
  level of, 46, 51

multiple and systematic approach, 46

measurement effectiveness principle, 48

measurement levels, 48
  interval, 179
  nominal, 42, 178
  ordinal, 49, 123
  ratio, 48
  categorical variables, 49
  continuous variables, 113
  downgrading, 42, 105
  upgrading, 42

measurement power, 177

measurement tools
  index, 42
  individual indicators, 122
  scale, 63

measures of association, 54
  coefficient of determination, 53
  contingency coefficient, 114
  correlation, 119
  correlation matrix, 131
  covariance, 120
  Cramer's V, 120
  Gamma, 119
  Lambda (λ), 131
  Pearson's γ, 116, 119, 131, 140
  Phi Coefficient, 115
  proportionate reduction of error (PRE), 118, 131
  Somers' D, 131
  Spearman's rho, 115
  tau-a, 117
  tau-b, 117
  tau-c, 117
  tau-y, 117

us to: (1) demonstrate that a conclusion logically follows from a set of premises, (2) derive new consequences of our premises that were not readily apparent simply by looking at the verbal formulation, and (3) detect logical flaws in an argument. Because so much of sociological reasoning is in verbal formulation, the consumer and the researcher cannot simply take for granted that a sociological argument is logically consistent, not self-contradictory, or not incomplete. Even the most sophisticated sociological theorists make mistakes which an alert and trained public can uncover and rectify. As an exercise, the reader might want to examine the following argument which was presented as an example of deductive theorizing:

1. If a society is to maintain its structural continuity, its members must conform to its norms.
2. Its members' conformity to norms is maintained by their expressing this horror collectively.
3. Their horror of nonconformity is maintained by their expressing this horror collectively.
4. The punishment of criminals, that is, nonconformists, is the means of expressing this horror collectively.
5. Therefore, a society that maintains its structural continuity is one in which criminals are punished. (Homans, 1964)

With just the simple tools that we have introduced in this chapter, it is possible to find logical problems with this argument (taken verbatim from chapter 25 of the *Handbook of Modern Sociology*).

If we paraphrase these statements, putting them in the present tense and in the "if...then" form, it will make our task easier:

1. If a society maintains its structural continuity, its members must conform to its norms.
2. If horror (of nonconformity) is expressed collectively, then members conform to its norms.
3. If horror of nonconformity is expressed collectively, then members maintain horror of nonconformity.
4. If criminals are punished, then horror (of nonconformity) is expressed collectively.
5. Therefore, if criminals are punished, then a society maintains its structural continuity.

Table 12.11 gives the symbolic translation of the simple statements. The careful reader will note that we have left the word *must* out of our translation of the original statement 1. The word *must* can be interpreted as "it is necessary that." With this interpretation, we can capture the meaning of the first statement by the way we formulate the conditional symbolically. In other words, we are suggesting

Table 12.11

Symbol	*Stands for*
p	Members conform to norms
q	Society maintains its structural continuity
r	Horror (of nonconformity) is expressed collectively
s	Members maintain horror of nonconformity
t	Criminals are punished

that the interpretation of the first statement is "only if members conform to its norms, then a society maintains its structural continuity," where "only if" expresses a necessary condition. The way we write this symbolically uses a negation: $\tilde{p} \rightarrow \tilde{q}$.[2] Now our symbolic representation of the argument is:

$$
\begin{aligned}
&1.\ \tilde{p} \rightarrow \tilde{q} \\
&2.\ r \rightarrow p \\
&3.\ r \rightarrow s \\
&4.\ t \rightarrow r
\end{aligned}
$$

therefore:

$$5.\ t \rightarrow q$$

The first thing to note about this argument is that statement 3 does not contribute to the argument since *s* occurs only in the consequent of statement 3 and nowhere else.

Now if statement 1 said $p \rightarrow q$, we would have an argument identical with our analysis of the theory of status characteristics, and the argument would look like:

$$
\begin{aligned}
&t \rightarrow r \\
&r \rightarrow p
\end{aligned}
$$

therefore: $t \rightarrow p$

and:
$$
\begin{aligned}
&t \rightarrow p \\
&p \rightarrow q
\end{aligned}
$$

therefore: $t \rightarrow q$

If that were the case, then the conclusion would follow from the premises. But statement 1 does not say that $p \rightarrow q$. This is the logical difficulty with the argument, because statement 1 ($\tilde{p} \rightarrow \tilde{q}$) is logically equivalent to $q \rightarrow p$. If the reader will go through the truth-table analysis of the four statements, using $q \rightarrow p$ as

---

2. The rationale for writing "only if *p*, then *q*" and "$\tilde{p} \rightarrow \tilde{q}$" becomes clear if it is spelled out. The "only if" statement means that for *q* to be true, *p* has to be true. If *p* is false, then $\tilde{p}$ is true, but *q* cannot then be true; hence $\tilde{p} \rightarrow \tilde{q}$.

statement 1, the reader will find that the argument does not follow. The difficulty arises because the conclusion depends upon $p{\rightarrow}q$, while an appropriate translation of the original statement 1 is $q{\rightarrow}p$, which is called the *converse* of $p{\rightarrow}q$. Indeed, many of the most common fallacies in reasoning arise from a confusion of a statement with its converse.

In this chapter, we have illustrated the power of logical tools and the importance of a logical analysis of sociological theory. We believe that this importance has not received enough attention among sociologists and that, indeed, logical analysis is prior to the empirical evaluation of ideas. If an argument is self-contradictory, it makes no sense to try to use evidence to evaluate the argument empirically. If an argument is incomplete, it is often difficult to bring evidence to bear without first teasing out the implicit assumptions that complete the argument. In the next chapter, we turn to the problem of empirically evaluating ideas.

SUGGESTED READINGS

A variety of introductory logic textbooks can be used to pursue the material in this chapter. Two useful texts are:

Suppes, Patrick. *Introduction to Logic.* New York: Van Nostrand, 1958.

Tarski, Alfred. *Introduction to Logic and to the Methodology of the Deductive Sciences.* New York: Oxford University Press, 1965.

THIRTEEN

The
Empirical
Evaluation
of Ideas

Collective evaluation using reason and evidence distinguishes sociological claims from matters of opinion. In the last chapter we presented a few simple techniques for the logical evaluation of sociological ideas. The next four chapters discuss the role of empirical research in the development and evaluation of sociological theories and knowledge claims; our discussion, however, will not deal with specific techniques but rather will focus on the issues involved in using evidence for the collective assessment of sociological claims. For the researcher, our analysis should provide a foundation for choosing among available options when conducting research; for the consumer, these chapters should illuminate some of the cautions that must be applied before accepting and using research results.

We will address fundamental questions of research strategy. We believe that one of the central purposes of research is to develop and choose among alternative explanatory theories. Empirical studies have the capacity to facilitate choices among a limited number of competing alternatives and therein lies their power. Our analysis in this chapter will show how we use an empirical study to evaluate alternative explanations. But in order that the reader may fully appreciate the significance of what we regard as the key role for empirical research, we must first clear away a number of misconceptions. In the process, we will take exception to some conventional wisdom, but what remains after "clearing away the underbrush" will provide a reliable foundation on which to build effective strategies for developing sociological knowledge.

The general conception of empirical research among lay people focuses on the single empirical study. A study, we often read in the media, discovers something or proves some claim. And when a knowledge claim is controversial, it is often dismissed as "unproven" so, to take two prominent examples, cigarettes have not been proven to cause cancer and evolution is a theory, not a proven fact. In chapter 5, we asserted that it is impossible to prove a knowledge claim empirically and that "proof" applies to logical analysis; causes of cancer and evolution share their unproven status with all other theories and knowledge claims. In this chapter, we will prove by logical analysis why we cannot empirically prove a knowledge claim. We will also indicate why empirical research is not particularly efficient for discovering something new; although discoveries sometimes occur in the course of an empirical investigation, there is no known *method of discovery*.

To clear away mistaken notions of empirical proof and misplaced emphasis on discovery will require an examination of both experimental and nonexperimental research. While our analysis of these has controversial aspects, we feel confident that most researchers would agree substantially with our conclusions.

In the course of the analysis, we will point out the limitations of the single isolated empirical study, limitations that even the best-designed, most rigorous, and most carefully conducted experiment cannot overcome. These limitations attest to the necessity of looking beyond the single study to a program of inter-related investigations. We will call such programs *cumulative research programs*. Viewing the single study in the context of a cumulative program suggests ways to increase its informative value, and chapter 14 will offer some proposals to that end. Chapter 15 will discuss general features of cumulative research programs, whereas chapter 16 will deal with stages of a cumulative program.

Before turning to the main analyses of this chapter, we need to distinguish three types of sociological research: exploratory, descriptive, and explanatory studies. We will devote most of our attention to explicating some aspects of explanatory research as this type will be the principal focus of our analysis.

EXPLANATORY RESEARCH—HYPOTHESIS TESTING

Sociologists do empirical research for a number of reasons. They carry out studies to generate ideas or to identify a phenomenon; they do empirical research to develop instruments and to assess the reliability and validity of their instruments. They engage in fact-finding investigations in order to characterize a social phenomenon or a social situation. Sociologists conduct tests of theories, knowledge claims, and sometimes singular propositions.

Writers on method conventionally categorize the different varieties of research in one of three basic categories. First, *exploratory studies* represent early attempts to gain knowledge about a phenomenon or a situation; these have little foundation of theory, prior knowledge claims, or prior observation on which to build and often use observations to sharpen vaguely formulated ideas. Second, *descriptive studies,* as the name suggests, intend to describe properties of a situation or phenomenon. A population census exemplifies a descriptive study since it characterizes the people of a particular place according to properties such as age, sex, and marital status. In our terminology, descriptive studies generate singular statements (observation statements) that are true of a given time-place situation.

The third category, *explanatory studies,* has a number of intuitively compelling, if somewhat vague, characterizations. These are studies that "go beyond mere description" and seek "understanding"; they intend to answer *why* questions about phenomena—for example, why is educational attainment a basis of prestige? Or *how* questions, such as "How do mass movements attain legitimacy?" In some usage, explanatory studies are studies that test hypotheses. Although we are sympathetic to these formulations, we believe that the concept needs further explication in order to be useful. Just as the idea of scientific explanation is more restricted than lay conceptions, the concept of *explanatory research* must be more restricted than these vague characterizations require.

Recalling the discussion of explanation in chapter 10, we can begin by restricting explanatory research to studies that intend to develop and/or evaluate one or more explanatory statements where such statements can be theories, knowledge claims, or singular propositions. (Although we have not discussed singular propositions as explanans, such explanations are particularly important in many applied or engineering studies where singular consequent statements are deduced from singular antecedent statements.) While this conceptualization needs further development, it provides more guidance than terms like "understanding" and it rules out some forms of hypothesis-testing studies—for example, those studies where a singular hypothesis is asserted flatly without any rationale or any linkage to other investigations.

Until now, we have avoided using the term, *hypothesis*. As chapter 4 noted, the word has many different usages in sociology and in the sciences in general. Since the concept will be useful in the analyses of the functions of empirical research, we now define hypothesis in a way that fits into our framework and yet is more or less faithful to the usages of sociologists who have considered problems of hypothesis testing:

An hypothesis is a predictive observation statement that is testable.

In other words, an hypothesis is (1) a singular statement that (2) predicts[1] a relationship between two or more indicators and (3) can be true or false. We now distinguish between hypotheses and observation statements: before one knows the state of affairs, a statement describing the set of observations—an hypothesis—can be true or false; when one already knows the state of affairs and formulates an observation statement to describe or summarize it, that statement can no longer be false.

For example, we can reformulate the first observation statement of chapter 5—which, if we stated it before knowing what the data look like, would be an hypothesis:

In Palo Alto, in November 1986, people who report an educational level "attending college" or higher will report that they voted more frequently than people who report an educational level "high school graduate" or lower.

---

1. Prediction is not limited to future observations. We can predict something about a past event provided we do not already know the result.

It is instructive to compare this with the statement in chapter 5. In the first place, this statement is formulated in terms of specific indicators referring to the questions on a questionnaire. Second, it is phrased in the future tense, emphasizing that it is a prediction of what will be observed when the data are analyzed. If, on analyzing the data, we can construct an observation statement equivalent to the hypothesis, then we can say the hypothesis has been supported. On the other hand, if we find the observation statement to be logically inconsistent with the hypothesis, we will say the hypothesis has been disconfirmed.

We can also formulate an hypothesis based on the theory of status characteristics and expectation states, presented in chapter 11. If (1) we assume that sex is a diffuse status characteristic, with male representing the high state and female representing the low state; (2) we let situation S be a task in which two people make judgments about a series of slides; and (3) we use "resisting change of judgment in the face of disagreement from the other person" as an indicator of position in the observed power-and-prestige order; then we can state the hypothesis: "In the proposed experiment, males will change their judgments less frequently than females."

We must spell out why we call most hypothesis testing, explanatory research. In our example, the hypothesis is an operationalization derived from the theory and three assumed initial conditions. The reader who recalls the discussion of explanation in chapter 10 will realize that, if a hypothesis is derived from a set of statements, that set of statements explains the hypothesis. Hence explanatory studies test explanations that logically account for hypotheses; the hypothesis is the explanandum, and the theory plus the set of initial conditions constitute the explanans. Where an hypothesis is supported, the supporting observation statement is equivalent to the hypothesis and so is also an explanandum. Of course, as we have defined the concept, a hypothesis need not be derived from a knowledge claim; it may simply be an observation statement from a previous study or a hunch generated from looking at a set of indicators and guessing their interrelationships. This type of hypothesis has a different status from those that operationalize knowledge claims and we exclude research testing these "nonderived" hypotheses from our category of explanatory studies.

In testing the hypothesis derived from status characteristic theory, suppose we observe that males actually change their judgments more frequently than females. That should say something about the truth of the theory. Unfortunately, it is not that simple.

It is clear that we want to use the consistency or inconsistency of the observation statement with the hypothesis to evaluate: the hypothesis; the knowledge claim from which the hypothesis was derived; and, if the knowledge claim is part of a theory, to evaluate the theory as well. But do these evaluations evaluate the empirical *truth* of the hypothesis, the knowledge claim, and/or the theory? When we asserted in chapter 5 that empirical proof of a knowledge claim was impossible, did we implicitly rule out empirical truth as a criterion? After all, if we cannot prove the empirical truth of a statement, of what use is the idea of empirical truth? Of course, observation statements can be empirically true, and, as we have formulated the concept, an hypothesis can be proven to be empirically true—at least, in principle—since it refers to a given study in a particular time and place.

Most of the time, however, our primary interest goes beyond the truth of the observation statements or hypotheses of the single study. We want to "generalize" from that truth to the truth of these statements in other situations and ultimately to their truth in all situations within their scope. But such "generalizing" is precisely what is ruled out, as we shall see after we reintroduce one key idea.

Nonuniqueness of Explanations

This conceptual tool is a reformulation of the principle stated on page 178 of chapter 10. It is easy to show that one theory or one knowledge claim (joined with one set of initial conditions) is not a unique explanation of an observation statement. In our hypothetical observation of men changing their judgments less than women, we might speculate that all the men were tall and all the women short. This speculation suggests that a theory relating height and resistance to influence would explain the observation statement in that event as an alternative to status characteristic theory. We could speculate on a number of ways in which the men and women of our study might have differed and construct an explanation based on each of these speculations. We term our reformulated principle, the nonuniqueness theorem:

Nonuniqueness Theorem:
Given any observation statement or set of observation statements, it is always possible to construct alternative explanations for the given statement or statements.

In general, given any conclusion of a logical argument, it is always possible to construct alternative premises from which that conclusion follows. If there are many explanations, they are all in one sense true explanations of the single instance; that is, they are all logically true in that the observation statement is deducible from all of them.

With our conception of explanatory research in mind and with the nonuniqueness theorem as an essential tool, we can now turn to an analysis of the limitations of empirical research. We must reiterate our view that a recognition of these limitations does not diminish the crucial importance of empirical studies, but does provide a foundation for understanding the proper use of empirical research in the creation, evaluation, and application of sociological knowledge.

WHAT EMPIRICAL RESEARCH CANNOT DO

In chapter 5, we asserted that there is no such thing as empirical proof of a knowledge claim. Now we must go beyond the assertion and demonstrate why this is the case and what its implications are for the conduct and evaluation of empirical research. We will analyze two cases. For Case 1, suppose a sociologist wants to test the following knowledge claim:

KC 1: Productivity of those who work in the physical presence of others is greater than productivity of those who work in physical isolation.

She formulates a set of scope conditions and decides to study a large company where some employees work on their individual tasks in the presence of others and some work also on individual tasks but in isolation from other people. She chooses to measure productivity using supervisor ratings on a "productivity" scale. She then states the following hypothesis[2]:

HYP 1: In company C during year Y, productivity of workers as measured by ratings of their supervisors will be greater for those who work in the physical presence of others than for those who do their work in physical isolation.

After conducting the research, she summarizes her results with the following observation statement:

OS 1: In company C during year Y, productivity of workers as measured by ratings of their supervisors was greater for those who worked in the physical presence of others than for those who did their work in physical isolation.

Since OS 1 is consistent with HYP 1 and HYP 1 is derived from KC 1, we can say that the study supports KC 1. No one would say that the study proves KC 1; after all, HYP 1 is only one instantiation of KC 1; we can conceive of many alternative sets of initial conditions which would lead to other instantiations. But suppose that sociologists have conducted many studies using different companies at different times and employing different indicators to measure productivity; for example, they have formulated and tested a number of different hypotheses that represent different derivations from KC 1 and varying initial conditions. We still cannot conclude that they have proven KC 1.

Proof—the guaranteed empirical truth—of KC 1 would require that we examine all instantiations of KC 1, those in the past, in the present, and in the future. Otherwise, no matter how many times the observation statement was consistent with the hypothesis there is no guarantee that the next study would not produce

---

2. Since the hypothesis is almost a direct translation, we omit statements of initial conditions.

an observation statement contradictory to the hypothesis.[3] Since it is clearly impossible to examine all instantiations, we must reject the possibility of empirical proof.

This conclusion seems straightforward; yet many who accept the result in principle violate its spirit in practice. If we approach the problem from a different starting point, we can show the difficulty with a common practice among some sociologists. For Case 2, suppose our sociologist does not begin with a knowledge claim, but in the course of studying a company makes a discovery which she summarizes as OS 1. Can this sociologist generalize OS 1 and argue that she has discovered a general principle and proven it by empirical research? Although, sadly, such things happen even in the scientific literature, they are totally unjustified. Let us see why.

One "generalization" of OS 1 arrives at a knowledge claim by dropping the place, time, and indicator specifications to arrive at KC 1. But if we reflect on the number of different "generalizations" that could be created from OS 1, we can appreciate the difficulties with generalizing from an empirical study. Just as we earlier generated a number of hypotheses instantiating KC 1, so we can formulate a larger number of "generalizations." We could "generalize" to all organizations over all time, to company *C* over time, to all companies like *C* at a given time, to all companies like *C* over time, and so on. Even a "modest generalization" like generalizing to company *C* over time gets us into trouble; while OS 1 may be true of company *C* this year, next year, and ten years from now, that doesn't preclude it being false in the eleventh year if, for example, company *C* introduces incentives that are known to increase productivity to those working in isolation and withholds them from those working in the physical presence of others. Since such "generalization" cannot anticipate all possible interferences, any number of intervening events could affect the truth of the statement in the future. Furthermore, how does one infer scope conditions from OS 1? Without scope restrictions, we cannot sort out future circumstances where the statement should be true from those where it should be false. The reader should recall the discussion of prophesy in chapter 6.

The "Problem of Induction"

The problem we have been discussing is known as the "problem of induction"; it is a classic philosophical problem, first analyzed in the eighteenth century by the philosopher David Hume. The problem is: *No set of rules exists that allows one to go from the truth of a singular premise to the truth of a universal statement.* Simply put, there is no way that the truth of a singular statement, or even the truth of a large number of singular statements, can guarantee the truth of a universal statement. In our Case 1, there is no logic that justifies inferring the

---

3. Some people would solve the problem of proof by saying that KC 1 is "probably true," but such an inference is not very helpful and equally prone to error. For a discussion of this point, see Popper, 1959, pp. 78–92.

Yule's Q coefficient, 117
methodology, 116
missing information, 117
  expected, 186
  restoration, 97
  unexpected, 97
missing values, 97
model fitting, 97
model specification, 96
multi-group comparison, 42
multidimensional scaling, 42
multidimensionality, 132
multifaceted research design, 161
multiple causation, 54, 153
multiple correlation, 81
multiple regression, 140
  nonlinearity, 143, 148
multivariate analysis of variance, 120,
  141, 171
mutual exclusiveness criterion, 142
neutrality, 147
nonlinear regression, 178
nonprobability sampling
  accidental, 41
  judgmental or purposive, 142
  quota, 62
  snowball, 62
normal distribution, 66
  Kurtosis, 62
  Skewness, 111
normal equations, 111
null hypothesis, 111
objectivity, 151
observables, 123
observation, 41
  announced, 38
  nonparticipant, 59, 75

participant, 75
one-way analysis of variance, 75
operationalization, 75
overreduction, 132
panel attrition, 37, 45, 46, 52
paradigms, 137
  critical, 68
  interpretive, 186
  positivist, 187
parameter, 187
parameter estimation, 186
parsimony principle, 123
part correlation, 128
partial correlation, 53
partial correlation coefficient, 140
partial regression coefficients, 139
path analysis, 141
  causal effect, 141
  direct effect, 150, 157
  indirect effect, 152
  total effect, 151
planning of research
  comprehensive papers, 151
  funding, 151
  oral examination, 10
  overview, 11
  research paper, 11
  study plan, 11
  term papers, 10
  theory paper, 8
  thesis/dissertation guide, 10
  thesis/dissertation proposal, 10
points of focus, 11
polynomial regression, 10
population parameter, 60
principal component analysis, 122
  component score, 126

success of all tests of hypotheses drawn from KC 1 from the success of any number of such tests. With our Case 2, no logic justifies deriving the general conclusion, KC 1, from singular premises like OS 1. Since Hume, many thinkers have attempted to develop an inductive logic, but none has succeeded.[4] The fact that the sun has risen every day since the beginning of time does not guarantee that the sun will rise tomorrow. The fact that our theory was supported in one study or in many studies does not foreclose the possibility that a new study would generate a disconfirming instance. Of course, when there are many supporting studies, scientists feel and behave differently from when there is only one study supporting a theory. But there is a difference between proof and feelings or beliefs. While in practice we may believe in our theory and use it, we have not proven it and must always regard it as provisional. We cannot overemphasize two points: (1) there is no valid inductive logic which enables us to prove generalizations from particular instances and (2) proof applies only to deductive arguments.

The absence of valid inductive logic—some would say the impossibility of an inductive logic—is the fundamental reason that we cannot empirically prove a knowledge claim or a theory. Since we cannot test the theory in all times and places where it applies, such proof would require a valid way to reason from a collection of singular statements to a set of universal statements. But there is no way to justify such reasoning intersubjectively.

These are the reasons beyond our assertion in chapter 5 that there is no such thing as a empirical proof. We speak of studies providing *support* or *supporting evidence* for a theory or knowledge claim. If an observation statement is inconsistent with the hypothesis, then we use the term *disconfirmation* and report that the study disconfirmed the theory or knowledge claim. When we examine the *confirmation status* of a theory, we are looking to summarize the overall support compared to disconfirmation from the empirical studies that have evaluated the theory. Again we must emphasize that these terminological distinctions do not represent excessive refinement nor are they semantic quibbles; they call attention to the provisional nature of scientific theories.

One more area of confusion surrounding "induction" deserves some comment. It is important to distinguish between justification as a logical process and idea generation as a psychological process. There is no question that we regularly think "inductively." We often arrive at universal statements by thinking about particular instances. How we do this is a problem for psychology. What needs to be emphasized is that such inductive leaps cannot be justified logically, but are acts of creative imagination. The statements that rise from our "inductive" thinking are conjectures that must be tested by means other than the psychological processes

---

4. Some people believe that statistical inference solves the problem of induction because statistical inference enables us to draw conclusions about the characteristics of a population from knowledge of the characteristics of a sample. Study of statistical inference makes it clear that the procedure is deductive, requiring assumptions about both the population and the sample and then deducing conclusions from them.

which generated them—the fact that we thought of these statements does not confirm their validity. To be sure, such conjectures are necessary acts, but they can in no way be regarded as having the same status as deductive arguments. Deductive arguments can be collectively justified. To put this another way, one person's inductions from particular instances need not bear any resemblance to another person's inductions.

Hence, recognizing the distinction between creating ideas and collectively justifying them leads to an inescapable conclusion: ideas that may be generated by a psychological process of induction need to be subjected to other tests before we can say the ideas have been collectively evaluated. In other words, when one creates KC 1 by generalizing from OS 1, KC 1 must be treated as a conjecture for future study. Thinking inductively, or generalizing, can generate ideas; but such ideas are proposals which remain to be collectively evaluated.

We must underscore one implication of the discussion to this point. Our argument maintains that the notion of establishing valid knowledge through generalizing from particular studies, although widely held, is untenable. No matter how well done a particular study or series of studies may be, the resulting observation statements cannot establish the truth of a universal knowledge claim. The practitioner who understands this argument will not ask whether a study done in Chicago can be generalized to San Francisco. The appropriate question for such a practitioner would be whether the ideas tested in a study in Chicago are applicable to the situation in San Francisco. The question can be answered by valid deductive argument if there are explicit, nontrivial scope statements, and does not require an inductive logic.

The scientist and the practitioner are quite appropriately concerned with the issue of generalization. But focusing on generalizing the results of a study puts this concern in the wrong place. We assert that the idea of generalizing can be appropriately applied to knowledge claims and theories, and we will explicate a meaningful concept of generalization—one that avoids the problem of induction—in chapter 15.

Returning to Case 2, we should emphasize that it is perfectly legitimate to formulate KC 1 as a possible explanation for OS 1 or, in other words, as a conjecture for future study. In so doing, we use an empirical study heuristically, that is, as an idea-generating activity. But one must clearly understand that OS 1 does not provide evidence for any conjecture generated after the fact. Because the explanation, KC 1, was constructed after we knew that OS 1 was true, OS 1 could not possibly contradict KC 1; if OS 1 was the only empirical result relevant to KC 1, then in this study KC 1 could not be wrong. The study, then, not only fails to prove KC 1—we have banned the idea of proof from the discussion of empirical outcomes—but it does not even provide supporting evidence. It has not increased the confirmation status of KC 1 at all. While KC 1 might turn out to be an important and useful idea, receiving empirical support from a large number of studies conducted over a wide range of conditions, only future investigations in which the possibility of disconfirmation exists will show that.

If Not Proof, What About Disproof?

It is a rule of logic that, if a conclusion is false, then the set of premises from which it is derived is also false. Can we not use this fact to define the principal function of empirical research as disproving theories and knowledge claims? Popper (1959) proposed a criterion of *falsifiability*, the possibility that observation statements can contradict knowledge claims, as the basis for distinguishing scientific knowledge claims from other kinds of knowledge claims. His criterion of *demarcation* is useful in marking off science from other intellectual activities and is the basis of the principle of testability: *To be empirically testable, scientific knowledge claims must be falsifiable.* Unfortunately, Popper's criterion, while successful as a rule of demarcation, does not solve our problem, since attempting to disprove a knowledge claim has similar difficulties to those that rule out proof.

First of all, if a hypothesis is falsified, the *set* of premises from which the hypothesis is derived is also falsified, but that set contains more than the theory or knowledge claim. In order to derive the hypothesis, a singular statement, from the universal statements of a theory, we need a number of linking statements, statements of initial conditions. The falsity of one or more of these initial conditions makes the set false. Hence it is perfectly consistent for a hypothesis to be false even if the theory is true as long as one or more initial conditions are false. Since sets of initial conditions are often quite complex, as we saw in chapter 5, it is not usually possible to locate which premise in the set generates the falsification.

Second, an hypothesis is one instantiation. What should be our position in the case where one hypothesis is falsified while other instantiations are supportive of a theory? Logically such a case is disproof of the theory when conjoined with the initial conditions, but if such disproof directs us to abandon the theory, then virtually every scientific theory would have to be abandoned. Since even the most widely supported theories have some falsified hypotheses, such disproof would be a poor guide to the way scientists should behave and we certainly would not want to make disproof the principal objective of empirical research. And in fact scientists rarely do empirical research to falsify their theories and seldom abandon their theories in the face of falsification. Typically, a theory or knowledge claim is abandoned when there is a better one to take its place.

IS PROOF POSSIBLE IN EXPERIMENTAL RESEARCH?

Some readers of this analysis might claim that we have "stacked the cards." They might argue that we find these limitations of empirical research because we have examined nonexperimental investigation and that, if we analyze rigorous experiments, we would reach very different conclusions. Other critics would claim, in addition, that there are types of nonexperimental research which enable researchers to make more powerful inferences than our analysis would allow. We believe that the limitations we have described are perfectly general and apply to empirical research, whether it is experimental or nonexperimental, whether it is more or less rigorous, whether it is quantitative or qualitative, whether it is historical or ahistorical. To demonstrate our contention we will apply our analysis to experimental studies and then note some differences between experimental and nonexperimental research. We will examine these issues in depth because we believe the implications of our analysis are critical to the development and evaluation of empirical research.

In sociology, considerable controversy surrounds the *experimental model*. Some people argue that the model is inapplicable in nonexperimental research, that much sociolgical research is nonexperimental, but that sociologists apply this model apishly (see Lieberson, 1985, pp.4-6). Others go so far as to dismiss all nonexperimental research as unscientific, and some use this argument to claim that sociology cannot be a science. Although there is some value to the arguments of these critics, there are many difficulties with them. In the first place, there are many experimental models rather than the one implied by the critics, so that it is not always clear what targets they have in mind. Second, while critics attack the use of the model, they invest the experimental method with magical powers in those circumstances where they believe it is applicable, powers of empirical proof. Contrary to these writers, no experimental model solves all problems of inference even when it is perfectly applicable.

Let us modify our previous example to fit with traditional discussions of experimental models. We will call this Case 3. A social psychologist wants to do an experiment to test KC 1 theorizing that the mere presence of other people in the absence of any interaction and any direct task assistance will result in higher productivity. Our investigator reformulates KC 1 and asserts scope and initial conditions:

KC 1*: Social facilitation increases productivity.

SC 1*: KC 1 applies to work situations where individuals are motivated to be productive.

SC 2: KC 1 applies to work situations where individuals have the necessary skill to perform the task.

IC 1: The mere presence of other people who do not interact with the individual, do not provide the individual with any task assistance, and do not exert any pressure to perform on the individual is an instance of social facilitation.

Our investigator decides to use proofreading as the task for the experiment and formulates three additional initial conditions:

IC 2: Speed and accuracy of proofreading are instances of productivity.

IC 3: Proofreading is an instance of a task for which college students have the necessary skills.

IC 4: Paying college students in this experiment according to their productivity will make the task an instance of one for which workers are motivated to be productive.

The experiment is designed to test the hypothesis:

H 1*: People who proofread in the presence of other people will complete more work and make fewer errors than people who proofread in a room by themselves.

Students in Introductory Social Psychology are randomly assigned to one of two experimental treatments; treatment C represents the presence of social facilitation and treatment $\tilde{C}$ represents its absence. All subjects work on the proofreading task and no communication is permitted in the C treatment. Unlike our example of the company study, the experiment takes place in a short time period under highly controlled conditions. Students come to the laboratory every hour from 9 A.M. to 10 P.M.; they are given instructions for the task and then placed in a room to work for one hour. Half the students work by themselves and the other half work in a room with three other people. The three others are confederates of the experimenter who have been trained to do their task at a fixed rate of speed and who have been instructed to take a break of five minutes after thirty minutes of work. In the C treatment, all subjects sit facing a different wall with their backs to one another. Data from the experiment support the hypothesis.

This example, except for its simplicity, comes close to an ideal experiment. Does it prove KC 1*? Or, using the terminology of causal inference, does it prove that there is a social facilitation effect? Because in this one study the observation statement is consistent with the hypothesis, we cannot guarantee that all such studies will have the same outcome. Experiments must operate in the same world as nonexperimental research—a world without a valid inductive logic. Even if the study were repeated a dozen times with the same result, we could not guarantee the outcome of the thirteenth repetition. In short, neither the procedures of this experiment nor the statistical models on which it is based provide a valid inductive logic. The reasoning which leads to this conclusion is identical to that we presented in analyzing nonexperimental research. Many writers will agree that proof is not possible, but then proceed in their analyses as if it were possible. Cook and Campbell (1979), for example, echoing the earlier work of Campbell and Stanley (1963), assert that "experiments probe but do not prove causal hypotheses" (p. 18). Apparently probe means "approximately prove" since these authors are concerned with *external validity*, which they define as the "approximate validity with which we can infer that the presumed causal relationship can be generalized to and across alternate measures . . . and across different types of persons, settings, and times." Their external validity, however, involves exactly the type of inductive reasoning which, in the absence of an inductive logic, cannot be intersubjectively justified.

But there is a more subtle view of experimental proof. Its adherents are concerned with making causal inferences such as, for example, inferring that working in the presence of others is the cause of increased productivity. Some of these researchers believe that if one does a proper experiment, one can prove the causal connection between an antecedent treatment and an outcome for the single instance of the experiment. Then if one can reproduce the experiment, researchers can act as if the causal connection were proven in general. We have avoided, and will continue to avoid, discussion of causality because the concept presents profound methodological and philosophical difficulties.

We can, however, deal with the issues involved. In our terms, the analogous problem of proof can be phrased as, "Given an hypothesis and a supporting observation statement, can one prove that a knowledge claim or theory that explains the hypothesis is true for the single instance of the empirical study?"[5] If we can and if we are able to reproduce the experiment, why is it not reasonable to act as if we have proven the knowledge claim?

Let us examine the second question first. Cook and Campbell (1979) write as follows:

> In choosing, however tentatively, a "true" theory, the scientist . . . operates in a practical way knowingly using an invalid deductive schema of the following sort:
> > If Newton's theory A is true, then it should be observed that the tides have period B, the path of Mars shape C, the trajectory of a cannonball form D.
> > Observation confirms B, C, and D.
> > Therefore, Newton's theory A is true. (p. 21-22)

This schema is invalid because it commits the logical fallacy of affirming the consequent.[6] Yet Cook and Campbell argue that the logical predicament can "usefully be interpreted in relevant ways, and indeed provides a framework in which our emphasis is on plausible rival hypotheses." While these authors may be able to develop an appropriate framework for empirical research from such a starting point, in general, it is not a good idea to base a methodology on a logical fallacy. Furthermore, this logical error has led many researchers to a mistaken view of what experiments can accomplish. The example quoted attempts to cloak inductive inference in an erroneous deductive argument. In other instances, researchers do not even bother; they simply use terms like "probably true" or "approximately true" and act as if they have demonstrated approximate or probable truth. But these qualifications are equally erroneous and misleading—for there is no way to infer approximate or probable truth either.

---

5. This question addresses issues involved in what writers such as Cook and Campbell refer to as the *internal validity* of an experiment.

6. As an exercise, the reader could apply truth table analysis from chapter 12 to see the difficulty with this line of reasoning.

To answer our second question, then, we assert that it is not reasonable to act as if a reproducible experiment has proven our theory. Furthermore, it is not necessary to act that way. It is not necessary to base a framework on such fallacies; rules and strategies can be developed which do not require a misuse of deductive reasoning nor a logic of induction.

What is required to answer the question of whether we can prove that our theory or knowledge claim is true for the single instance? If H 1* is supported in our experiment, and KC 1* together with IC 1 through IC 4 explain H 1*, can we show that the explanation is true for this one time and place without committing the error of affirming the consequent? The answer is "yes, in principle, but. . . ." Applying the nonuniqueness theorem introduced at the beginning of this chapter indicates that, whatever the hypothesis, there are alternative possible explanations of it. We claim that some of these alternative explanations are consistent with the truth of H 1* *and* the falsity of KC 1*. If we could show that all of these alternative explanations were false, then the alternative explanations that remained would be only those that were consistent with the truth of KC 1* and that would prove the truth of KC 1* in the single instance. Some writers believe that experiments allow us to do this, but unfortunately, no experimental method yet devised can achieve even this limited objective.

Some Experimental Models

To understand why experiments cannot prove the truth of a theory or knowledge claim even for the single instance, we need to examine some general models that are the basis of experimental research. The experimental example in our Case 3 employs a particular experimental model that involves the comparison of two "treated" samples (actually a "treated" with an untreated sample). Suppose OS 1*, the average number of pages proofread by students in treatment C is greater than in $\tilde{C}$, describes the outcome of the experiment. While that result is consistent with H 1*, it could be consistent with either the truth or falsity of KC 1*. Assume KC 1* is false, that is, social facilitation does not increase productivity. We could still obtain the experimental outcome characterized by OS 1* if the students in treatment C were, on the average, better proofreaders than the students in $\tilde{C}$. If we insure that the two treatments are equal in average proofreading skill, then the statement, "KC 2: Differential skill at a task increases task productivity," cannot explain OS 1*. We have therefore ruled out *one* of the alternatives consistent with the truth of H 1* and the falsity of KC 1*. Most experimental models employ and extend this logic.

The model underlying Case 3 assumes that the two samples are equivalent with respect to the outcome variable before the treatment. By insuring that C and $\tilde{C}$ were equivalent in average proofreading skill, we presumably meet this requirement of the model. In addition, the model presumes the absence of factors other than the treatment that might explain the outcome. Insuring the equivalence of

C and ~*C* with respect to factors other than the treatment that might explain the outcome, whether the knowledge claim is true or false, is known as instituting controls. The problem of proving that KC 1* is true for this instance, then, requires, for a given knowledge claim, that we control all factors that represent alternative explanations to KC 1* (or at least those that can explain the outcome when KC 1* is false). But in practice that is impossible. In this respect, experimental and nonexperimental research are alike; neither provides a method of proof.

Experiments, where it is possible to conduct experiments, do have some advantages over nonexperimental research. Experiments, for example, allow greater control of factors that might represent alternative explanations, but no experiment—or series of experiments—enables us to eliminate all alternative explanations. Hence, they do not enable us to prove the truth of a knowledge claim. Those who believe that the experimental method is a method of proof have not thought the matter through. Some of these people even go further and claim that the experimental method is not only a method of proof but also a method of discovery; they believe that experiments allow us to discover which factor explains the outcome and to prove the truth of that explanation. Such beliefs rest on fundamental errors, but, in fairness to these researchers, we should note that these errors have a long intellectual history.

To gain an appreciation of what experiments can and cannot do, we must understand the issues. Many of the current experimental models have their roots in John Stuart Mill's methods of experimental inquiry and contemporary models draw heavily on the work of the statistician, R. A. Fisher. Cook and Campbell, for example, "deliberately adopt an outmoded position derived from Mill's inductivist canons..." (1979, p. 1) and modify it with randomization techniques drawn from Fisher. "Random assignment is the great *ceteris paribus*—that is, other things being equal—of causal inference" (p. 5). But Mill's canons cannot be used inductively and random assignment is more limited than these authors recognize. Both Mill's canons and randomization can play significant roles in empirical research, but their proper use depends on a clear understanding of their limitations. We will first analyze one of Mill's methods and after that examine randomization.

Over fifty years ago, Morris R. Cohen and Ernest Nagel (1934) demonstrated that Mill's canons were neither methods of discovery nor methods of proof. Their analysis was both elegant and cogent and has not been improved upon in the intervening years. Our discussion owes much to their critique of Mill's Method of Difference.

Mill's statement of the Method of Difference is:

> If an instance in which the phenomenon under investigation occurs and an instance in which it does not occur, have every circumstance in common save one, that one occurring in the former; the circumstances in which alone the two instances differ, is the effect, or the cause, or an indispensable part of the cause (Mill, quoted in Cohen and Nagel, 1934, p. 256).

Our Case 3 may help understand Mill's nineteenth century English. The two "treatments" represent two instances, one where performance was "high" and one where performance was not. The two treatments are alike in time spent on the task, the number of men and women in each, the fact that both did proof-reading, both involved no interaction with others and no pressure from others to produce; they differ in that one received the treatment of other people present and the other did not. These then are the circumstances in common save the one in which they differ.

In table 13.1, we present the Method of Difference abstractly. Row I represents the group which received the treatment C and in which the phenomenon, outcome E, was present; Row II represents the group that did not receive C and in which E was absent. A and B stand for the circumstances the two instances have in common.

Table 13.1 Mill's Method of Difference

	Controls		Treatment	Outcome
I	A	B	C	E
II	A	B	$\tilde{C}$	$\tilde{E}$

Is Mill's Method of Difference a method of proof? If higher productivity occurs in the sample where people worked in groups, have we proven KC 1* is the explanation, that social facilitation results in higher productivity? If KC 1* is the only possible explanation of H 1*, then we would be justified in making the generalization: Whenever H 1* is true, then KC 1* explains it. Note that we still have no basis for generalizing H 1*. But even our conditional statement cannot be justified; the nonuniqueness theorem precludes KC 1* as the only possible explanation. Suppose, however, that we knew all the alternative possible KCs that could explain H 1* and further, that the initial conditions that link these alternatives to H 1* were all false. Then we could prove for this instance that the presence of others and not something else produced the improvement. We can only justify such an inference if the As and Bs in our diagram represent all the factors in which the two samples could differ and we can also rule out the inference that difference between the two was accidental, that is, due to chance. Statistical procedures allow us to deal with accidental differences, so let us assume that we can rule out chance as the explanation of the results of our experiment; but can we assume that the As and Bs represent all of the alternative KCs that could explain H 1*?

Our two samples can be characterized by a large number of variables: gender composition, average age of subjects, average year in college of subjects, and personalities of the people in each sample, to mention only a few. Is it ever reasonable to assume (1) that we know all the possible explanations for either the support of H 1* or its disconfirmation and (2) that our two samples are equal on all factors in that very large set of variables? In terms of table 13.1, aren't there always unknown or unstated circumstances X, Y, and Z on which the two instances differ? Furthermore, it does not help matters to say that the two instances do not need to be equated on all variables but only those variables that are relevant to outcomes, E and $\bar{E}$. Unless we know the total list of factors that affect the outcomes, we can never assert that the two instances are equivalent with respect to all relevant factors. Since we do not know this list nor do we know all the variables on which the two instances may differ, we never know that we have not overlooked one or more relevant factors. We therefore cannot rule out the possibility that X, Y, and Z represent alternative KCs that explain H 1* when KC 1* is false, and hence cannot prove that KC 1* is true in even the single instance of our experiment. The Method of Difference, then, does not accomplish the limited proof of truth in the single instance.

A similar argument demonstrates the failure of the Method of Difference as a method of discovery. The investigator may find two instances, one in which E is present and one with E absent,[7] but nothing in the method indicates how to sort all the factors that might characterize the two circumstances into those that are relevant and those that are not. Furthermore, the method does not indicate how we should analyze complex factors, separating the As and Bs from the Cs. In short, the use of the method depends on an *a priori* analysis and the formulation of an *a priori* hypothesis, and thus is not a method for discovering explanations.

This analysis does not intend to suggest that the Method of Difference is useless or that we should not do experiments. However, before we can consider how the Method of Difference and experimental methods in general can be appropriately used, we must take our analysis further and deal with some objections that can be raised to what we have presented so far.

Randomization—No Solution

Some critics might suggest that we have set up a straw man that we could easily knock down. They might claim that the analysis of Mill's Method of Difference ignores the modification of this model that resulted from the contribution of R.A. Fisher and other statisticians. They argue, as the *ceteris paribus* comment of Cook and Campbell implies, that randomization provides a procedure that allows us to equate our samples on all factors, even those of which we are unaware. As usually used in social science discussions, randomization means randomly assigning subjects to experimental treatments, and random assignment is what distinguishes experiments from quasi-experiments in these discussions. As the terminology implies, quasi-experiments are only approximations to the real thing and, according to some writers, somewhat less scientific. But we contend that randomization does not solve the problem of proof; neither quasi-experiments nor experiments based on the most sophisticated statistical models are methods of proof or methods of discovery. And the assertion that experiments are more scientific or that quasi-experiments and nonexperiments are unscientific is insupportable. We need to look

---

7. Although this discussion deals with dichotomies—present and absent, same or different, it can easily be extended to situations where the As, Bs, Cs, and Es are continuous variables.

extraction, 153, 173
Probability or random sampling
  cluster, 156
  multistage, 155
  simple, 66
  stratified, 67
  systematic, 63
probability sampling, 66
program design, 63
program implementation, 61
public access databases, 82
publication, 82
  book, 73
  camera-ready copy (CRC), 197
  copyrighted materials, 201
  doctoral dissertation, 204
  journal article, 204
  master's thesis, 201
  rewriting, 201
  thesis/dissertation, 201
  timing, 201
Q technique, 242
qualitative methods, 202
qualitative research, 153
quality control, 85
quantification, 80
quasi-experimental design, 100
questionnaire, 47, 90
  administered, 70, 71
  mailed, 77
questionnaire design, 77
  logical check, 77
questionnaire survey, 77
  interview, 98, 100
  Mailed, 83
questions, 88
  closed-ended, 88

open-ended, 89
R technique, 90
random assignment, 90
random numbers, 153
randomization, 72
rank relation, 63
recoding, 72, 135
  collapsing and categorization, 49
  dichotomization, 96
  dummy variables, 99
reduction, 99
reduction of data, 99
redundancy, 79
regression, 105
regression analysis
  statistical redundancy, 53
  stepwise methods, 184
  regression coefficient, 142
  regression line, 141
regression equation, 121
  nonstandardized, 142
  standardized, 121
relational research or analysis, 120,
  141
relative effects, 34, 176
relative scope of research
  case study, 34, 177
  census, 143
  sample study, 59
reliability, 59
  split-half method, 59
  test-retest method, 130
replacement sampling, 54
Request for Proposal (RFP), 54
research
  alternative approaches, 79
  courses, 68

elsewhere to find the methodological implications of the differences between experimental and nonexperimental research. First, however, we must demonstrate that our analysis of the Method of Difference also applies to experiments involving randomization.

Let us expand the scenario of Case 3, our social facilitation experiment. Suppose our sociologist had an eager research assistant who decided to analyze the data further and as a result came up with two additional findings. First, he noticed that time of day made a difference, and so he decided to compare "alone" with "in group" subjects, holding time of day constant. Dividing the day into morning, afternoon, and evening sessions, the analyst found that in the morning, people alone performed better; in the afternoon, people alone and people in groups performed equally well; in the evening, people in groups performed better. In short, the results of the experiment depended on the time of day. The overall difference came about because students were more available in the evening; many more subjects took part in evening sessions.

The second finding resulted from probing the data further. Our research assistant discovered that a greater number of sophomores were assigned to the "in group" treatment than were assigned to the "alone" treatment and this difference was most pronounced in the evening sessions. That imbalance is not at all inconsistent with random assignment—if you flip a coin ten times and repeat the ten flips a number of times, once in a while you will get nine, or even ten, heads by chance. But suppose it is true that sophomores are more eager workers than other students; the experimental result may be due to that alone and unrelated to the experimental treatment.

These examples may be a bit bizarre, but they are certainly not impossible and they illustrate the critical points we wish to make. If those who believe that random assignment makes Mill's Method of Difference viable by equating instances on all the unknown factors (the Xs, Ys, and Zs) are right, then we should be able to justify the inference that working in groups improve an individual's performance. But our examples indicate why random assignment does not resolve the issues.

The finding that time of day makes a difference illustrates one general problem that random assignment of individuals to treatments does not affect at all. Factors that are not characteristics of individuals may affect the outcome either directly or because the treatment factor operates differently at each value of the factor in question. When the treatment behaves differently in the morning, afternoon, and evening sessions, we have what statisticians call *interaction effects*. If the investigator is aware of potential interaction effects, he or she can design the experiment to deal with them, but that does not solve the problem we started with. In order to justify the inference that working in a group improved performance, we have to rule out the possibility that some *unknown* factor was responsible for the improved performance. Our first example illustrates the fact that random assignment does nothing to equate instances with respect to factors that are not characteristics of individuals which may produce either direct or interaction effects.[8]

Random assignment of individuals to treatments is not supposed to affect factors which are not characteristics of individuals, but the procedure is supposed to equate treatments on both known and unknown characteristics of the experimental subjects. Yet our second example illustrates failure there too—more sophomores were randomly assigned to the "in a group" treatment. As we noted, by chance that can happen. Unfortunately, in some quarters, random assignment becomes a magic wand to be waved at everything instead of used carefully, self-consciously, and analytically. If one analyzes random assignment using principles from probability theory, two results emerge. First, for any given factor, the probability that two treatments differ on this factor approaches zero *in the long run*. Experiments, however, are not done in the long run and so there is always a non-zero probability that the experimental treatments differ substantially on the given factor. The second, and more paradoxical, result follows from the first: The greater the number of unknown factors that may be involved in an experiment, the greater the likelihood that experimental treatments differ substantially on at least one of these factors.[9]

This brings us back to the original problem. To justify the inference that working in groups improves performance requires the elimination of all competing inferences that explain the outcome when KC 1* is false, but random assignment does not make that possible. In the short run, it is almost guaranteed that there is some alternative inference that is equally well-supported by the experiment.

Our purpose is not to demolish experimental methods nor to suggest that random assignment is an inappropriate procedure. Rather, it is to provide a foundation for a proper appreciation of the place of random assignment in particular and experimental methods in general. Such an understanding will also support the view that there is *not* a radical discontinuity between experimental and nonexperimental methods. What we have demonstrated thus far is that random assignment does not rescue Mill's Method of Difference as an inductive canon; even when used in conjunction with randomization and statistical reasoning, it is neither a method of proof nor a method of discovery. Furthermore, this assertion applies not only to the Method of Difference but also to experiments in general:

Experiments are neither methods of discovery nor methods of proof.

While genuine discoveries sometimes occur as a result of experimentation, no method known to science can produce discoveries on demand. And proof that

---

8. The reader might object that the interaction effect of the example still shows that the experimental treatment had an effect, if only in the evening sessions. But making an inference about a conditional effect is not the inference that motivated the original experiment. To preview what is ahead: experiments help evaluate the experimenter's analysis of the phenomenon; hence the proposition the experiment is designed to evaluate is critical, and different propositions require different experimental designs.

9. If we assume the treatments' differing on one factor does not affect their differing on another, then a simple analogy demonstrates the point. Suppose we have an urn with nine white balls and one black ball and a trial consists of drawing one ball from the urn and then putting it back. On each trial, we have one chance in ten of drawing a black ball, but as the number of trials increases, we will eventually draw a black ball.

a treatment is invariantly related to an effect entails requirements that no experiment can possibly meet, as we saw in our earlier discussion. Hence the widely held but unexamined belief that experiments are superior to nonexperimental research, because they allow proof or discovery, will not stand up to close scrutiny.

For those people who cannot live without proof, the consequences of our analysis are quite unsatisfactory, but that is the nature of the research situation. And it is the case despite all the claims—even those made by scientists themselves—that "research has proven. . ." For people who are able to cope with the ambiguities of this world, experiments despite their inherent limitations—and there are some additional limitations that we have not considered—are extremely powerful tools.

WHAT EXPERIMENTS CAN DO

What value, then, does the Method of Difference have? If we view our factors A, B, and C of table 13.1 as representing alternative explanations, we can delineate a clear and important way to use this and other experimental models. Recall the nonuniqueness principle—whatever explanation we construct, someone can always come up with an alternative. (Since we regard an explanation of the result as the most important inference we can make from an experiment, it is clear that we can always make alternative inferences.) The nonuniqueness principle, however, does *not* imply that all explanations are equivalent. Some are better than others. An important use of Mill's Method of Difference and of experimental models in general is to assist us in evaluating alternative explanations.

Another fundamental principle underlying this book is that research is only as good as the analysis on which it is based. When we observe an effect, we want to analyze possible explanations for the effect. When we have a treatment factor, we want to analyze its consequences and the conditions under which those consequences occur; in other words, we want to propose statements about our treatment as a possible explanation for observation statements describing these consequences. If the analysis enables us to reproduce the effect or effects in times and places of our own choosing, then we are likely to consider the analysis sound. In one sense, an experiment is a demonstration of the ability to produce an effect. Sometimes experiments are carried out without much analysis behind them: "Let's see what happens if we apply treatment T to object O." Such experiments are not very informative; if object O changes, the experimenter does not know what factors in addition to T, or instead of T, contributed to the changes. Similarly, if O does not change, he does not know that T is ineffective, since some other factors may have interfered with the operation of T. Hence, while some experiments represent "fishing expeditions" with only minimal analysis as a foundation, our discussion will focus on experiments based on serious analyses of the problem.

Experiments serve their most important function by providing a basis for choosing among alternative explanations. Consider the method of difference again. Now let *A, B,* and *C* in table 13.1 represent factors that are contained in alternative explanations of *E*. Once again, let *E* be high(er) productivity in proofreading (relative to *Ē*). Now suppose we have three candidate explanations for the observed experimental result:

1. Social facilitation increases productivity.
2. People working with familiar material are more productive than people working with unfamiliar material.

   H 2 Experimental subjects will proofread text dealing with familiar topics faster and more accurately than text dealing with unfamiliar topics.
3. Women are more meticulous than men and thus are more productive on those tasks requiring meticulousness.

   H 3 Women will proofread faster and more accurately than men.

We omit the scope statements and initial conditions that allow us to deduce H 2 and H 3 from their corresponding KC statements. Since the process is the same as illustrated for KC 1*, the reader should have little problem in formulating these statements for KC 2 and KC 3.[10] Now we set up our experiment to evaluate KC 1 versus KC 2 and KC 3. We attempt to construct a situation in which KC 2 and KC 3 could *not* explain the outcome whereas KC 1* possibly could. If such an experiment came out as expected then KC 1* would be preferred over the others as an explanation of the result. We use the following principle:

No factor can explain both an outcome and its opposite.

Once again, our treatment factor is the presence (*C*) or absence (*Ē*) of others working in the same room and we use the same procedures to produce it that we described earlier. To insure tht KC 2 cannot explain the outcome, we choose text dealing with topics that are unfamiliar to our subjects and use the same text in both the *C* and *Ē* samples. Furthermore, we interview subjects afterward and discard from our analysis any who say they are familiar with the topics covered in the text. Thus both samples are equated on factor *A*. To equate them on the gender composition of the treatment, factor *B*, we use either only men or only women or we insure that an equal number of both men and women are present in each treatment sample.[11] If the "in group" sample has high productivity (*E*) and the "alone" sample has low productivity (*Ē*) and we have conducted the experiment

---

10. We should also note that KC 2 and KC 3 could have explained the result of the original experiment, if subjects in the "in group" treatment had by chance been working on familiar material, while those in the "alone" treatment were not, or if there were more women in the "in group" situation and more men in the "alone" treatment. This is another illustration of factors potentially present that, unknown to the researcher, could explain the result.

11. Of course, we could do a more complex study and evaluate KC 1 and other explanations for males and females separately. That would not introduce any new principles and would complicate this discussion.

appropriately,[12] have valid measures of the relevant factors, and have not obtained an accidental result, then, applying the principle, A and B cannot explain both E and $\bar{E}$. Provisionally, then, we prefer KC 1* to KC 2 or KC 3 as the explanation of the outcome. To put it another way, we have eliminated explanations involving factors A and B. The Method of Difference, then, is a *method of elimination*, providing of course that E and $\bar{E}$ are reproducible results.

We have described the use of the Method of Difference to eliminate explanations based on factors A and B when the experiment "worked," that is, when the "in group" treatment results in E and the alone treatment $\bar{E}$. What happens when the experiment fails, when both treatments result in E (or both in $\bar{E}$)? A principle based on Mill's Method of Agreement provides a guideline:

No factor can explain an outcome satisfactorily that is not common to all occurrences of that outcome.

Using this principle, the treatment factor (C) is not present for the two occurrences of E and so explanations involving this factor are eliminated. Of course, the requirement that the experimental outcome be reproducible also applies to this case.

Experiments, then, allow us to choose among alternative explanations by eliminating those that one of our two principles rule out on logical grounds. When we use a model based on the Method of Difference and the result is E in one sample and $\bar{E}$ in the other, we eliminate explanations based on A and B but do not eliminate explanations based on C. We provisionally accept C explanations because we have tested them and have *not* eliminated them. If our experiment produces E (or $\bar{E}$) both when C is present and when C is not, then, applying the second principle, we eliminate explanations based on C, but provisionally retain those based on A and B. Of course, the failure of an experiment designed to produce E and $\bar{E}$ does not tell us much about A and B, but, since we equated our samples on A and B, there was some reason to believe that factors A and B might be involved in explaining an E, $\bar{E}$ outcome. An E, E (or $\bar{E}$, $\bar{E}$) outcome provides no reason to alter this original belief, and so A and B explanations are still viable. To be sure, the "no difference" outcome is not as informative as the outcome where the samples differed; in the E, $\bar{E}$ case we eliminated all the alternative explanations connected with those factors on which we equated the two samples (A and B stand for many factors) whereas in the "no difference" case we only eliminated explanations based on the treatment factor C. Furthermore, the A and B factors on which we equated the samples may represent quite unrelated ideas so that both

---

12. To fully spell out "appropriately" would take a book by itself. What we intend to convey here is the idea that the experimenter carried out the experiment in such a way that positive and negative outcomes were both possible, that is, the experimental techniques did not force a particular outcome for reasons quite irrelevant to the treatment factor. For example, many social-psychological experiments have been criticized for producing artifactual results by creating pressures on the subject to conform to the wishes and expectations of the experimenter. This is known as *experimenter effect* and, when it is a factor and is not dealt with, the experiment has not been conducted appropriately.

A and *B* cannot contribute to the development of an interrelated set of knowledge claims. In our example, we are not much farther ahead if we know that both "familiarity with the material" and gender may be viable explanations of productivity. Further experiments opposing *A* factors to *B* factors where one or the other is varied (treated like *C* of the previous case) would be required. Of course, doing such studies would depend on the theoretical relevance of, or substantive interest in, *A* and *B*.

We conclude from this analysis that the major function of explanatory research is to evaluate one or more explanations against a *limited set* of alternatives. The models we have discussed apply to both experimental and nonexperimental research; the Method of Difference can also be employed in nonexperimental research using statistical procedures that produce comparisons on *C* where *A* and *B* are held constant. Thus experimental and nonexperimental research can employ the same logic, at least in terms of this model, and both share the aim of eliminating some explanations while provisionally retaining those that are not eliminated. Research is designed to provide the means for comparing alternative explanations and eliminating some of those compared. Procedures such as sample selection, random assignment, controlling variables by experimental intervention or with statistical techniques, and documentation of sources in historical research all facilitate the elimination of alternatives. Furthermore, good practice requires a demonstration of reproducibility for all provisional elimination as well as provisional acceptance.

We commented earlier that experiments have some advantages over nonexperimental research. Our analysis enables us to spell out three respects in which experiments are more efficient:

1. Experiments are more easily reproducible than nonexperimental studies.
2. Experiments allow the researcher to eliminate more alternative explanations than nonexperimental research.
3. Experiments enable the researcher to separate and isolate factors that are not separable in nature and thus to make comparisons that are not possible where the factors are confounded.

These efficiencies suggest that when a problem is equally amenable to experimental or nonexperimental investigation, researchers are more likely to do experiments. However, they cannot support the view that experiments are totally distinct from other research procedures. The same logical justifications and the same problems of inference underlie both experimental and nonexperimental research.

To be sure, there are many phenomena for which it is either premature or impossible to do experiments; some ideas are not amenable to experimental study because they cannot be operationalizd in a laboratory. Processes that must operate over a long time span to have an effect represent an important class of problems that researchers cannot investigate experimentally. In addition, for some problems ethical considerations rule out experimental research. Nonexperimental research in these cases is perfectly appropriate scientific procedure.

There are also processes, however, for which experiments are particularly suitable—for example, the processes that require actors to be consciously aware of the factors that are operating. In choosing to do experimental or nonexperimental research, the researcher must weigh the advantages and disadvantages of each type of research for the particular set of ideas under investigation.

LIMITED VALUE OF THE SINGLE STUDY

We have shown that one cannot generalize from a single study and that one cannot prove that an explanation is true even for the instance of the single study. One can compare alternative explanations in a single empirical study. However, demonstrating the reproducibility of the provisional elimination or acceptance of an explanation requires at least a second study. Furthermore, our conclusion that an explanation is evaluated *relative* to a limited set of alternatives provides another strong argument for the view that a single study is not very informative.

It should be obvious that we can compare only a limited number of alternatives in a single study. While we are designing an investigation, we can usually think of more alternative explanations than those we choose to test; hence, we always make choices and these choices require scientific judgment and analysis. Such judgments and choices are not at all compatible with mechanical applications of things like the Method of Differences. In addition, different investigators dealing with the same problem, the same theory or knowledge claim, and the same resources are still likely to make different choices. Thus there is no "right" study, no "best" study, and not even an "optimal" study for a given problem.

Finally, we must abandon as untenable the widely held notion which is such a key element of the mythology of science, the idea of a "crucial experiment" that in one stroke decides the fate of an idea. Even for the purpose of choosing explanation C over explanations A and B, one experiment is not sufficient. At a minimum, the researcher must demonstrate that the experimental outcome is reproducible and, as we have emphasized, this cannot be done in a single experiment. Reproducibility is so important that, in some disciplines, journals will not publish research that has not been replicated.[13] In addition to the requirement of reproducibility, however, another consideration rules out the crucial experiment. The hypotheses, procedures, and measures of an experiment represent only one set of instantiations of the concepts contained in explanations A, B, and C. An experiment that stands alone without a context of supporting research cannot address the reliability and validity of the procedures and measures employed. If we

---

13. Social science journals have not adopted this policy and replication remains a controversial matter. As a student, I was taught that, from a statistical point of view, exactly replicating an experiment is equivalent to increasing the number of subjects. While that is true, too many people draw the false conclusion that it is unnecessary to replicate an experiment if the results are statistically significant. Beyond increasing the number of subjects, replication is one way to demonstrate reproducibility, and that cannot be demonstrated merely by enlarging the original experiment.

recognize that our objective goes beyond comparing the particular hypothesis represented in C to those represented by A and B to a more general evaluation of the explanatory ideas involved, then we can appreciate the inadequacies of the ''crucial experiment'' approach. Only a series of experiments can serve our objectives; in other words, evaluating our ideas requires a program of research.

The fact that there is no crucial study or no optimal study does not imply that all studies contribute equally to the development and evaluation of theories and knowledge claims. Although we believe that discussions of research and training of researchers have put too much emphasis on the design, execution, and analysis of the single study, we certainly agree that researchers and consumers cannot be indifferent to the quality of the single study. Precisely because single studies can only evaluate limited numbers of alternative explanations, the investigator must pay particular attention to choosing which alternatives to evaluate, to formulating hypotheses, and instituting appropriate controls in order to evaluate them. Because integrating single studies into research programs enhances their value, investigators have special responsibilities to build on prior research and to plan research that will follow up.

This discussion suggests a new criterion for the evaluation of research: the informative value of a study. We noted in the last section why a ''no difference outcome'' $(E, E$ or $\bar{E}, \bar{E})$ was less informative than an experiment in which treatment variation produced differential effects. The number of alternative explanations that can be eliminated is an important aspect of this criterion and the criterion applies equally well to experimental and nonexperimental research. Moreover, in this standard, we want to incorporate the fact that alternative explanations can have differential theoretical relevance or widely varying substantive interest. In keeping with our orientation to the production and evaluation of knowledge claims, we can ask how informative the research is to a set of ideas and particularly to the significant ideas in that set. In the next chapter, we will discuss ways to improve the informative value of empirical research as a foundation for the examination of research programs in chapters 15 and 16.

CONCLUSIONS

The analysis in this chapter has produced four conclusions: (1) Empirical research does not discover or prove explanations. (2) Mill's Method of Differences and randomization do not solve the problem of induction—there is no logic that justifies inferring universal conclusions from singular premises. (3) Using the Method of Differences, we can evaluate alternative explanations by eliminating those which cannot explain the empirical outcome of the study. (4) A single study can evaluate a very limited number of alternative explanations and thus single studies by themselves are of limited value.

From the last conclusion it follows that we must pay careful attention to the alternative explanations we wish to evaluate. It is even possible to ''stack the cards'' in favor of a particular explanation by choosing absurd alternatives and designing the study accordingly. Furthermore, as the nonuniqueness theorem asserts, for a given empirical result many explanations would fit equally well, and thus it is always possible to negatively evaluate a particular explanation. These considerations put an added burden on the analysis underlying the empirical study. And, since "stacking the cards" represents one inappropriate way to conduct research, we also require some constraints on those candidate explanations that enter the competition for empirical evaluation.

SUGGESTED READINGS

Cook, Thomas D., and Donald T. Campbell. *Quasi-Experimentation: Design and Analysis Issues for Field Settings.* Chicago, Ill.: Rand McNally, 1979. Ch. 1 and 2.

These authors present a somewhat apologetic version of the conventional view of experimental and nonexperimental research. They provide a philosophy of science base for their position, but are perhaps overly sensitive to Anti-Positivist arguments.

Lieberson, Stanley. *Making It Count: The Improvement of Social Research and Theory.* Berkeley: University of California Press, 1985.

Lieberson presents an unconventional critique of many aspects of social research. Both his critique and his proposals for improvement represent a somewhat different perspective from the position presented in this chapter. See particularly pages 4–6.

FOURTEEN

Making Empirical Research More Informative

While we have demonstrated that a single empirical study does not provide a sufficient basis for accepting or rejecting a theory or knowledge claim, the fact remains that the single study is the basic unit in our thinking about research. Both the lay public and the research community focus primarily on the outcome of one investigation. Although we believe that there is too much emphasis on the design and conduct of single studies in the training of sociologists, nevertheless, we must consider how to improve the single study.

Despite our emphasis on research programs made up of cumulative series of studies, we recognize that such programs have discrete steps and these steps are usually single studies. One of the most important features of a research program, however, is that the studies are connected, not isolated. Furthermore, how studies are linked to one another in a program generates a set of issues for research methodology that we have only begun to address. In the next two chapters we will examine some of these issues and develop some suggestions for the pursuit and evaluation of research programs. Since the single study may be the beginning or at least an integral element of a cumulative series of studies, however, we must examine a few critical issues in the planning and execution of single empirical investigations.

Once again we will focus the analysis on explanatory research, but many of the same issues and the same conclusions apply to exploratory or descriptive studies as well. In the last chapter we argued that the purpose of explanatory research is to allow the investigator to choose among alternative explanations by eliminating some and retaining provisionally the explanation or explanations that were not eliminated. Since a single study can test only a limited number of alternatives, the informative value of that study depends on careful analysis of the set of alternative explanations to be evaluated.

The first step, then, in planning a study involves examining the set of possible explanations that could account for the outcome of the investigtion. Some studies focus on the thing to be explained, the explanandum—for example, attempting to understand why people working in groups do better at a given type of task than people working alone. Other studies focus on evaluating a theory or knowledge

degree, 86
design, 246
funding, 13
grant, 242
guide, 29
in higher education, 15
learning, 242
limitation, 241
linear model, 14
preparation, 194
proposal, 179
publication, 243
purpose, 244, 7, 10
textbooks, 26
thesis, 19
topic, 242
research design
   cross-sectional, 242
   first-hand information, 15
   human subjects, 67
   longitudinal, 18
   secondary sources, 18
research document, 68
research framework, 18
research hypothesis, 36, 40, 45, 180
research languages, 34
research logic
   deductive, 41, 123
   inductive, 98
   generalization, 241
research methods, 39, 45
research phobia, 39, 45
research population, 186
research preparation
   pilot projects, 60
   reading, 9
   research assistantship, 248

specialization, 247
   typing skills, 244, 18, 22
research problem, 226, 18, 22
research proposal, 9
   guidelines, 244
research purpose, 31, 35
research questions, 86, 180
   narrowing down, 88
   topics, 35
   uncertainties and risks, 28, 31, 33, 35, 41, 189
research report, 28, 30
   audience, 28
   conclusions, 33
   content, 181, 189
   discussion, 181, 193
   purpose, 181
research skills
   data collection, 194
   data management and analysis, 193
   instrumentation, 181
   interpreting results, 245
   organization of the research team, 245
   pilot study, 245
   planning, 246
   qualitative, 245
   quality control, 245
   quantitative, 245
   sampling, 247
   statistics, 245
   use of computers, 247
   writing research report, 245
Research students, 247
research topic, 245
response rate, 246
sample size, 241, 7

claim as an explanans and derive a prediction which the theory explains; for example, Blau's theory discussed in Chapter 10 might allow the following prediction: "As universities grow in size, the number of administrators increases more rapidly than the number of faculty members." Historical data collected from a sample of universities could then be used to evaluate Blau's theory as an explanation. In both types of studies, maximizing informative value requires an *a priori* analysis of the set of alternative explanations.

Apprentice researchers are taught about the importance of having an hypothesis in explanatory research and they are also introduced to the idea of comparing their hypothesis with one alternative. This alternative is that chance explains the outcome of the study. But chance represents only one alternative explanation, and writers frequently comment that chance represents a weak alternative explanation. Even so, methodological writing and research training do not emphasize the importance of considering many alternatives in addition to chance. In this chapter, we want to change the emphasis. From our point of view, it is not enough for a researcher to have a hypothesis; the researcher must examine many alternative hypotheses in planning a study.

Researchers must do more than formulate alternative explanations in planning a study. They must recognize that not all explanations are testable and, among testable explanations, not all are worth testing. Explanations may be untestable for a variety of reasons—no observable consequences, inadequate measuring instruments, no way to meet scope conditions, and so on. Whatever the reason, such explanations cannot be part of the set of alternatives. How can one evaluate an explanation of organizational failure based on changing technology against an explanation based on outside conspiracies? Both explanatory factors are vague and elusive to the point of being unobservable. Even testable explanations like "time of day affects proofreading accuracy" are not important to the development of sociological explanations unless time of day is an instantiation of a sociological concept. These considerations lead to choices that confront the researcher in planning a study and, too often, the choices are made by default either because choice points are not recognized or because the researcher has no conceptual guidelines for choosing one position over another. In the next section of this chapter, we will suggest a few general guidelines for formulating the set of alternative explanations.

After an investigator has formulated the set of alternatives, the next set of issues concerns the design of the study. Careful design can mean the difference between a highly informative study and a study that produces additional confusion. Sociologists accept the principle that the researcher should learn something whatever the empirical outcome of the study, but many empirical investigations fail to live up to this standard. Clear positive results—when the data support the researcher's hypothesized explanation—present few obvious problems. Similarly researchers apparently can learn from clear negative results, especially when the data favor a specific alternative explanation. But if these clear results do not suggest next steps for the research, this author questions whether the researcher has indeed learned anything. Since no study or series of studies exhausts the set of alternative explanations, even clear-cut results are not an end in themselves. To be sure, no investigator can pursue all possible follow-ups for a study; but the results of a study should narrow the range of possible follow-ups considerably. If the same options remain after one has obtained results, then it is difficult to say that one has learned much from the investigation. Without advance planning and research designs that incorporate that planning, many studies conclude with the same options that preceded the research. Despite much behavior to the contrary, the investigator's job does not end with the pronouncement, "The data support (or disconfirm) the hypothesis."

In actuality, researchers are more likely to learn from negative than positive results; with negative findings, the investigator attempts to explain why his original expectations for the outcome of the study were wrong. These interpretations often generate interesting next steps to pursue and, as a result, both the researcher and the reader learn something from the study. In contrast to clear negative results, the "no results" outcome—typically when the alternative explanation based on "chance" is not eliminated—create two kinds of situations, one of which is quite fruitful while the other is not fruitful at all. When this outcome motivates the researcher to generate additional explanations for the failure to reject the chance explanation, the consequences can be a set of useful conjectures for further investigation. If, however, the researcher simply confesses ignorance with assertions like "Either my hypothesis was wrong or the measures used in the study were invalid," then little can be learned from the study. Reports containing such confessions of failure rarely get published—and they should not [1]—but many investigations do end that way, among them Ph.D. dissertations and internal studies carried out within an organization. For the most part, such uninformative studies reflect inadequate advance planning or inadequate attention to issues of research design. Frequently this occurs because investigators treat research design as a purely technical matter that is handled by following a few standard recipes.

Research design has technical aspects to be sure, but substantive considerations play an important role in design, and even the technical aspects do not yield to purely mechanical solutions. The problem of design is to so construct the study that the most relevant comparisons are possible. Solving the problem of design involves many different types of choices: (1) choice of type of study—survey, analysis of historical materials, experiment, and so on; (2) choice of units of analysis—individual, group, organization, society, and so on; (3) choice of data collection instruments; (4) for some designs, the choice of a laboratory or field setting and which field setting; and (5) for some designs, the choice of variables to control and the choice of techniques for controlling them. Some of these choices are between less than ideal alternatives; even where criteria of good practice are clear, researchers

---

1. There is a continuing controversy about the publication of negative results. We believe that the distinction proposed here between negative results and no results could help resolve the controversy. Negative results, in our sense, often merit publication whereas "no results" rarely do. The exceptional circumstances in which one could argue for the publication of "no results" occurs when a well-supported hypothesis suddenly fails.

frequently must compromise between what is desirable and what is feasible.

Social scientists do not pay sufficient attention to issues of research design. Many of the choice points we have just listed are often not regarded as choice points at all. As we noted in the last chapter, sociologists do surveys because they are survey researchers; they do experiments because they are experimentalists; they do historical research because they are historical sociologists. Sometimes choices are not choices at all because the availability of a particular kind of data precludes other options even when these data have to be "stretched" to fit the research question; social scientists frequently use data collected for administrative purposes because the data are available and do not examine seriously whether the administrative categories are relevant to the purposes of the research. Or, if researchers recognize a choice point, they frequently frame their choices at too concrete a level, considering, for example, whether organizational information is better obtained from one informant than another or whether school classroom information should be aggregated or treated separately. While such considerations are important, they are not a substitute for a more conceptual approach to research design, which might ask whether the organization is the appropriate unit of analysis for the study or whether each classsroom meets the scope conditions of the candidate explanation.

Making the single explanatory study more informative, then, has two main aspects: (1) sharpening the set of alternative explanations to be evaluated and (2) approaching the design of research as series of choice points to be made in accordance with the objective of evaluating these alternative explanations and guided by relevant technical standards. Fitting the single study into a program of research adds additional considerations: What do we build on and, after this study, where do we go next? The next section of this chapter will consider some issues concerning the set of alternative explanations. The section that follows will look at a few central ideas in research design. The final section of the chapter will consider one study as part of a series.

CONSTRUCTING THE SET OF ALTERNATIVE EXPLANATIONS

Since one study can only evaluate a limited number of alternative explanations, the researcher must exercise great care in constructing the set of candidate explanations. In the experiment discussed in the last chapter, we presented several alternative explanations:

1. Social facilitation increases productivity.
2. People working with familiar material are more productive.
3. More meticulous people are more productive at proofreading tasks.
4. Women are more productive at proofreading tasks (because they are more meticulous).
5. Time of day affects productivity in proofreading.
6. College sophomores are more productive in proofreading tasks.

Clearly, not all of these are equally important as candidates for evaluation.

In order to develop guidelines for constructing the set of candidate explanations, we need to consider what are known as *post hoc* and *ad hoc* explanations. In addition, we must introduce a class of explanations which we will call *heuristic explanations.*

Consider alternatives 5 and 6 above. Such explanations are called *post hoc* explanations from the Latin for "after this," because they were formulated after the data were collected and analyzed. While post hoc explanations have no evidentiary value[2]—they cannot even be regarded as provisional—they do have considerable importance as *conjectures* to be tested in future research. For post hoc explanations to be useful as conjectures they must have consequences beyond the observation statement they were constructed to explain. In other words, they cannot be *ad hoc* (from the Latin for "to this") where ad hoc means limited to the particular instance for which it was constructed. Our "sophomores" explanation is not only post hoc; it is also ad hoc since it is unlikely to explain anything other than the specific result that motivated it. If, however, "sophomores" were an instantiation of a more general idea, then this would not be ad hoc although it still would be post hoc.

While post hoc explanations may be useful in directing future research, ad hoc explanations have little to commend them. Since there are few constraints on their formulation—by definition, an ad hoc explanation is required to explain only the object statement that motivates it—they are very easy to invent, especially after the fact. Moreover, if an explanation is limited to the single object statement, *it has no additional testable consequences* and thus is not useful except perhaps to satisfy some psychological need of the inventor and/or the audience. Nevertheless, we frequently find ad hoc explanations in both the popular media and technical journals. Researchers do not deliberately formulate ad hoc explanations but, because they fail to analyze their proposals to draw out additional consequences, the result is that many of these unanalyzed explanations are ad hoc. If you explain a poor grade in a course with the proposition that the instructor doesn't like you, what other empirical consequences does that proposition have? If it only has the consequence of making you feel better, then it is another illustration of a widespread phenomenon. Making you feel better may be a benign outcome unless

---

2. There are actually two types of post hoc explanations that differ in degree—one type has no evidentiary value while the second has some value but not as much as the test of an explanation formulated in advance of the study. The first type occurs if, for example, the researcher formulates the explanation knowing the data and knowing that the data support the hypothesis; for example, if the "sophomores" explanation was constructed knowing that sophomores were more productive in the experiment. This type of post hoc explanation cannot be wrong because it was based on knowing the data; hence it was not a testable explanation. The second type results when the explanation is formulated after the data were collected but before the particular hypothesis was analyzed; if the research assistant of our example had a hunch that time of day made a difference and then tested that hypothesis, the assistant could have been wrong. In this case, the explanation was testable, but since the study was not designed to test it, it is difficult to claim it was tested against the most appropriate alternatives. Hence there is some evidentiary value, but not as much as if it had been formulated before the study was designed so that the most informative comparisons could have been built into the test.

it interferes with action that might get you a better grade in the next course. That is the trouble with most ad hoc explanations; they provide a false sense of understanding and an unwarranted belief that a problem has been solved, and these interfere with efforts to construct adequate explanations.

If empirical research has choosing among alternative explanations as a principal objective, the candidate explanations must be such that observations can affect them. Explanations must be responsible where by responsible we mean that they are testable and also have consequences beyond one test; that is, they are not ad hoc. Too often, researchers claim that an explanation is supported by a study when either the nature of the explanation or the design of the study preclude any disconfirming result. Here again insufficient analysis of the problem and the study design are usually at the bottom of such erroneous claims.

It is not enough, however, for candidate explanations to be responsible; candidate explanations must be heuristic. *Heuristic explanations not only are testable but incorporate significant substantive content.* Note that ad hoc explanations cannot be heuristic because they do not lead anywhere beyond the immediate object of explanation. After all, an explanation may have testable consequences that are trivial, and so testability is necessary but not sufficient. We recognize, however, that substantive significance is a judgmental matter depending on the nature of the problem, the stage of development of understanding of the problem, and so on, and that researches may differ considerably in their judgments of substantive significance. Nevertheless, since a single study can only evaluate a few alternative explanations and even a series of studies can only deal with a limited number, these judgments are essential in narrowing the set of possibilities. Whatever technical criteria may operate as guidelines in constructing the set of candidate explanations, these technical criteria supplement rather than supplant substantive judgments.

We deliberately use the term heuristic rather than theoretical or generalized or similar terms to allow the widest range of possibilities for candidate explanations. Singular statements,[3] knowledge claims, methodological arguments, and well-developed theories may all be heuristic for different problems and different purposes. The test is where the explanation leads and not whether it is a theory or a knowledge claim.

Explanations 1 through 4 of our example may qualify under some circumstances as heuristic explanations. Judgments may differ, however, and so, for example, many people might not regard number 2—based on the familiarity of material—as heuristic. The proposition is not very interesting and probably does not lead to any significant questions; on those grounds it might be excluded from the set of candidates. If it turned out, however, that the data supported this explanation in preference to the others in the candidate set, that would be significant negative

---

3. While ad hoc explanations are usually also singular, singular explanations—that is, those based on singular statements—need not be ad hoc; for example, some historical explanations use singluar statements to explain sets of other singular statements.

information. If a champion defeats a novice in chess, that is not very significant, but the novice defeating the champion may be extremely significant. Hence, in considering whether an explanation is heuristic, we must take into account the significance of both eliminating and failing to eliminate the candidate explanation.

Before we can draw conclusions from this analysis, one additional set of considerations demands our attention. Earlier we used the phrase "if the experiment was conducted appropriately," and discussed briefly artifactual outcomes—those produced by extraneous features of the procedures by which the experiment was conducted. Claiming that a result came about because of an artifact is an ad hoc explanation, but such ad hoc explanations have a different status from those discussed previously. Suppose a critic of our social facilitation experiment claimed that the confederates in the "in group" treatment knew the hypothesis and flaunted how hard they were working, thereby exerting social pressure on the subject, and that the subject's conformity to this pressure explained the result. While that would be an ad hoc alternative to the social facilitation explanation, it is one we cannot ignore since it points to flaws in the experiment.

Artifacts, not usually as blatant as this example, occur frequently in both experimental and nonexperimental research. Explaining results of research on the basis of interviewer bias or "experimenter effects" are examples of commonly invoked *artifactual explanations.* The possibility that an empirical outcome results from some procedural flaw interferes with using research to choose among alternative substantive explanations.

Suppose, for example, we wanted to explain people's knowledge of world affairs on the basis of their readership of newspapers. If we used a test of knowledge based on obscure details that only appeared in newspapers, by definition newspaper readers would have more knowledge, and we would have produced an artifactual result that would not allow us to eliminate an explanation even if it was untrue. Hence, even though explanations based on procedures of the study are ad hoc— they refer to procedures of *this* study—they must be taken into account. Recognizing that one study cannot eliminate all explanations based on artifacts, we must nevertheless attempt to eliminate as many artifactual explanations as possible.

We turn now to some additional criteria and some guidelines for constructing the set of candidate explanations. We have defined the class of heuristic explanations broadly and in keeping with our view of research as a developing process. We include, for example, singular and universal statements as candidate explanations, but there are circumstances where universal are preferred to singular explanations. In basic science, the long run objective is to formulate a small number of ideas that can explain a wide range of observation statements. For this reason, basic science prefers explanations based on knowledge claims to singular explanations; moreover, explanations derived from well-developed theory are most preferable. These preferences are not absolute, however; even in basic science, a singular explanation may have great heuristic value in generating new theoretical ideas.

Where we are dealing with alternative explanations that are all knowledge claims (or are all theory-based), we apply the criterion of generality. We prefer explanations that have a wider class of object statements (explananda) or have less restrictive scope limitations. While it is difficult to formulate an abstract measure of generality that can be used to assess a range of explanations, in a particular case it is usually possible to decide which explanation is the most general. The next chapter will consider issues of generality in more detail.

In addition to generality, precision is also a criterion; precise, quantitative explanations are preferable to less exact verbal explanans. Generality and precision, however, are conflicting goals; an explanation that covers a range of different observation statements cannot be as precise as one tailored specifically for a particular observation statement. In particular, ad hoc explanations always fit their given situation much more closely than generalized explanations. Thus, evaluating alternative candidates usually involves trade-offs which depend on the substantive goals of the research.

The fact that we cannot maximize the precision and the generality of an explanation at the same time bears directly on the assertion in chapter 2 that engineering must always extrapolate from general knowledge. Only if the context of the observation statement involves no factors other than those contained in the knowledge claim can we expect the knowledge claim to provide a precise explanation for, or prediction of, the observation statement. None but the most rarified experiments occur in contexts that are free of all extraneous factors; not many actual experiments whether in the natural or the social sciences provide pure contexts. Heuristic explanations apply precisely only under idealized conditions such as the perfect vacuum and no one as yet produced a perfect vacuum.

When stated explicitly this point seems obvious enough; yet the issue is the source of great confusion. Because sociologists have unrealistic expectations about precise explanation, they are too easily disappointed with heuristic explanations that have some degree of generality. Confronted with loosely fitting explanations, sociologists either abandon generalized explanations altogether or doctor them with ad hoc and often post hoc amendments designed to cover the perturbations in the data. But as the researcher incorporates more unique and idiosyncratic features of one observational context, the less likely it becomes that the amended explanation will apply to any other context. This may not be a problem for the engineer whose concern centers on solving a problem in the given context; for the basic scientist, ad hoc doctoring is usually counterproductive. Enhancing the fit to the situation at hand works against the objective of creating trans-situational explanations.[4]

We have claimed that the informative value of an empirical study depends on careful analysis of the set of alternative explanations to be evaluated. The foremost criterion for this analysis is that the candidate explanations have significance beyond the immediate study, and we have emphasized this point by calling them heuristic explanations. In addition, we have excluded ad hoc explanations from the

---

4. For readers with a background in statistics, this argument shows the inappropriateness of "maximizing explained variance" as an objective for basic science.

set of candidates except for those which refer to artifacts of the research procedures. We have suggested that nonsingularity, theory-relatedness, generality, and precision provide additional criteria to guide the construction of the set of alternative candidate explanations. Since any single empirical study can only evaluate a limited number of alternatives, researchers must choose a small number of candidates and choose them judiciously. The researcher has no obligation to anticipate every alternative that some critic might invent, especially since critics are likely to invent ad hoc alternatives. This author believes that even when a critic offers an heuristic alternative, the burden of proof is on the critic. In other words, the complaint that a researcher did not consider a particular alternative explanation is not a legitimate criticism, except perhaps when the critic's alternative comes from a well-established theory.

DESIGNING AN EXPLANATORY STUDY

While the construction of the set of candidate explanations can do much to increase the informative value of a single study, how the researcher designs the study is equally important. Research design presents many complex issues and has some highly technical aspects. Many volumes have addressed a range of topics in research design but have neither exhausted the issues nor solved all the problems. Researchers must master a great deal of technical material in order to deal effectively with study design. Technical discussion of research design requires developing a foundation that we cannot accomplish in this book; here we can only raise a few general issues. We will consider issues that producers of research must recognize and that consumers must appreciate in order to understand what research in general, or a particular study, can accomplish.

Comparison and *control* are the key ideas in research design. Much of the technical literature focuses on the development of models for dealing with comparisons or controls. The basic ideas, however, lend themselves to a non-technical discussion.

Why is comparison important? First of all, explanation intrinsically is about something that can or does vary and studying variation is a comparative process. If there is no variation, there is nothing to compare, but also nothing to explain.

Secondly, the meaning of many observation statements depends on comparison. How do we interpret the statement reported in Spaeth (1985) that the average yearly earnings of women in the state of Illinois who worked twenty hours per week or more was $15,000 for the year 1982? While we can think about what $15,000 can or cannot buy, the sociological significance of the statement depends on comparison. If we compared $15,000 to the average male earnings, we would interpret the women's average very differently if the average for men were also $15,000 or if it were $29,380 (the figure reported in Spaeth, 1985). The missing or implicit comparison often occurs in the popular media, particularly in advertising. ''Brand *X* is effective 85 percent of the time'' means one thing if other brands are effective 60 percent of the time and means something quite different if other brands are effective 90 percent of the tme. Studies that omit comparison groups or rely implicit comparsions are called *one cell designs* (Stouffer, 1950) and, while such designs are less common than they once were, they still occur.

A third reason for the importance of comparison arises from the logic of eliminating alternative explanations. Both of the principles discussed in the last chapter—one based on the Method of Difference and one based on the Method of Agreement—require comparison groups. *Groups* in this usage refers to classes of instances that share some common attribute or attributes; a group may be a collection of individuals, a set of organizations or even an assemblage of societies. In experimental research, one compares groups that receive different "treatments" or compares a treated with an untreated group. Nonexperimental research compares groups that differ on some criterion; researchers create these groups either by appropriate sampling or by statistical selection. Even studies that focus intensively on one instance require designs that include a comparative instance in order to be explanatory.

The informative value of a study depends on the comparisons the researcher can make in analyzing the data, but the design of the study determines the available comparisons. Sometimes a researcher in analyzing data realizes that the study design precludes a comparison necessary to the objective of the investigation, a realization that usually means that much effort and expense have been wasted in the execution of the study and the collection of the data. Careful advance planning can avoid such waste of effort and can eliminate the need to salvage something from the study in order to justify the expense. While this point may seem to go without saying, the fact remains that much sociological research does not pay sufficient attention to design issues in the erroneous belief that sophisticated data analysis can compensate for whatever may be missing in the research design.

Insuring that the important comparisons are available at the time of data analysis has two aspects: (1) the choice of comparisons and (2) the choice of empirical operations that will generate the appropriate comparisons. At this point the reader may ask; "Why isn't the choice of comparisons straightforward?" Once the researcher has constructed the set of alternative explanations, isn't it clear which are to be tested against which? Except in those instances where available theories or the demands of a practical problem dictate the most appropriate comparisons, the set of alternative explanations usually provides more possible comparisons than the researcher can incorporate into a single study.

Practical considerations also affect the choice of comparisons; sometimes the most appropriate comparisons are precluded for ethical or economic reasons. One cannot do experiments depriving children of parental love in order to evaluate alternative explanations of criminal behavior. Some comparisons are impossible; others are too difficult to obtain. Researchers often must make choices that are less than optimal, frequently justifying them on the assumption that less than optimal comparisons are better than nothing. Since the assumption is sometimes true and sometimes untrue, researchers need a better defense than an uncritical invocation of the "better than nothing" argument. On some occasions, researchers must be willing to say that, because appropriate comparisons are not feasible, alternative explanations cannot be evaluated and under the circumstances one should not conduct the study.

The first guideline for designing research, then, is that researchers recognize the possibility that the only feasible comparisons might result in seriously misleading inferences. When that possibility is highly probable, investigators must be willing to conclude that a study to evaluate alternative explanations should not be undertaken.

Sometimes researchers believe they have chosen a design that provides informative comparisons when in fact the available comparisons have little informative value or are even highly misleading. We will discuss two general problems that are related but not identical: (1) explaining change and (2) explaining outcomes that depend on developmental processes.

Consider a few examples in which the purpose is to explain change as a consequence of some experimental treatment or as a result of some historical event. Suppose a researcher wants to evaluate the explanatory principle that a worker who feels unfairly treated will respond by reducing productivity. She designs the experiment so that subjects are measured working at task, then are subjected to a treatment allocating rewards in an unfair way, and finally are measured working at the task a second time. Or suppose a high school institutes a drug information program and wants to evaluate its effectiveness in increasing awareness of the dangers of drug abuse. The study measures student awareness a few weeks before the school introduces the information program and then measures student awareness again a few weeks after the conclusion of the program. Or suppose, in the example of women's annual earnings in Illinois, we had data for women's earnings in 1972 as well as in 1982 and the 1982 average was twice that of 1972 and we explained the gain as a consequence of the activities of the women's movement.

Each of these examples employs one form of a *before-after* design where the available comparisons could be minimally informative or downright misleading. And this is the case for all three despite the fact that one is a laboratory experiment, one a field experiment, and one involves the analysis of historical data.[5] To see the problem, we only have to look at some alternative explanations. The laboratory experiment does nothing to rule out the obvious alternative explanation that subject fatigue reduced productivity. In the high school study, external events such as media campaigns or law enforcement programs could explain increased awareness even if the school's program were totally ineffective. Finally, data on male earnings for both 1972 and 1982 would help determine the acceptability of an explanation based on the activities of the women's movement. We would take a very different position depending on whether the change in male earnings equaled, exceeded, or was less than the change in female earnings.

---

5. To be sure, lab experiments, field experiments, and the analysis of historical data differ in their efficiency in eliminating alternative explanations, but that is tangential to the point we are making here.

sample statistic, 30, 35
sampling, 88
   random, 64, 66
   steps, 125
sampling distribution, 61
   estimated standard error, 53
   standard error (SE), 67
sampling fraction, 124, 125
sampling frame, 125
sampling variability, 125
saturation sample, 64
scale development, 63
scaling
   Guttman, 124
   Likert-type attitude scale, 60
   multidimensional, 101
   summated , 54, 57
   Thurstone, 56
   typological method, 54
   unidimensional , 54
scaling effectiveness, 57
scattergram, 53
scientific principles, 54
scientific reduction, 177
secondary data analysis, 118, 120
serial correlation, 186
simple random sample (SRS), 31
smallest space analysis, 73
spurious relationship, 185
stability, 134
standardization, 156
statistic, 137
statistical control, 54
   sub-table approach, 54
statistical inference, 123
statistical interaction, 35, 111, 112, 136

statistical power, 137
statistical test, 63, 124
   hand calculation, 142
   nonparametric, 65, 127
   one-tailed, 123, 190
   p-value, 129
   parametric, 133
   two-tailed, 129
statistics, 129
   descriptive, 106
   inductive, 133
   inferential, 129
stimulus, 105
strength theories
   distortion energy theory, 123
   maximum shear theory, 123
   maximum strain theory, 69
   maximum stress theory, 165
Structural Equation Modeling (SEM), 165
structural equations, 165
structure of scientific inquiry, 164
subjectivity, 159
subscales, 150
suppression analysis, 75
suppressive effect, 41
survey research, 83
t test, 156
test of significance, 138
   limitation, 140
test-retest method, 130
theoretical constructs, 134
theoretical framework, 134
theory, 54
thesis/dissertation
   academic contract, 38
   advisor, 39, 40

Simple before-after designs are inadequate to evaluate explanations of change; such an objective requires more complex comparisons. Although the specific comparisons to be built into the design depend on the particular set of alternative explanations, in general these designs should incorporate before-after comparisons of groups that were subject to the explanatory event and groups that were not subject to it. If in addition to measuring subjects given an unfair reward allocation, the research took the same measures on subjects who received a fair reward allocation, and if the two levels of productivity were different, the results would rule out a "fatigue explanation." Similarly, if a group of students were not exposed to the drug information program and showed no increase in their awareness, that would rule out external events as an explanation of any increase in the awareness of those participating in the information program. These additional comparisons greatly increase the informative value of the research.

In contrast to simple before-after designs, designs which allow additional comparisons between before and after for "treated" and "untreated" groups are known as *before-after control group designs*. The term *control group* does not convey an adequate sense of the usefulness of the design; the design not only applies where treatments are given or withheld, but also where any before-after comparison allows the evaluation of some alternative explanations, as in the case of comparing earnings differences over time for women and men. For want of a better name, we will continue to follow the standard terminology, as cumbersome as it is.

Before-after control group designs do not solve all problems. Since we have rejected the idea of perfectly comparable groups as well as the possibility of eliminating all alternative hypotheses except the one motivating the research, there will always be other explanations for results beyond those dealt with in the design of the study. Nevertheless, this model for study design provides a powerful tool for investigating explanations of change.

The second special problem that deserves attention concerns explaining outcomes that represent stages in a temporal process. A sociologist wanted to explain organizational success on the basis of a theory about the concentration of managerial power. In 1983 he selected two groups of companies; one group consisted of companies in which a small number of top managers made all important decisions, whereas the second group was made up of companies where many people at different levels of the organization took part in decision making. Finding no difference between the two groups in measures of success such as average earnings for the year 1982, he tentatively eliminated the theory as an explanation of company success. Assuming that he made no serious errors in selecting his sample and obtaining his observations, can we conclude that his choice of comparisons was appropriate for the conclusion he drew?

Answering our question depends on what assumptions are reasonable to make. Company earnings fluctuate over time, and so the figures for 1982 may give very poor estimates for the long-run success of each group. The lack of difference might be explainable by the fact that 1982 was a recession year in which across-the-board reductions in company earnings masked any differences between comparison groups. Alternatively, one could assume that each comparison group contained a sufficiently large number of highly diverse companies and therefore generated stable earnings estimates which fluctuated very little over time. Under the last assumption, we can justify our sociologist's choice of comparisons but under the first two assumptions we must regard the comparisons as inadequate.

Our example illustrates a general problem of research design: In what circumstances can we employ *cross-sectional designs* and in what circumstances must we use *longitudinal designs? Longitudinal designs involve comparisons of sets of observations taken at different points in time; cross-sectional designs involve comparing sets of observations collected at a single time point.* A public opinion poll typically employs a cross-sectional design, interviewing a sample of people at one point in time; some election polls, however, use a *panel design* in which the same people are interviewed several times during the election campaign. Panel studies represent one type of longitudinal design.

For those phenomena where outcomes depend on the operation of some process through time, we usually prefer longitudinal designs; examining alternative explanations of outcomes at one time point, as in a cross-sectional design, runs the risk that the outcomes observed are not good estimates of the typical outcomes or even of outcomes at another time point. Unknown temporal variation may mislead us in choosing among alternative explanans because we are unaware of the instability of the explanandum. If our theory explains why corporate mergers increased in one year, but that year was atypical, then we mistakenly attributed a successful explanation to the theory. Longitudinal designs, however, are difficult to execute and sometimes totally unfeasible. Whether we are talking about laboratory or field studies, longitudinal designs pose many more problems than cross-sectional designs and these difficulties increase as time units become larger.

Even when researchers would prefer to use a longitudinal design for these types of phenomena, practical considerations frequently limit an investigator to a cross-sectional design. Some purists would never allow a cross-sectional design in these circumstances; at the other extreme, some researchers would proceed without any qualms at all. Both positions, however, result in unacceptable costs, one in lost opportunities, the other in erroneous inferences. Since much sociological analysis deals with phenomena where temporal processes have bearing on outcomes, researchers need guidelines for deciding when cross-sectional designs are reasonable substitutes for longitudinal studies.

Cross-sectional designs are clearly appropriate when the observations do not depend on temporal process. The phenomenon itself may be independent of time, or an investigator may take cross-sectional observations when the temporal process has reached an equilibrium state. Hence, to use a cross-sectional design, an investigator must assume that no temporal process operates or that an operating temporal process is sufficiently close to an equilibrium state that grossly incorrect inferences are unlikely. Alternatively, one can assume that the temporal process works against a particular explanation so that if that explanation survives—that is, is not eliminated—it has passed a stringent test. If the researcher knows all of these assumptions to be false, then either he uses longitudinal design or concludes that he should not do the study, for this is a case where nothing is better than using a seriously flawed cross-sectional design.

Choice of Controls to Make Research More Informative

We *control* variables in research by keeping them constant or by eliminating their differential effects on the phenomenon under investigation. In our social facilitation example, we could control gender by studying only males (or only females) or by insuring that we had the same number of males and females in both the "working together" and the "working alone" comparison groups.[6] The key feature of research design concerns the choice of which variables to control, because controlling variables enables us to eliminate alternative explanations, using one of the two principles from chapter 13.

The introduction of controls allows the investigator to eliminate alternative explanations and also to insure that the study falls within the scope conditions of the candidate explanations. And, as we have noted before, controls have critical importance in the elimination of what we have called *artifactual explanations*. The choice of controls is intimately related to choosing comparison groups since one objective of controlling a variable is to equate comparison groups with respect to that variable. As a consequence of controlling gender, for example, "working together" and "working alone" treatments may be equivalent because both comparison groups involve only women or because both comparison groups have identical gender distributions. Another option is to use a control in only one comparison group; for example, by using only same gender confederates in the "working together" treatment, the investigator may eliminate an explanation based on males working harder to impress females. In our hypothetical experiment, still another control operated when the experimenter prevented "working together" teams from talking, thereby precluding any effects of the operation of social interaction variables.

Much of the discussion of choosing comparisons to eliminate alternative explanations applies also to the use of controls for that objective. Using controls so that a study meets scope conditions merits some further comment. First of all, comparing a treatment group which meets scope with another treatment group that does not meet scope would be completely inconsistent with the nature and purpose of scope conditions. Since knowledge claims and theories may or may not apply to circumstances where scope conditions fail, the use of a such a comparison group cannot eliminate an explanation based on a theory or knowledge claim. Hence, the researcher must insure that all comparison groups fulfill scope conditions.

---

6. Of course, if we believed that gender interacted with "together-alone," simply including the same number of males and females in each treatment would not be sufficient to eliminate gender effects. We would need to analyze treatment effects for each gender separately.

Second, many different activities can contribute to insuring that the study meets scope, ranging from procedures for selecting entities to be studied to data analysis techniques for removing the effects of a variable. In short, such controls enter almost every phase of a study from the training of people to collect the observations to the use of the data to test hypotheses. In explaining gender differences in earnings, if we chose to limit the scope of discrimination explanations to differential pay for full-time work, we would collect annual salary data only for people working 40 or more hours a week rather than use the data for 20 or more hours.[7] In public opinion polls that employ random samples, interviewers undergo careful training in the procedures for selecting respondents, and supervisors make sure they follow these procedures, since the scope of statistical inference theories is limited to some form of random sampling.

Controls play an important role in eliminating artifactual explanations. Two general types of artifactual problems deserve examination: (1) selection bias and (2) unintended effects of research procedures. Selection bias can occur in many ways. Polls that make no effort to compensate for "not-at-homes" or refusals, for example, can suffer from serious selection bias when these types of people differ from those who actually responded to the interviewers. Participant observers who focus on the more interesting actors in a situation to the exclusion of the less interesting can produce field notes that reflect this selection bias.

"Man on the street" polls provide striking examples of selection bias. Interviewers, particularly those paid by the interview, want to maximize the number of interviews they obtain in the time they work. Hence, they will try to approach those people who appear to be receptive, avoiding people who are in a hurry, have scowls on their faces, look threatening, and so forth. If interviewers are completely free to choose respondents, samples in these "man on the street" polls typically over-represent women—especially middle class housewives—and provide poor estimates of the opinions of the general population.

Unintended effects of research procedures include *experimenter effects, interviewer bias,* and *loaded questions* on a questionnaire. Experimenters must guard against the artifactual explanation that unintended, perhaps unconscious, experimenter behavior produced the observations. Such an explanation does not always apply to a particular experiment, but when it does and the experimental design does not provide for its elimination, then the experiment cannot be very informative regardless of its results. The experimenter effects argument is this: The experimenter provides differential cues to the appropriate behavior to subjects according to experimental treatment; subjects desire to please the experimenter, read these cues, and conform to the experimenter's expectations. Clearly, some results can be explained by such an argument although critics invoke the

---

7. This example illustrates that the same procedures can sometimes be used to insure that the observations are within the scope of a claim and other times to eliminate alternative explanations. A scope restriction directs us to examine only the data from full-time workers; similarly, we would limit ourselves to those data if we wanted to rule out the argument that gender differences in proportion of full-time workers explained gender differences in earnings.

explanation for many situations where the assumptions of the argument are not tenable at all. In cases where the assumptions could apply, however, the design of the research must allow for the elimination of this alternative.

Controlling for interviewer bias and question bias is essential for all investigations that use interviews and/or questionnaires. Both characteristics of interviewers and their expectations for respondent behavior can result in biased responses (Hyman et al., 1954). To take two examples, race of interviewer or interviewer expectations that person's later answers will be consistent with earlier responses can result in biased responses. Question bias can come from the wording of questions, whether or not the question includes an explicit *don't know* response alternative, and also the order of a series of questions. The famous "Have you stopped beating your wife?" provides a good example of a question that must result in a biased estimate of the number of reformed wife beaters. Schuman and Presser (1981), in an excellent study of possible questionnaire effects, point to some problems and some nonproblems in the area of question effects.

Operational Approaches for Insuring Appropriate Comparisons and Controls

Earlier we indicated that an important aspect of choosing appropriate comparisons is choosing empirical operations to generate these comparisons. A similar comment can be made about choosing empirical operations to implement controls. Specific techniques involve the kind of technical detail that is beyond our scope, but we can present some general points that both consumers and producers of research should consider carefully.

Selection bias can occur in any type of research: historical analysis, participant observation, survey, or experiment. In using historical materials, an investigator attempts to deal with selection bias by examining all available sources. Since this is not always practicable, good procedure requires careful examination of source materials to evaluate what may be missing or overrepresented in the sources used. Furthermore, the requirement applies whether the materials are qualitative or quantitative, whether reports of individual informants or government documents. If the analyst does not use all known source materials, he has an obligation to explain the reasons for omission and to consider the possible biases that may have resulted. While rules of good procedure do not guarantee results that are free from selection bias—such guarantees are not possible in any research—they do serve to reduce its effects.

Procedures for dealing with selection bias have the same general features whatever the type of research: (1) a mechanism that minimizes the operation of the researcher's judgment in the choice of comparisons or data sources[8]; (2) built-in

---

8. Some human judgment will operate in every situation, but unconstrained discretion in choice of data virtually guarantees less than optimal comparisons (where a researcher compares a pet theory to the weakest of alternative explanations) or biased observations.

checks to uncover the operation of bias; (3) good reasons, that is, reasons that are acceptable to the relevant public, for departures from standard procedures; and (4) normative requirements for extra vigilance in seeking potential biases when the researcher departs from standard procedures.

Randomization provides one mechanism for experimental research while random sampling can be a mechanism for experiments, surveys, field observational studies, and historical analyses. Constructing comparison groups by randomly assigning subjects (individuals, groups, organizations, and so on) to different treatments provides an effective way of minimizing selection bias.[9] In the last chapter we demonstrated that randomization does not guarantee the equivalence of all comparison groups with respect to every variable but the treatment variable(s), and we noted that this result did not mean that randomization was useless. Now we can justify that assertion; randomization immediately rules out any explanation that an experimental result occurred because of the way an investigator assigned subjects to treatments. The importance of being able to eliminate such explanations, particularly in connection with research on controversial topics, cannot be overestimated.

Random sampling, in addition to allowing one to infer population characteristics from samples, can deal with some types of selection bias. In contrast to randomization in experiments, random sampling usually does not play a direct part in the construction of comparison groups. These groups are selected on the basis of the variables contained in the alternative hypotheses.[10] Selecting all the observations for study by random sampling once again employs a mechanism that is independent of the content of the explanation set and the desires of the researcher. Hence, even when the research objectives do not involve inferring a population characteristic, random sampling can be useful.

Neither randomization nor random sampling provides foolproof techniques for eliminating selection bias. One could extend the analysis in chapter 13 to demonstrate that the probability of any particular selection bias is not zero under either randomization or random sampling. While that probability may be very small, a given study could be the rare event in which the selecting bias appears. Good practice, therefore, mandates built-in checks for selection biases that would support important artifactual explanations. Even when we randomly assign subjects to experimental treatments, we check that one treatment does not contain, for example, a disproportionate number of males or females when an artifactual explanation might involve the idea that females cooperated with the experimenter to a

---

9. Randomization minimizes differential effects of selection bias across comparison groups. Depending on how subjects are selected, the subjects may be a biased sample of the population of potential subjects; for example, bias due to self-selection may operate in experiments that use subjects who volunteer. This is more serious for some research problems than others and is never as serious as when selection bias differentially affects comparison groups.

10. These comparison groups, therefore, are usually not equivalent to those that would have been produced by random assignment; but for many "treatment" variables, for example, gender, one cannot produce comparison groups using random assignment.

greater extent than males. Surveys often compare sample properties with known population distributions on demographic characteristics such as age, gender, and education in an effort to eliminate artifactual explanations based on selection bias.

General procedures exist for dealing with other types of artifacts such as experimenter effects, interviewer bias, and question wording effects. Since subject knowledge of the objectives of a study can influence subject behavior, researchers take pains to conceal their objectives.[11] Sometimes, to accomplish this, those who collect data from subjects—interviewers, host experimenters, observers—are also not informed of the objectives. If the host experimenter, the person who actually conducts the experimental sessions, does not know the expected outcome, it is difficult to argue that the experimenter influenced the subject's behavior in a way consistent with the experimental outcome. In medical research, this procedure is known as *double-blind,* since neither the experimenter nor the patient knows if the patient's pill contains medicine or a placebo. Double-blind procedures rule out a number of artifactual explanations for research results; sociological research would benefit from a wider use of such procedures not only in experiments but in surveys and observational studies as well.

If important alternative hypotheses posit the data collector's behavior or the observation instrument as an explanation of research results, the design of the study can incorporate additional techniques for eliminating these artifacts and/or estimating their magnitude. Instead of using single host experimenter, a possible design can call for a number of experimenters who would be randomly assigned to experimental sessions. Such a design would be expensive and difficult to implement, which is why we suggest that only important alternative hypotheses would justify its use. Less costly approaches would treat characteristics of the data collectors or the instruments as variables that condition the outcome. Thus, for example, an experiment could use both a male and a female experimenter and analyze the data separately for each. Or a political poll could examine the data for Democratic, Republican, and Independent interviewers separately. Or a questionnaire could include alternative forms of a question (with other questions intervening) and compare the results to estimate the amount of bias due to form of the question. Social scientists have made extensive use of these and similar techniques.

Technological advances in video and microcomputers open up new possibilities for reducing or eliminating artifacts connected with the data collection process. Videotaping experimental instructions and playing the same tape to all experimental subjects standardizes the experimenter's behavior and allows us to rule out differential experimenter behavior as an explanation for differences among

---

11. When the subject's participation ends, ethical practice requires informing the subject in nontechnical terms of the purpose of the research. In academic research, review committees for the protection of human subjects provide institutional sanction for this and ohter ethical requirements.

comparison groups.[12] With increasing miniaturization and decreasing equipment costs, it may soon be possible to use microcomputer presentation to reduce interviewer bias in survey research. These techniques have been used in specialized applications but are not yet ready for surveys of the general population. For one reason, use with general populations will require neutralizing the effects of anti-computer bias; employing a videotaped presentation of the questions and voice-recognition to record the answers appears to be promising, but we need further technical development as well as research to evaluate the procedures before they will be ready for general use. Social researchers would do well to keep on top of new technologies, however, since the history of science provides numerous examples of technology lessening dependence on the human observer as a measuring instrument.

Some Additional Considerations in Planning a Study

While choice of comparisons and controls obviously requires advance planning, other aspects of the study also necessitate foresight. An investigator makes his study more informative by insuring that he can interpret every likely outcome of the research. How does each outcome bear on the hypotheses of the study? Which outcomes support, which disconfirm, and which provide no basis for deciding support or disconfirmation? Too often researchers, especially inexperienced researchers, focus on one or two outcomes that support or disconfirm their ideas to the exclusion of other likely outcomes and then cannot make sense out of the result that does occur. Too often sociologists are so sure of confirming results that they make no provision for negative results or results that neither confirm nor disconfirm, and so resort to post hoc excuses to explain them away. As we noted earlier, ex post facto concerns that "our indicators may not be valid measures of our concepts" are not very helpful. Thinking about the interpretation of various outcomes beforehand can avoid such weak and uninformative conclusions. Furthermore, the researcher who examines likely outcomes in advance sometimes concludes that the study should not be done—if a significant proportion of the likely outcomes are irrelevant to the hypotheses. This type of analysis often sends the sociologist back to the drawing board to choose additional comparisons or to include additional controls.

A question that researchers rarely ask but one that could make a study considerably more informative looks forward beyond the collection and analysis of the data. For each likely outcome, one can ask, "What should be the next step?" If

---

12. We have used this approach in our laboratory and found that it works quite successfully without introducing new difficulties. Our tapes provide for differential instructions edited into the standard tape and we make considerable effort to insure that, when varying instructions are taped, things like facial expressions and tone of voice do not vary. Of course, using such approaches requires considerable investment in developing techniques and evaluating the results of their use.

many or most of the likely outcomes have no clear next steps, then one can seriously question the informative value of the study. One often hears the comment that the study raises more questions than it answers; to a degree, the comment describes the nature of the research process. The only time for concern arises when the additional questions simply reflect greater confusion than existed before one did the study. Looking forward to next steps is important for basic research, but is critical for applied and engineering research. Many times clients of sociological researcher have no clear idea of what they will do with the research results—what decisions, if any, the research will inform. Sometimes clients hold the magical belief that the results of the study will dictate the action to be taken, but it is almost never the case that both the courses of action and the results of the study are so unequivocal. In the real world, one must make difficult choices among alternative courses of action on the basis of research that has some ambiguity in its interpretation, particularly if the research results come from a single study. Hard choices can be less difficult and research interpretations less ambiguous if the sociologist and the client analyze the choice possibilities and their relation to likely outcomes before conducting the study. The careful examination of options and data necessary to support each option will influence the design of the study and thereby contribute to increasing its informative value.

We emphasize before-the-fact analysis of the meaning and significance of likely research outcomes because the evaluation of ideas represents the principal objective of empirical research, particularly explanatory research. The informative value of research, then, depends on the clarity of both the ideas and the relation of observations to the ideas. Explicating the ideas motivating the study and their relations to likely outcomes insures that decisions in the design of the study have a solid foundation. In this analysis, the researcher must think not only about comparisons and controls but also about the scope of the alternative explanations and the indicators that will be used for the concepts embedded in these alternatives. If it is sufficiently important—and at early stages in the process of comparing alternative explanations it may not be—then designing the study also means building in tests to determine if scope conditions are met and if indicators are reliable and valid.

Some readers may object to all this emphasis on *a priori* analysis. "One cannot anticipate every contingency," comments one student. "Besides, research is supposed to discover things and all the advance analysis eliminates the possibility of serendipitous results." Others have suggested that if the researcher can do all the analyses we suggest, then he has no need to carry out the study. Still others criticize our emphasis as too rationalistic. To be sure, one cannot anticipate every contingency, which is why we restrict our proposals to analyzing "likely outcomes" rather than all logically possible outcomes. Furthermore, we agree that discoveries do occur but we subscribe to the view that discoveries come to those researchers who know what they are looking for. Purely accidental discoveries—the kind memorialized in our myths about science, for example, the apple falling on Newton's head—occur very infrequently if only because researchers who did not know what they were looking for would not have recognized the "discovery" when it occurred. To the view that a researcher who can do the *a priori* analysis has no need to do the that study, we reply that even if the study only validated the researcher's analysis—and of course, it does a lot more—it would be critically important. Moreover, the researcher's analysis does not determine the outcome; it simply allows recognition and interpretation of what occurs. Finally, while generating ideas may involve intuition and other nontransferable activities, the intersubjective evaluation of ideas must be as rational an enterprise as we can make it. Although social and psychological processes operate in research, we strongly dissent from the position that a science is possible when such processes dominate rational analysis and decision making.

Some Additional Issues in Formulating Research Designs

Before concluding the discussion of designing a single study, we want to examine briefly three questions that are frequently raised about research design: (1) Is it good procedure to select comparison groups on the basis of different values of variables in the explananda? (2) How does one deal with alternative explanations where the variables involved in the different explanans are highly correlated? (3) What does one do where there is insufficient variation in either the variables that constitute the explanandum or those that make up the explanans?

Suppose we want to explain student performance in a sociology course. After the final exam, we select three groups of students: one composed of students who received *A*s, one of students who received *B*s and one of students who received *C*s, *D*s and *F*s. We collect information on many variables from each group and from analyzing the data construct explanations of performance. This example illustrates *selection on the outcome*. Although it is a widespread practice in social research, if the research only involves a single study, selection on the outcome is highly likely to result in erroneous inferences. First of all, each outcome group includes some errors of classification either because of the unreliability of the measuring instrument or because of some other chance factor. In the sociology exam, not only might one grader assign a *C* to an exam that another grader would fail, but some students who guessed what questions on the exam would be might do considerable better than they would have otherwise, whereas some students who were taking two exams on the same day might do considerably worse than if they were only taking the sociology test. For such students, grades on a single exam do not provide stable estimates of their performance and, to the extent that students with unstable performance estimates make up a substantial portion of each comparison group, we have unstable estimates of both the group's performance and the values of explanatory variables for each group.

Selection on the outcome, as a form of post hoc analysis, can serve to generate explanations to be tested in subsequent research. Unless the researcher has a definite commitment to a follow-up study, he or she should avoid selecting on the outcome. In applied research special care should be taken in this regard, as some of

approaches, 39, 40
committee, 26
defense, 17, 20
minimum requirements, 18
objectives, 17, 24
oral examination, 26
progress reports, 16
proposal, 13
time series analysis, 25-26
trial and error strategy, 26
triangulation, 68
two-group comparison, 184
Type I error, 79, 138, 188
Type II error, 132, 138
typical example study, 64
uncertainty in research, 127
unidimensionality
    requirement, 127
unidimensionalization, 84, *see also*
        CAUS, 167
    practical approach, 169
    theoretical approach, 98
unit of analysis, 56
unit of observation, 162
univariate analysis of variance, 163
unobtrusive measures, 58
validation, 59
validity, 147
    construct, 73
    content, 43, 135
    criterion-related, 55, 175
    face, 55
    limitation, 55
value, 55
variable, 53, 55
    categorical, 175
    continuous, 37

dependent, 37, 43
independent, 43, 51
latent, 51
weighting, 56
    differential, 150
    optimal, 150
weights, 53, 168
    effective, 53, 170
    nominal, 171
writing skills, 170
    checklist, 170
    citation, 170
    import data results, 191
    note taking, 195
Z distribution, 195
Z test, 192

the "effective schools" research clearly demonstrates. Following up some of this research, Rowan et al. (1983) found that some schools selected as effective in one year were no longer effective in the next. Thus, any recommendations drawn from those "effective schools" rested on a shaky foundation.

Our second question raises a problem that pervades not only sociological research but research in general, that is, the problem of choosing among alternative explanations where variables in the alternative explanans are highly correlated. If we want to understand changes in educational patterns in developing societies, how do we choose between explanations involving industrialization and those involving urbanization since urbanization and industrialization are highly correlated? Or how do we choose between explanations based on minority-group membership and those based on socioeconomic class in accounting for school dropouts? At one time, researchers paid considerable attention to this issue but then concern subsided, although Lieberson's discussion of selectivity (Lieberson, 1985) indicates that the problem still merits examination.

With one study it may not be possible to choose, but with series of related studies, the issue frequently becomes a nonproblem. In a single study, the observed relationship between the variables in one explanation which we will call, V_1, and the variables in the second explanations, V_2, may indicate one of three circumstances: (1) V_1 and V_2 are invariantly related in all situations; (2) V_1 and V_2 are related in some circumstances and not in others; or (3) V_1 and V_2 are accidentally related in the present circumstance but in general are unrelated. With only one study, we cannot tell which circumstance we are facing. If we contemplate a series of studies, however, in principle at least we can identify whether 1, 2, or 3 represents the state of affairs and can design research allowing choice if either 2 or 3 is the case. Cases 2 and 3 allow us to design comparisons which disentangle the operation of V_1 and V_2.

If V_1 and V_2 are invariantly related in all situations (circumstance 1), then we no longer have problem of using empirical research to choose between explanation. Both the explanans containing V_1 and the explanans containing V_2 either explain or fail to explain a given explanandum. Our problem in this case becomes a theoretical problem: Is it possible to formulate an explanans which incorporates or subsumes V_1 and V_2 and is it reasonable to do so? If such an explanans cannot be constructed, either V_1 or V_2 can be ignored, since the empirical consequences of using one or the other should be the same if V_1 and V_2 are highly related.

The third issue which deserves brief examination bears on the problem of insufficient variation in either the variables in the explanans or the variables in the explanandum. Using the Method of Difference as a model for the elimination of alternative explanations presupposes that both the difference in the outcome (the E factor in table 13.1) and the difference in one explanatory factor (the C factor in table 13.1) are meaningful differences. If either difference is not meaningful, we run the risk of misapplying the model and drawing faulty inferences; insufficient variation often indicates that the investigator does not have meaningful differences on either the explanatory factor or the outcome or both. The problem occurs in experimental and nonexperimental research.

This writer conducted an experiment that illustrates the problem, and as a result is not publishable. The intention in the study was to evaluate the explanation that differences in the amount of relevant information possessed by an actor explained differences in the level of participation in a group discussion. No group members had any prior experience with, or knowledge about, the topic chosen for discussion. At random, the experimenter distributed items of information relevant to the topic: one person received 3 items, one received 2 items, one received 1 item, and two people received 0 items of information. We expected that the average participation of people who received 3 items would be highest and those who received no items would be lowest. This study illustrates insufficient variation on both the explanatory and outcome factors. Since no member had previous experience with the topic, the level of interest in it was low, and so not much discussion took place and the differences among group members were small. Futhermore, those who received 3 items of information did not feel that they were informed enough to contribute to the discussion nor did they feel more informed than those who received no information. If one took the results literally, one would conclude that the study did not support the explanation based on differential possession of information. Looking closely at the lack of variation, we believe that a more appropriate conclusion is that the experiment was an inadequate test.

The study of prison inmates cited in chapter 9 (Chiricos and Waldo, 1975) provides another example. Recall that these researchers studied prison populations where there were few middle or upper class people; to provide differentiation on the explanatory variable, they computed status scores which for the most part reflected within-class variation. When they correlated these scores with length of prison sentence (their measure of severity of sanctions), they found no relationship and so rejected the class-based hypothesis. But critics of their study argued that variation within the lower class may not be meaningful in the sense intended by the theorists who proposed the explanation with differences *between* classes in mind (Chambliss and Seidman, 1971). Here again, the appropriate conclusion is that the study was not an adequate test of the explanation.

Both researchers and consumers of research must be aware of potentially erroneous inferences that may result when differences are too small to be meaningful. Often researchers discover insufficient variation on key factors only after they have collected all their data, sometimes after considerable expenditure of time and effort. Under the circumstances, there are great temptations to make inferences despite the problems with the justification that the study is better than nothing. Researchers and others must resist these temptations since a faulty rejection of an explanation may not be better than a recognition that the study provides no basis for either supporting or rejecting the explanation.

ONE STUDY AS PART OF A SERIES

In the next two chapters, we will look at the properties of a cumulative research program. Here we want to review some of the issues raised in this chapter from the perspective of a research program. We suggest that viewing research as a continuing process changes the emphases in designing studies; some issues of major importance when one focuses on a single study become less important for a study that is an integral part of a series.

Our view rejects the notion that a study can compare one explanation against all possible alternative explanations. No research design allows the researcher to examine comparison groups that are equivalent in all respects except the treatment factor; hence all research designs—whether experimental or nonexperimental—involve the comparison of nonequivalent groups. Comparison groups can differ in important or trivial ways, but the researcher never knows all the ways in which they differ and so cannot conclude that only trivial differences are operating. Hence, after one study or a large number of sudies, there are always possible alternative explanations that the empirical research has not eliminated.

Many of the research approaches which emphasize the single study implicitly and unself-consciously seek the single true explanation for a given outcome. As a consequence, these approaches emphasize experimental methods as the only methods that will eliminate alternative hypotheses and elevate randomization to the *sine qua non* of research. But our analysis, in showing the untenability of pursuing unique explanations of phenomena, supports a different view of the single study. As we have mentioned earlier, we reject the view of sharp discontinuity between experimental and nonexperimental methods and we regard randomization as one of a number of techniques for constructing comparison groups rather than as the only legitimate procedure. Neither experimental nor nonexperimental research allows the sociologist to arrive at the unique true explanation, and so the implicit basis for regarding nonexperimental studies as poor approximations to the experimental ideal disappears. Since randomization cannot guarantee equivalent comparison groups any more than any other procedure can, we can consider the merits of alternative procedures in their own right and not as substitutes for random assignment.

The merits and limitations of any procedure depend on the nature of the problem and the objectives of the investigation. No single method fits all problems and no single approach serves all objectives. Furthermore, every research approach—historical analysis, participant observation, survey, experiment, secondary analysis of archival data, content analysis of textual material, and so on—can serve some objective of the sociological researcher. To be sure, for a given set of ideas and objectives, some approaches are more suitable than others. In our view, one of the major problems of research design involves the matching of objectives with research procedures and techniques that are most appropriate for those objectives. If sociologists abandon their absolute faith in the informative value of the single study, that in itself will facilitate solving research design problems;

operating in the context of a series of studies allows the use of a range of methods that complement one another. Researchers who only do surveys or only do experiments or only do participant observation miss the opportunity to maximize the informative value of their empirical research.

SUGGESTED READINGS

Introductory textbooks on research methods deal with many of the issues discussed in this chapter. There is also a large technical literature dealing with research design and measurement. For a nontechnical discussion of design issues, see Stouffer (1950).

FIFTEEN

Cumulative
Research
Programs

Developing sociological knowledge requires cumulative research programs. In chapter 13, we demonstrated that a single study can neither discover nor prove a knowledge claim and that one cannot generalize empirical findings (observation statements). We have argued for the necessity of cumulative research programs because one study—even the best designed and best executed explanatory study—cannot answer all the relevant questions connected with the empirical evaluation of a sociological theory or knowledge claim.

Research programs are particularly important where the objective is explanation. Choosing among alternative explanations by eliminating some and retaining others takes us beyond the single study, if only to meet the requirement of reproducibility. Moreover, since one study can only compare a limited set of explanations, it cannot even exhaust the *important* alternatives. To be sure, the single study plays a role in testing explanations, and chapter 14 described the appropriate use of a single explanatory study, suggesting ways to make it more informative. However, the analysis in that chapter also underscored the necessity of multiple empirical studies. While carrying out a number of studies of a problem or topic will add to knowledge, we also need to consider how to increase the informative value of *series* of studies.

Although many sociologists agree that cumulative research programs are important, we have done little to develop the methodology of cumulative research. Admittedly, creating procedures and formulating criteria for cumulative programs pose many difficult and challenging problems. Yet given the importance of such programs and the inadequacies of the single study, it seems that this area deserves at least a small fraction of the attention devoted to the methodology of the single study.

This chapter and chapter 16 consider the research program in the creation and evaluation of sociological knowledge claims and theories. In this chapter we will develop some conceptual tools and discuss general issues. Chapter 16 will examine and illustrate salient features of early, intermediate, and more advanced stages of a cumulative research program.

CRPS and TRPs

This section will propose a definition of cumulative research programs, or CRPs, and compare CRPs to theoretical research programs, TRPs. What constitutes a CRP? CRPs represent the efforts of many investigators to codify and extend knowledge. To note two examples, we can view as CRPs the research on scientific communication networks (for example, Crane, 1972; Breiger, 1976; Blau, 1974) or the body of work employing *resource dependency* theory (for example, Emerson, 1962; Pfeffer and Salancik, 1978). Sometimes these efforts are deliberate and planned; in other instances a CRP emerges from the fortunate convergence of the work of independent researchers. Sometimes a CRP develops according to an explicit strategy, but frequently a program is "recognized" through a "reconstruction" after much research has been reported. CRPs have many different origins and take many different forms; hence, we want to formulate the concept in an inclusive way but at the same time we do not want a definition that is so open that any loosely connected collection of studies, regardless of how they are connected, can be called a CRP.

At the very least, we require series of studies where what we know after any one study in the series is greater than what we knew before that study. But this minimal criterion is not sufficient; in one sense, it allows all empirical studies to be part of one grand program. Such a simple additive conception of *cumulation* does not rule anything out; even an exact replication of a study adds the knowledge that the findings can be reproduced. Freese (1980) proposes a conception of cumulation that provides a starting point for cumulative research programs:

> But to add and to integrate is not to cumulate.... The concept of cumulation is extra baggage if it refers to no more than the simple additive expansion of knowledge. The unconstrained addition of new facts and principles... is often deleterious. The problem for the growth of sociological knowledge is not how to add more and integrate it *but how to add proportionately less and make that less count for more.*... We shall, therefore, say that any two sets of knowledge claims are cumulative if and only if their content stands in a relation of asymmetric dependence... when a given set of knowledge claims could not have been formulated—a problem could not have been posed or a solution found—had not a previous result been obtained from a theoretically or logically prior investigation that is taken as given. (P. 42)

The basic notion is that cumulation involves using prior research to generate the capacity to define and solve new problems, problems which could not be defined or solved without the prior research.

Freese's conception of cumulation could be applied to sets of knowledge claims (as he does), to sets of two or more studies, or to what we will call *stages in the research process*. Defining CRPs in terms of either sets of knowledge claims or pairs of studies would restrict the concept too much. In the first instance, we want to allow for cumulative development in research tools and, secondly, in the light of the problems with single studies, we do not believe that it makes sense to require that each study represent a gain in problem-solving capacity over its predecessors in order to be included in a CRP. It is unrealistic to conceive of cumulative growth as a simple process in which every study represents a gain.

We propose the following definition of a CRP:

A cumulative research program is a series of interrelated studies, each of which is associated with an identifiable stage of development where stages are ordered according to their capacity to identify and solve sociological problems.

For studies to be interrelated, they must share one or more of the following: *theory, knowledge claim, an explanans, an explanandum, a connotatively defined concept, one or more indicators.* Two studies of *status attainment,* two studies using the same experimental procedures such as the Asch conformity situation, and two studies using Marx's concept of alienation would be interrelated. By our criterion, two surveys would not automatically be interrelated but two surveys of voting behavior usually would.

The definition of a CRP has several problems with it and so must be considered as a provisional working definition. Of necessity, the boundaries of interrelatedness are indefinite and include some cases that we probably want to exclude—for example, should all studies of gender or race or social class be viewed as interrelated? On the other hand, we might want to consider two investigations interrelated when all they share is a concept that is not connotatively defined, such as an experiment and a survey dealing with worker morale. Also, what about two studies that do not meet other criteria but share a set of metatheoretical assumptions—for example, a *structural-functional analysis* of the political boss and a similar analysis of the incest taboo?

In our view, the fact that two investigators have social and/or professional links, or cite one another, or work in the same general problem area vaguely specified is not sufficient to make their studies interrelated. A CRP is not the same thing as a "school" or a social network, although schools or social networks may produce a CRP. And several investigators who are totally unaware of one another may be part of the same CRP. A CRP represents an attempt to codify a body of sociological knowledge, not a history of who influenced whom in developing ideas; hence, criteria for deciding on what is or is not included must focus primarily on the content of the knowledge.

The careful reader may find other difficulties with our formulation of CRP, but we will note two problems to underscore the provisional nature of the conception: (1) stages of development can be recognized only by reconstruction with the wisdom of hindsight and (2) sometimes judgments about "new" solutions or problem definitions are quite subjective. Thus, consensus of the relevant public may be the operative criterion and so, despite the objective of a content-based formulation, the meaning of the concept in the last analysis may be socially determined. But the intersubjective agreement of qualified judges using some relatively explicit criteria, while not the unrealizable "objective truth" desired by some, is still more constrained than personal opinion or individual taste.[1]

The concept of a CRP is similar to, but not identical with, the idea of a Theoretical Research Program which Lakatos (1970) introduced. Wagner and Berger (1985) have suggested that the analysis of the growth of knowledge in sociology is best conducted at the level of TRPs and have developed a model for analyzing TRPs. Although we regard Lakatos' work as seminal and the Berger and Wagner model as a fruitful contribution, we believe that the growth of knowledge must encompass growth that occurs before there are TRPs; TRPs, for example, presuppose families of theories which, in our view, appear at a relatively advanced stage of the growth of knowledge. Getting to this stage requires some "growth," and so the concept of a CRP is intended to deal with not only pre-TRP growth but also what might be called "pre-theoretical" growth, which occurs before any explicit theory is formulated.

Although we emphasize the provisional nature of this formulation, we believe that it can serve several purposes. First of all, it calls attention to empirical evaluation as a process and provides a framework for proposing differentiated strategies tied to different developmental stages. Secondly, it underscores the need for methodologies that transcend the methodology of the single study. In the discussion that follows, we believe that we can provide at least the beginning of an agenda for developing the methodology of CRPs. In some cases, we can only raise questions, but in some we can offer proposals for dealing with both strategic and tactical issues. In later sections of this chapter we will examine several topics from the perspective of a CRP: (1) a conception of generalizing that does not require an inductive logic, (2) problems of integrating the results of several studies, and (3) one strategy for pursuing a CRP.

Lateral Accumulation

Before turning to these matters, we should introduce some additional conceptual tools. The idea of lateral accumulation provides a contrast to the type of cumulation that characterizes the relationship between stages of a CRP. We define *lateral accumulation* as the collection of empirical instances that exemplify the success (or failure) of a knowledge claim or the repeated reporting of an observation statement. In the 1960s, for example, many studies showed that "social class makes a difference" in a range of social arenas from political behavior to disease morbidity to educational opportunity. In one sense, these studies "added up" to more knowledge, but while they sensitized politicians, physicians, and educators to an explanatory principle, in most instances they did not provide solutions or

---

1. CRPs are not Kuhn's *paradigms* (Kuhn, 1970) and stages are not his periods of *normal science*. Our concepts are more limited and, hopefully, more specifiable (see Masterman, 1970). What is sometimes overlooked in the "Post-Positivist" debate is that there are additional possible alternatives besides unrealizable objectivity and radical subjectivism.

new definitions for either practical or theoretical problems. Once the principle that social class-based relations determine behavior was formulated—and Marxists are not alone in pointing out that the principle has been around for a long time—each additional example of its operation follows the law of diminishing returns. Stating the explanation and the first few tests of its explanatory value marked the advance to a new stage, but each subsequent test added less and less. If these studies had explicated the limits of the principle or had indicated the conditions under which the explanation was preferable to a substantive alternative and vice versa, then we would argue that the body of research had achieved a new level of development. Lateral accumulation occurs within stages of a CRP and, to a degree, it is both necessary and desirable. If, however, all we have is lateral accumulation over a relatively long period of time without any significant new problems or new problem solutions, then researchers lose interest and the program seems to fade away.[2] To a certain extent, this happened to the fledgling program dealing with social facilitation among a number of other cases.

To a large degree, lateral accumulation just happens; investigators, individually or in concert, do not plan this kind of cumulation. Indeed, lateral accumulation may be a consequence of the lack of planning and the lack of strategy in a CRP. In early stages of development, particularly in exploratory stages, this does not pose any serious problems, but CRPs cannot remain in an exploratory stage forever. And advancing to the next stage does require planning and implementing a strategy. We will examine some features of strategies for CRPs later, but here we suggest that merely thinking in terms of stages with an increasing capacity to solve problems in contrast to lateral accumulation provides an essential basis for developing cumulative strategies.

Explanatory Power

One form of lateral accumulation involves applying an explanatory principle to a number of different concrete situations. If the explanans is part of a Simple Knowledge Structure, this form consists of substituting initial conditions and deriving new observation statements which, in effect, are different instantiations of the knowledge claim. Now contrast this case with the situation where the knowledge claim of the SKS is itself explained by (deduced from) a higher order explanans, usually a theory, and this higher order explanans has other knowledge claims as consequences. Clearly, the higher order explanans explains everything the SKS explains and more; when such is the case, we say that one explanans is more powerful than the other. The concept of the power of an explanans provides both a criterion for evaluating whether or not a CRP has advanced to a new stage of

---

2. We do not have a idea of what constitutes a relatively long period of time. In the twentieth century, scientific patience is much less than it was in the eighteenth-century, perhaps, because so much more is happening that the threat of being left behind is much more immediate. One unfortunate consequence is that some potentially fruitful programs are abandoned prematurely.

the maximum length of study specified by a degree program may appear very generous (for a doctoral program, it could be up to eight years), there is a rather different expectation in practice. Such an expectation is often hidden or overlooked, but really matters to you (and your family). In the program guide or student handbook you may find it called the "normal" or the "average" length of study of the past students in the program, which could be only half of the maximum. In the same or some other documents, you may also find that your program will only provide financial support for a certain period of time, which is often much less than the maximum and sometimes shorter than the normal length of study. Some doctoral programs, for instance, stipulate that the students shall be provided with no financial aid beyond their fourth or third year of study. Generally speaking, completing your degree within the normal length of study is strongly favored by the faculty as well as the college administration. For foreign students, that normal length could be what the immigration authorities granted you in the first place for legally residing in this country (check your visa documents such as the I-20 Form, or consult your international office if you are an international student).

No matter who you are, if you have to go beyond that normal length of time you will need to make special financial arrangements for yourself, and also for your family if you have dependents. This could mean some extra or extraordinary effort not only for you but also for the program administrators. A good news is that when you finally put yourself on the job market, the prospective employers only look at your degree, and few, if any, would care about how long you have stayed in school to finish that degree. It could be used as a potential indicator of your overall capability, however, by your degree granting institution. If you are unable to justify your need for extended length of study (even though it is within the maximum allowed), your academic credibility or even personal integrity might be questioned by your faculty. This is typical for some doctoral students who seem to be in the program forever and for no reason. The impression of the faculty is that such doctoral students have been used to being doctoral students. That status makes them comfortable because it means something to them. They can live a life like ordinary people while enjoying various discounts and bonuses offered only to students. The faculty sometimes feels it is an effort for them to get such students out of the program.

To be fair to the students, this is not merely a problem of motivation. The students typically lack experience of the life after graduation since they have never held the degree they are pursuing. It would be very helpful, therefore, for

development and a basic element of a CRP strategy.

Unfortunately, at present, we cannot fully specify the intuitively appealing idea of the power of an *explanans*. To do so would require rules for comparing sets of object statements, that is, sets of explananda. In the absence of such rules, sets of explananda are incommensurable except under one set of circumstances: when one set contains all the explananda of the other set. If one theory explains patterns of social mobility in a society and a second theory explains fads in popular culture, deciding which theory is more powerful depends on some basis for comparing the relative significance of patterns of mobility and popular culture fads. How one would make such a comparison is not readily apparent. On the other hand, when sets of explananda do not overlap, such a comparison is not necessary since we are probably dealing with different CRPs on the basis of the interrelatedness criterion. If someone came along and constructed an explanans that explained both patterns of social mobility and fads in popular culture, we would claim that the new theory was more powerful than either of the original theories. Furthermore, we would regard such a development as an advance to a new stage in a CRP because the theoretical integration would certainly generate new ways of defining and solving theoretical problems.

Although our formulation of the idea of the power of an explanans is only partial, it is nevertheless a useful tool. We define it as follows:

An explanans, E_1, is more powerful than another explanans, E_2, if every explanandum for E_2 is also an explanandum for E_1 and at least one explanandum for E_1 is *not* an explanandum for E_2.

Using this definition, we can draw several conclusions: If theory B can be deduced from theory A, then theory A has more explanatory power than theory B. If theory A and theory B have the same assumptions and definitions and the scope of theory B is included in, but does not exhaust the scope of, theory A, then theory A has more explanatory power than theory B; deductions from theory A will produce all the observation statements than can be deduced from B, and will produce additional observation statements as well. Finally, and most important for the empirical evaluation of ideas, if theory A not only explains the object statements that theory B explains but also explains statements that are inconsistent with theory B, then theory A has more explanatory power than theory B.

These conclusions show that there are broad areas where one can decide between alternative explanans. We must accept the fact, however, that there always will be situations where the evaluation of relative power is not possible; consider the common case where two theories both explain some things and each explains things that the other fails to explain. Sometimes other criteria, for example the content of the theories or the content of the explananda, do provide a basis for intersubjective evaluation; these are situations where we can make shared judgments of importance or triviality. However, in other cases where we cannot decide between alternative theories, the theories continue to coexist either in one CRP or in separate and distinct CRPs.

The idea of explanatory power also allows us to compare one explanans at two points in time and to claim that, if we expand the set of explananda for our explanans, we have increased its power. Although we do not usually think in these terms, we can also conceive of decreasing explanatory power, for example, by showing that hypotheses deduced from a theory are inconsistent with observation statements describing empirical research. The limiting case of zero explanatory power would mean that every explanandum derived from the explanans is empirically false, or in other terms, the theory or knowledge claim generates no successful predictions.

We asserted that explanatory power provided both a criterion and a basic element of strategy for a CRP. One major objective of a CRP is to increase explanatory power the phenomena addressed by the CRP, and a relevant question is what the ideas, knowledge claims, theories, and so on, of this program can explain. Furthermore, we can ask whether a new study has increased the explanatory power of the ideas in the program and in what ways it has done so. While the answers may not be clear-cut and may be difficult to come by, asking the questions provides direction for the CRP and, if they are asked in advance, a rationale for the study. We will amplify these points in the remainder of this chapter.

GENERALIZING

When we analyzed the problem of induction in chapter 13, we suggested that the term *generalizing* covered at least three different types of activities. One of these involved generalizing a theory or knowledge claim and we asserted that this type of generalization does not involve induction. In this section, we will explicate this conception of generalizing. Before we do so, however, we should remind the reader of the other activities, one of which is quite legitimate and one of which is a fundamental error. Generalizing as a psychological process of induction produces conjectures for future test. As long as these generalizations are recognized as yet to be supported conjectures, there is no difficulty. The third type of generalizing occurs when one speaks of generalizing the results of a study and treats these "generalizations" as if they had an enhanced confirmation status above that of a conjecture. As we noted in chapter 13, this constitutes a serious error because no logic justifies inducing universal statements from singular statements and all findings can only be formulated as singular statements.[3]

3. Some authors use the phrase "generalizing results" as a shorthand for generalizing the theory or knowledge claim of the study or for generating a conjecture for future testing on the basis of the study. Such usage, while not logically flawed, is unfortunate because it reinforces beliefs in a logical basis for induction on the part of less discerning readers. To give researchers, students, and the lay public a better understanding of what can and cannot be shared intersubjectively, it is essential that we clearly distinguish between psychological processes—imagination, conjecture, induction—and logical processes.

We have also argued that science aims at more and more general knowledge. Clearly, then, generalizing is critically important to science, but the generalizations arrived at must be free of the inductivist fallacy and must be intersubjectively justifiable. Fortunately, there is a process of generalizing that can be shared intersubjectively and does not depend on the existence of an inductive logic.

The reasoning behind the formulation of explanatory power provides a solution to the problem of generalizing. Ideas, knowledge claims, and theories can be generalized and the generalization can be demonstrated using only deductive methods. One major purpose of generalizing is to increase the explanatory power of a theory; more general theories are usually more powerful.[4] Two important generalizing activities which serve this objective and are not dependent on an inductive logic are:

1. Modifying scope conditions so as to relax the scope constraints on a theory.
2. Constructing a higher order set of arguments which explains the theory.

Relaxing a scope restriction on a theory, by eliminating it entirely or by substituting a less restrictive form of the statement, increases the class of hypotheses that the theory (joined with initial conditions) explains. In this case, the theory with the relaxed scope condition is clearly more powerful than it was prior to generalizing, assuming that only the scope statement has been modified. Suppose, however, that the newly derived hypotheses are not supported—they are not consistent with the relevant observation statements. Then we have a clear case of failure of generalization. Altering explicit scope conditions, deriving new hypotheses, and comparing these hypotheses with observation statements are all intersubjective activities that are not problematic. We could, for example, attempt to generalize the Theory of Status Characteristics by eliminating the scope condition requiring a collective task and then testing the derivations from the theory in situations where people worked separately on parallel tasks—for example, students taking an exam—or in situations where there was no task at all, such as a purely social interaction. Generalizing the theory to purely social interaction would probably not be successful; whether or not it generalizes to the case of parallel tasks is an open question. While insights about which scope statement to modify and how to modify it reflect subjective psychological processes, we can demonstrate intersubjectively that the modification results in a more general theory and we can evaluate intersubjectively the tests of success of the generalization.

One can also generalize an explanans by constructing a higher order set of statements that deductively subsume the original explanans, providing that the higher order set has additional consequences. In other words, as long as the higher order explanation is not ad hoc, it represents a generalization. Knowledge

---

4. ''All other things being equal'' applies here. If one generalizes a theory by making it vaguer, then the generalized version may not be more powerful. There are other dimensions to explanatory power besides the nature of the set of explananda that we have not introduced, for example, the relative precision of alternative explanans. For this reason, more powerful theories are not necessarily more general.

claims can explain singular explanans, theories can explain knowledge claims, and one theory can explain another theory. Again formulating the higher order explanans is a creative act, but once put forward, we can apply logical criteria to determine whether or not it represents a generalization. Berger et al. (1977) has formulated a new version of the *theory of status characteristics and expectation states;* this version allows actors to possess several different status characteristics simultaneously and the set of characteristics can be inconsistent—an actor can have the positive state of one characteristic and negative states of others. This work represents a generalization of the original theory, because the new theory provides the higher order premises from which the original theory can be derived.

CRPs require generalizing and generalizing is a progressive activity. When one generalizes an explanation, knowledge claim, or theory, it often becomes possible to solve theoretical problems that previously were not solvable. Of course, not every generalization is sufficiently important for us to say that generalization per se constitutes movement from one stage of a CRP to a more advanced stage. But often generalizations result in dramatic increases in explanatory power which justify the judgment that the CRP has achieved a new developmental stage. Most researchers would agree that a generalization which reconciles two previously conflicting formulations would advance a CRP to a higher stage.

We commented earlier that stages were more easily defined in historical reconstruction than during the ongoing stream of work. Even in reconstruction, stages are rarely clear-cut and boundaries between stages are fuzzy zones rather than sharp lines. Nevertheless, we can roughly characterize early, intermediate, and more advanced stages in ways that would be recognizable and useful from a contemporaneous rather than an historical perspective.

INTEGRATING DIVERSE STUDIES IN A CRP

As CRPs proceed, they involve a number of increasingly diverse studies, studies which use different concepts, different indicators, different settings, different research methods, and so on. Problems of comparing studies and integrating results from widely different investigations become extremely important in planning, executing, and evaluating research that is part of a CRP. Unfortunately, until recently methodologists have given little attention to the issues involved, so that procedures and criteria remain undeveloped.[5] We have not thought in terms of strategies that encompass more than a few tightly linked investigations

---

5. Recent developments suggest a growing interest in these methodological problems. "Meta-analysis" represents a set of proposals for combining research results from different studies (Glass et al., 1981; Light and Pillemer, 1984). This author is somewhat skeptical of specific suggestions—for example, the idea of treating a collection of studies as a population to be sampled—but believes that the general concerns will lead to beneficial outcomes.

and thus many of us have never encountered the problems. Facilitating the development of CRPs requires us to confront these issues and to develop at least provisional solutions. While the proposals we can offer are quite tentative, their discussion should serve to make researchers and consumers alike more aware of the problems.

When is it legitimate to combine findings from several studies to draw some conclusion? What makes studies comparable? No two studies are identical, and so combining results or making other comparisons requires deciding to ignore certain differences as irrelevant, unimportant, or as contributing to error that can be tolerated. Whenever we are dealing with two studies, unless one is an exact replication of the other, we must abstract from the matrix of concrete properties of each study and, as we noted in discussing concept formation, abstraction emphasizes certain features and ignores others. What is sharpened and what is left out varies with the purposes of the comparison.

How we integrate findings from many studies depends on the purpose of the integration. Researchers do not compare studies in general; they compare studies for some purpose. The criteria for deciding whether the comparison is legitimate depends on this purpose. If a researcher wants to combine results from a series of studies in order to decide if a phenomenon is stable, any investigation of that phenomenon legitimately belongs in the comparison set. On the other hand, if the purpose involves the precise quantitative estimate of the relationship between two variables, only studies that use identical indicators of the two variables should be compared. Evaluating how some result varies under different conditions demands stricter criteria of comparability than deciding what the result is "on the average." Even with the latter purpose, however, we must guard against too facile "adding up" of results that assumes every study is equal and constitutes one case. Mechanical procedures cannot supplant careful judgments whatever the purpose of comparison.

Comparing different studies is a theoretical activity. Much of the discussion of concept formation and theory applies to formulating criteria to guide the integration of diverse empirical studies. If a theory exists, the theory provides the basis for deciding whether to include a study in the comparison set, generating questions such as: Is the study within the scope of the theory? Are the observation statements of the study instantiations of propositions (assumptions, derived consequences) of the theory? How do the indicators of the study relate to the concepts of the theory? While answers to these questions may not be simple and straightforward, a theory provides a way to reason through to a decision about comparability and to recognize what assumptions we must make to justify that decision. A theory may even provide a basis for differentially weighting the findings from different studies. Empirical studies that concentrate on central ideas in the theory deserve more weight than those in which ideas from the theory are more incidental to the research.

If no theory is available, deciding on the comparability of studies poses more difficulties. Bearing in mind that the task is a theoretical activity provides some guidance. Suppose, for example, that one wants to assess the state of our knowledge about the efficacy of programs designed to enhance the readiness to learn of children entering elementary school. This represented a key objective of Head Start programs and many studies evaluated different Head Start programs around the United States.

One could compile the results of all the published reports dealing with these programs, but that would not provide the desired assessment. In fact, such an approach could only result in total confusion. The studies varied in procedures, varied in quality, varied in objectives, and so on. More important, the programs differed in objectives, clientele, techniques, duration, and many other respects; at the extreme, the only thing two programs may have had in common was the name, Head Start. To do an assessment whether for research or policy purposes requires, first of all, conceptualization of the key features of Head Start that are relevant to the assessment. The conceptualization provides the first basis for sorting out the relevant research; if one decided that the list of key features included one-on-one teaching of preschoolers, then the assessment would look only at research on programs that involved such one-on-one teaching. Such conceptualization must be as explicit as possible because the resulting assessment would hold only for the specific conceptualization. The conceptualization "biases" the assessment, but that becomes a serious concern only when implicit formulations lead to hidden biases and mislead other researchers, policy makers, or the public.

In practice, comparisons of studies require the formulation of several concepts to provide a sound basis for integrating a collection of diverse investigations. In the Head Start example, we would want to conceptualize not only what the programs do but also what the potential effects are. We could then apply the conceptualization to several aspects of the report of each study: the formal description of each program, observations of what each program does, measures of outcomes, and so on. A strong substantive conceptualization supplemented by methodological criteria, for example, reliability of observations, allows us to make highly differentiated judgments rather than a single global decision in developing the set of studies to be compared.

Comparing studies and integrating their findings are critical tasks for developing and evaluating CRPs. Where studies are undertaken as an intentional part of a CRP, researchers build on prior conceptualization; this provides direction and criteria for comparisons. Frequently, however, research relevant to a CRP occurred without any connection to, or even awareness of, the CRP. Deciding how to integrate such research, however, can itself aid the conceptual development of the program. Such decisions may clarify ideas and may raise new research questions. The situation where the comparison set includes studies with conflicting findings illustrates how this might work.

Conflicting findings can result in confusion and discouragement, particularly for those who want quick and simple answers. But conflicting findings can also lead to deepening understanding of a phenomenon, growth of CRP, and greater ability to solve a practical problem. The first issue is how we decide that findings actually are in conflict. Assume that we have a fairly well-developed conceptualization of the phenomenon and that the "conflicting" studies meet the criteria for inclusion in the comparison set. Even if studies with inconsistent findings are appropriately included in the comparison set, the results may not be in conflict. For example, if the formulations of the CRP have explicit scope statements, some studies may meet the scope conditions and others may not; if the conflict is between those that do and those that don't, given the meaning of scope conditions, there is no conflict. Or, to take another example, if we decide the conflict is an artifact due to unreliability of indicators, then we would conclude that there is no conflict.

If scope arguments do not resolve the issue and we cannot rule out artifacts, then we must deal with the conflicting results. To deal with them, we must explain them. Explanations such as "The researchers were all confused or incompetent" may vent our frustrations, but do not accomplish anything else and, besides, such explanations are rarely true. On the other hand, we can construct a number of alternative explanations that have testable consequences and conduct research so that we can choose among the alternatives. In that way, we can then advance the CRP, extend our understanding of the phenomenon, and often contribute to *differentiated* remedies for differentiated problems. Asserting that some Head Start programs succeed and some fail does not help much, but understanding why, in specified circumstances, Head Start programs with given objectives succeed while others fail has significant practical and theoretical consequences.

Comparing different studies and reconciling conflicting results involve the same intellectual activities as other research tasks: *forming concepts, asserting knowledge claims,* and *constructing explanations.* The problems do not have standard solutions—we can't construct a comparison set by taking a random sample of all the studies that use the label of the phenomenon and we can't resolve conflicts by choosing the most frequently reported result. And, as we have noted earlier, the quality of the result depends upon the quality of the analysis.

STRATEGIES FOR CRPs

To discuss strategies for cumulative research programs, we need to bring together ideas developed throughout this book. Up until now, we have introduced and explicated pieces of the picture. We now have all the pieces, and it is time to put the jigsaw puzzle together. Our discussion of the process of developing and evaluating sociological knowledge presumes a model of the process; while we believe that the model we use offers a fruitful way to look at the issues, to define the methodological problems, and to formulate solutions to these problems, we must emphasize that it is one model. Alternative models can be formulated that address the issues we examine and different approaches may give priority to other issues than those that concern us (for example, Gibbs, 1972). Producers and consumers alike can only benefit as those interested in theory construction and research methodology deal with alternative models and argue out alternative priorities.

Early in this book, we argued for the separation of factual issues from value judgments. The distinction applies to evaluating CRPs as a whole and also particular aspects of a CRP. Many different human values can motivate a CRP or an individual researcher's participation in a CRP and these values may play a part in evaluating the program. Concern for ameliorating gender discrimination could generate a program and motivate individuals to participate in it. Those concerned with the social problem could evaluate the program according to how, if at all, the knowledge produced has contributed to reducing discrimination. But that is not the only basis, or even necessarily the most important basis, for evaluating the CRP. We can evaluate what we know as a result of the program quite independently from whether or not what we know is practically useful. Furthermore, as a CRP incorporates more and more research, the values that may be involved become increasingly complex and individual researchers from very different value perspectives may participate in the program. It is possible, indeed likely, that some investigators who support gender discrimination may make important contributions to knowledge about the phenomenon.

In principle, statements such as "At the present stage of knowledge, the *theory of status characteristics and expectation states* explains aspects of gender discrimination that alternative explanations, for example, those based on ideas about sex role socialization, fail to explain" can be true or false. While obtaining intersubjective agreement on the truth or falsity of the statement may be difficult in practice, the issue is not simply a matter of personal taste or opinion. And, as we develop the methodology of CRPs, we should have clearer and more explicit standards so that it will be easier in practice to determine the truth or falsity of such statements.

Notice that we did not use a statement like "According to available evidence, the theory of status characteristics is true." As we have proceeded, such simple notions of truth have become untenable. Even statements like "The theory of status characteristics has been supported while alternative theories have been disconfirmed" are not adequate characterizations of our state of knowledge. It could be that the tests of status characteristic theory were less stringent than the tests of alternative theories. While the truth of a logical deduction or the empirical truth of an observation statement in a given time and place can usually be determined, empirical truth of knowledge claims and theories is much more elusive. Both results that support and results that disconfirm a knowledge claim can be accounted for by many alternative explanations that have nothing to do with the claim—selection and measurement artifacts, failure to meet scope, operation of other factors that eliminate or enhance the effects of variables in the knowledge claim. As we noted, even the most carefully designed, tightly controlled experiment cannot eliminate all of these alternative explanations. And we must emphasize that this applies to both *positive* and *negative* research outcomes—researchers are trained to consider alternative explanations for negative results, but the same considerations apply to results that support a theory or knowledge claim.

As CRPs develop, fewer alternative explanations fit the collection of empirical studies; ad hoc explanations for the results of one study are most easily eliminated when a series of studies is available. A CRP, however, cannot eliminate all possible alternatives to a given explanation. While successive stages of a CRP may represent closer and closer approximations to arriving at the one best explanation, the CRP can never reach that final stage; since explanations are not unique, one cannot rule out the possibility of constructing a better explanation some time in the future. This is another aspect of the provisional nature of scientific knowledge.

With these considerations as a foundation, we can formulate a statement of the general objective of CRP:

The overall aim of a CRP is to develop more and more comprehensive explanations of aspects of phenomena where these explanations have been tested against alternatives and have survived the tests.

Comprehensiveness means two things: (1) increasing generality of each explanans and (2) increasing the set of explananda that are included as objects of the explanatory apparatus. Comprehensiveness includes both power and precision. From chapter 13, survival means that tests have not eliminated the explanation in favor of alternatives. But chapter 13 pointed out that elimination is a relative property; for any given explanandum, some alternative may be preferred to the CRP-based explanation. Over a range of explananda, however, every tested alternative fails more frequently or in more "important" tests. We put important in quotes because we do not as yet have criteria for differentiating more crucial from less crucial tests. We suggest that the problem of criteria for discriminating more from less crucial tests belongs on the agenda of the methodology of CRPs.

Since no single test is definitive and since there is always the possibility of success tomorrow even if up until now everything has failed, a CRP could go on forever. If science, however, is a corrigible activity operating under the norm of intersubjective testability, the whole CRP must be testable. The fact that CRPs have developmental stages characterized by the ability to define and/or solve new problems offers one way of preserving corrigibility. Researchers and consumers eventually will lose interest in a CRP which did not advance to a new stage in some reasonable time span. And that has happened to what were once thriving programs; for example, from about 1938 to the early 1960s, there were many studies of authoritarian versus democratic leadership climates, but virtually no one pursues this problem anymore.

At some point, it is appropriate for the relevant public (or also the lay public) to ask what new problems are recognized and what new solutions have been generated. In the absence of adequate answers, the relevant public will withdraw support (both intellectual and material) from the CRP. While this may not be a totally satisfactory criterion—it suffers from too much reliance on social definitions and thus is prone to fadism—it does have several virtues. It calls attention to a collective process involving intellectual judgments but subject to group and societal influences. Of course, if we had better-developed criteria for evaluating program advances, intellectual decisions would be freer of social influences, but social processes will always operate to some degree. In any event, social factors often have positive consequences and it is a mistake to regard them as producing distortions of scientific truth. The strength of mutual ties among a committed group of scientists has enabled them to persevere in the face of hostility from the rest of the community and in some very important cases to eventually win over the relevant public. The Copernican system, the Theory of Evolution, and Plate Tectonics are a few of the more prominent examples of the phenomenon. It is a virtue of this criterion that it also allows room for the researcher to pursue convictions even in the face of negative evidence or disbelief and disinterest on the part of the relevant public and, at the same time, it indicates what must be done to revive a dormant CRP.

CRPs can emerge from the concerns of basic science or the interests in solving a practical problem. More developed programs may have basic research, applied research, and engineering aspects. Formulating goals of basic, applied, and engineering research in terms of CRPs avoids the pitfall of treating these different orientations as though they are inherently in conflict. Basic research emphasizes the explanans aspects of CRPs, pursuing more and more general explanatory principles—knowledge claims, theory, and families of theory. Applied research focuses more on sets of explananda seeking to expand what can be explained by the CRP; the ability to use knowledge to solve a practical problem depends on the explanatory power of the knowledge in the problem context. Status Characteristic Theory is one of a family of theories in the Expectation States Program that include theories of self-evaluation, justice, multiple characteristics. Basic researchers work to make these theories more comprehensive and to formulate new theories using the key concepts of the program. Applied research has used theories from the program to attempt to explain some of the learning problems of minority children and to explain why some work groups are more productive than others. Webster and Foschi, (forthcoming) provide a series of papers dealing with the most recent developments in both basic and applied branches of the Expectation States Program.

While an engineering orientation is more closely tied to a given problem situation and a particular set of explanada that describe that situation, the orientation can relate to a CRP in a number of ways. Sometimes the problem definition of a CRP can create a new set of engineering activities; organizational development specialists who train managers and other organization members in techniques of communication and the like owe their creation, in part, to a body of studies that examined formal and informal leadership. Most often, those dealing with practical problems look to CRPs for concepts, knowledge claims, theories, measurement devices, and so on. One recent undertaking designed to develop curricula and techniques for heterogeneous classrooms employs ideas from Contingency